Understanding and Managing Threats to the Environment in South Eastern Europe

NATO Science for Peace and Security Series

This Series presents the results of scientific meetings supported under the NATO Programme: Science for Peace and Security (SPS).

The NATO SPS Programme supports meetings in the following Key Priority areas: (1) Defence Against Terrorism; (2) Countering other Threats to Security and (3) NATO, Partner and Mediterranean Dialogue Country Priorities. The types of meeting supported are generally "Advanced Study Institutes" and "Advanced Research Workshops". The NATO SPS Series collects together the results of these meetings. The meetings are coorganized by scientists from NATO countries and scientists from NATO's "Partner" or "Mediterranean Dialogue" countries. The observations and recommendations made at the meetings, as well as the contents of the volumes in the Series, reflect those of participants and contributors only; they should not necessarily be regarded as reflecting NATO views or policy.

Advanced Study Institutes (ASI) are high-level tutorial courses intended to convey the latest developments in a subject to an advanced-level audience

Advanced Research Workshops (ARW) are expert meetings where an intense but informal exchange of views at the frontiers of a subject aims at identifying directions for future action

Following a transformation of the programme in 2006 the Series has been re-named and re-organised. Recent volumes on topics not related to security, which result from meetings supported under the programme earlier, may be found in the NATO Science Series.

The Series is published by IOS Press, Amsterdam, and Springer, Dordrecht, in conjunction with the NATO Emerging Security Challenges Division.

Sub-Series

A.	Chemistry and Biology	Springer
B.	Physics and Biophysics	Springer
C.	Environmental Security	Springer
D.	Information and Communication Security	IOS Press
E.	Human and Societal Dynamics	IOS Press

http://www.nato.int/science
http://www.springer.com
http://www.iospress.nl

Series C: Environmental Security

Understanding and Managing Threats to the Environment in South Eastern Europe

edited by

Gorazd Meško
University of Maribor
Faculty of Criminal Justice and Security
Ljubljana, Slovenia

Dejana Dimitrijević
University of Belgrade
Faculty of Security Studies
Belgrade, Serbia

and

Charles B. Fields
Eastern Kentucky University
College of Justice and Safety
Richmond, Kentucky, USA

Published in Cooperation with NATO Emerging Security Challenges Division

Proceedings of the NATO Advanced Research Workshop on
Managing Global Environmental Threats to Air, Water and Soil –
Examples from South Eastern Europe
Ljubljana, Slovenia
28–30 June 2010

Library of Congress Control Number: 2011920992

ISBN 978-94-007-0613-2 (PB)
ISBN 978-94-007-0610-1 (HB)
ISBN 978-94-007-0611-8 (e-book)

Published by Springer,
P.O. Box 17, 3300 AA Dordrecht, The Netherlands.

www.springer.com

Printed on acid-free paper

All Rights Reserved
© Springer Science + Business Media B.V. 2011
No part of this work may be reproduced, stored in a retrieval system, or transmitted
in any form-or by any means, electronic, mechanical, photocopying, microfilming,
recording or otherwise, without written permission from the Publisher, with the
exception of any material supplied specifically for the purpose of being entered and
executed on a computer system, for exclusive use by the purchaser of the work.

Foreword

The transnational nature of environmental crime is nowhere more apparent than in the case of South Eastern Europe. What happens in one country will usually, and in many cases, inevitably, affect people, ecosystems and animals well beyond that country's borders. In this region, we are all interconnected and, regardless of specific political, social and cultural differences, we all have obligations to be good neighbours. Our general wellbeing, our fate, rests in the hands of those who share borders, rivers, winds and transportation networks with us, as much as in what we ourselves do or not do in regards to the environment.

The recent toxic sludge incident in Hungary provides a tragic illustration of what regional partnership really means. A thick red torrent of sludge burst from a reservoir at a metals plant 100 km south of Budapest in early October 2010. At least seven people died as a result of the sludge surge, some went missing and over one hundred persons were physically injured as the toxic substance flowed into nearby villages and towns. The toxic sludge reached the Danube River several days later, from where it could flow into six other European countries before reaching the Black Sea: Croatia, Serbia, Romania, Bulgaria, Ukraine and Moldova. An ecological and social disaster for Hungary thus simultaneously poses an environmental threat to surrounding countries, and the human inhabitants, ecosystems and animal life of these.

How to prevent, interpret and respond to such events is part of the mandate of those criminologists who have an interest in analysing the threats to environmental wellbeing. From the perspective of what I call eco-global criminology, the core issues attending to this type of criminological work include a concern with the ecological, the transnational and the environmentally harmful. Transgressions against humans, ecosystems and nonhuman animals provide the substantive focus for such work. But it is the conventional methodological tools of mainstream criminology as well as the conceptual innovations of green criminology that make this work so interesting, exciting and urgent.

Taken in its entirety, this book demonstrates how criminological knowledge, techniques and science can be utilised in the study of and response to environmental threats. From situational crime prevention through to problem-solving investigation techniques and environmental risk assessment, the concepts, methods and applications of criminology are increasingly being mobilised to study immediate and longer-term threats to environmental well-being across South Eastern Europe. The threats are many, and the potential impacts are profound.

Transnational environmental harm is always located somewhere. That is, while risk and harm can be analysed in terms of movements and transference from one place to another, it is nonetheless imperative that threats to the environment be put into specific regional and national contexts. This is important for several reasons. First, environmental threats originate in particular factories, farms, firms, industries and localities. Second, the political and policy context within which threats to the environment emerge is shaped by the nature of and interplay between local, national, regional and international laws and conventions.

What happens at the local level counts. What happens at the local level is likewise implicated in decisions and processes that transcend the merely local, given the complex ties and connections between businesses, governments, workers and activists. Yet, it is at the level of the nation-state that clear differences in approach and in problems is perhaps most easily discerned.

From an analytical perspective, it is essential that examination of environmental threats take into account specific national context. Living in Serbia or the Ukraine or Slovenia makes a big difference in terms of what the specific environmental risks and harms are, and how authorities and others may respond to these. Criminologists have to be aware of national differences (and similarities) in the course of investigating environmental threats. They also have to draw upon cross-national comparisons and global developments in order to provide critical scrutiny of practices and policies that offer the best option in addressing specific issues relating to environmental threats.

The world is watching what is happening in South Eastern Europe, not only because of the devastating events that have recently taken place in Hungary, but because, ultimately, it is in all of our interests to do so. How eco-crime and eco-risk is dealt with in this region, how threats to the environment are conceptualised and responded to, and how criminological policy and practice can be fashioned to address contemporary issues in the best possible way are matters of great importance to anyone and everyone with an interest in ecological justice and environmental wellbeing.

It is with this in mind that we say that – regardless of physical and historical affiliation – we all live near Chernobyl, we all swim in the Danube, we all breathe the air of Ljubljana. South Eastern Europe is, metaphorically and in many other interconnected ways, part of our global home. Accordingly, what happens anywhere in our house is and should be of concern to all of us.

The future of the planet is increasingly linked to the environment and how we collectively deal with threats to the environment. Looking over the horizon means developing strong analytical tools and positive strategic interventions that will ensure that our children and our children's children have a decent home in which to live and grow. As this book indicates, many issues demand our attention in many different parts of the world including South Eastern

Europe specifically. Moreover, the past, present and future form a continuum of analysis that can expose deep seated problems as well as suggest specific means of addressing these. The crucial thing is that we do indeed turn our gaze to that which in the twenty-first century is most important to our future wellbeing – understanding and managing threats to the environment.

Rob White
School of Sociology and Social Work
University of Tasmania, Australia

Contents

1 Introduction to Understanding and Managing Threats to the Environment in South Eastern Europe 1
Gorazd Meško, Dejana Dimitrijević, and Charles B. Fields

2 Slovenian Environmental Policy Analysis: From Institutional Declarations to Instrumental Legal Regulation 11
Andrej Sotlar, Bojan Tičar, and Bernarda Tominc

3 Situational Crime-Prevention Measures to Environmental Threats 41
Gorazd Meško, Klemen Bančič, Katja Eman, and Charles B. Fields

4 The Role of Economic Instruments in Eco-Crime Prevention 69
Radmilo V. Pešić

5 International Waste Trafficking: Preliminary Explorations 79
Ana Klenovšek and Gorazd Meško

6 Primary Categories and Symbiotic Green Crimes in Bosnia and Herzegovina 101
Elmedin Muratbegović and Haris Guso

7 Usage of Special Investigation Measures in Detecting Environmental Crime: International and Macedonian Perspective 123
Marina Malis Sazdovska

8 Solving Problems Related to Environmental Crime Investigations 135
Bojan Dobovšek and Robert Praček

ix

Contents

9 The Benefits from Using Professionally Developed Models of Possible Hazardous Materials Accident Scenarios in Crime Scene Investigations 151
Damir Kulišić

10 Environmental-Security Aspects of Explosion in the Ammunition Storage: Opportunity for Policy Making 187
Zoran Keković

11 Nanotechnology: The Need for the Implementation of the Precautionary Approach beyond the EU 205
Dejana Dimitrijević

12 Management of Spring Zones of Surface Water: The Prevention of Ecological Risks on the Example of Serbia and South Eastern Europe 225
Miroljub Milinčić and Tijana Đorđević

13 Environmental Risks to Air, Water and Soil Due to the Coal Mining Process 251
Ivica Ristović

14 Solid Municipal Wastes in Ukraine: A Case Study of Environmental Threats and Management Problems of the Chernivtsi Dump Area 265
Igor Winkler and Grygoriy Zharykov

15 Environmental Risk Factors in Connection with Hospital Laundry Effluent 279
Sonja Šostar-Turk and Sabina Fijan

16 Risk Assessment of Chemicals in Food for Public Health Protection 293
Elizabeta Mičović, Mario Gorenjak, Gorazd Meško, and Avrelija Cencič

17 Possibilities of Risk Quantification in the System of Save-for-Health Food Production 311
Midhat Jašić and Dejana Dimitrijević

18 Solutions to Threats and Risks for the National Security of Slovenia 327
Teodora Ivanuša, Matjaž Mulej, and Iztok Podbregar

Contents

19 Uncertainty in Quantitative Analysis of Risks Impacting Human Security in Relation to Environmental Threats 349

Katarína Kampová and Tomáš Loveček

20 Environmental Conflict Analysis 365

Katarína Jelšovská

About Editors and Authors 379

Index 387

Chapter 1
Introduction to Understanding and Managing Threats to the Environment in South Eastern Europe

Gorazd Meško, Dejana Dimitrijević, and Charles B. Fields

This monograph presents a selection of reflections on environmental issues in South-Eastern Europe from diverse contemporary scientific disciplines. The chapters present a variety of crucial issues regarding definitions and aspects of national environmental policies, situational prevention of threats to the environment, economic instruments in prevention of crimes against the environment, international waste trafficking, definitions and aspects of environmental (green) criminology, detecting and investigating of crimes against the environment, investigation of crimes against the environment, post-disaster management of explosions, prevention of ecological risks in relation to spring zones of surface waters, nanotechnology, threats to air, water and soil in relation to mining, management of dump areas, risk factors of hospital wastewater, environment protection and food safety from perspectives of public health, risk analysis, safe food production, application of dialectical systems theory in problem solving following environmental disasters, environmental risks and human security, and environmental conflict analysis. The authors present diverse perspectives and come from South Eastern and Eastern Europe, Western Europe, and the United States.

In the second chapter, Andrej Sotlar, Bojan Tičar and Bernarda Tominc identify and analyse the existence of a comprehensive environmental and security policy (or its fragments) and security systems in the fields of environmental crime and protection of the environment. Furthermore, the administrative- and criminal-legal

---

G. Meško (✉)
Faculty of Criminal Justice and Security, Institute of Security Strategies,
University of Maribor, Kotnikova 8, 1000 Ljubljana, Slovenia
e-mail: gorazd.mesko@fvv.uni-mb.si

D. Dimitrijević
Department of Environmental Protection, Faculty of Security Studies,
University of Belgrade, Gospodara Vučića 50, 11000, Belgrade, Serbia
e-mail: ddejana@eunet.rs

C.B. Fields
Department of Criminal Justice and Police Studies, College of Justice and Safety,
Eastern Kentucky University, Lancaster Avenue 521, 40475 Richmond, KY, USA
e-mail: chuck.fields@EKU.EDU

G. Meško et al. (eds.), *Understanding and Managing Threats to the Environment in South Eastern Europe*, NATO Science for Peace and Security Series C: Environmental Security, DOI 10.1007/978-94-007-0611-8_1, © Springer Science+Business Media B.V. 2011

regime of environment protection in the Republic of Slovenia (RS) related to their membership in the European Union (EU) is analysed. The study is based on the analysis of the existence and content of strategic documents (declarations, programs, strategies) and the implementing documents (regulations and codes) that define the organisational and functional aspects of state bodies, including those of the national security system, and which should address the intentional and unintentional threats to the environment The study is also based on a comparative and descriptive analysis of the *de lege lata* legal system of environment protection in the RS as a member of the EU. The study shows that there have been many sectoral (line) and cross (interdepartmental) approaches and policies (e.g. policy of protecting and conserving the natural environment and space; policy of protection against natural and other disasters; national environmental action programme, etc.) dealing with the question of environment protection (and thus society) in a wider sense. The authors emphasize that it is still difficult to identify a single, comprehensive and consistent "environmental security policy" that would address a wide variety of sources and types of threats to the environment. In this regard, the instrumental-legal regulation is divided into two major areas: (1) pollution control and remediation, and (2) resource conservation and management. Instrumental-legal regulation is often media-limited; it pertains only to a single environmental medium, such as air, water, soil, etc., and it controls both emissions of pollutants into the medium, as well as liability for exceeding permitted emissions and responsibility for a clean-up.

In Chap. 3, Gorazd Meško, Klemen Bančič, Katja Eman and Charles B. Fields present situational crime-prevention measures in the field of environmental protection both from the Slovenian as well as international perspectives. Crimes against nature are far-reaching, dangerous and complex, and the efforts to detect and deter environmental crimes should be as effective as possible. The purpose of this chapter is a presentation and application of the possible forms of Clark's model of situational crime prevention techniques in the field of environmental crime. The proposals are based on the results of the review of detected environmental threats in Slovenia and the consequences that arise from this connection. Situational crime-prevention has, in combination with rising of people's awareness on crime and other forms of threats, proved to be quite an effective form of prevention. These are four types of measures: (1) aggravation of access, (2) increasing the risk of the offender, (3) reducing the potential for damage (reduction of perpetrator's profit) and, (4) elimination of the excuses that someone did not know that a specific action is unacceptable, prohibited or illegal. These situational crime-preventive aspects of responding to environmental threats represent a novelty in the Slovenian practice of crime prevention methods in the environment protection field. The results of the debate can be used in the planning of crime-preventive measures with regard to environmental threats. Furthermore, these measures represent a very useful beginning for all further research and development of programs, resolutions and other documents about responding to environmental threats.

Radmilo Pešić, in Chap. 4, discusses the relationship between designs of environmental policy instruments and eco-crime prevention activities. In that context, the author extends the Emery-Watson Model by introducing a distinction between direct and indirect benefits of non-compliance. The extended model also assumes that the benefits are not static, but result from a series of events in a time sequence, right up to the moment of the detection of the crime. In conclusion, it has been stated that an essential precondition for a good environmental policy is to reduce the expected present value of non-compliance. Other conclusions assert that the higher the eco-standards imposed by a regulator, the higher the direct economic benefits of non-compliance. Furthermore, the longer the period of violation, the higher the sums of benefits obtained from non-compliance and the lower the net present value of the expected penalty. Consequently, a proposition for an effective environmental policy mix that prevents eco-crime has been put forward with country specific recommendations tailored for the Republic of Serbia.

In Chap. 5, Ana Klenovšek and Gorazd Meško discuss the problem of international waste trafficking, one of the "dark" sides of technological development in the late twentieth century. The first section frames illegal waste trafficking within the field of green criminology and cross-national environmental crime. The chapter defines waste and the classification of waste in different categories and consequences of illicit waste dumping. Reasons for international waste trafficking and the methods used are presented and discussed, as well as characteristics of destination countries, phases of trafficking and the traffickers themselves. In addition, international legislation is the essential element of social control activities, taking into consideration the Basel, Bamako and Lomé IV Conventions. Domestic waste laws of some countries are also examined. To gain better insights into the seriousness of the problem, some of the most notorious cases worldwide are summarized and international illicit waste trafficking combating efforts are outlined.

Elmedin Muratbegović and Haris Guso, in Chap. 6, highlight the problem of environmental degradation in Bosnia and Herzegovina from a criminological perspective. The authors define two types of "Green Crimes" – primary and secondary. The four "primary categories" of green crimes in which the environment becomes degraded through human action – all have become the subject of legislative efforts in recent years. These new categories are as follows: (1) crimes of air pollution; (2) crimes of deforestation; (3) crimes relating to the decline of certain species and against animal rights; and (4) crimes of water pollution. The secondary categories of green crimes can be referred to as symbiotic green crimes. These are defined as crimes that deal with ignoring rules that seek to regulate environmental disasters. They include state violence against opposition groups, hazardous waste and organized crime. The 'green criminological thought' is still taboo in many South-Eastern European countries, and there are many legal and policy-related issues raised by green crimes. The primary problem is that those crimes are not considered important by the institutions of formal social control. Case studies from Bosnia and Herzegovina are presented and illustrate examples of crimes against the environment which are currently taking place there.

In Chap. 7, Marina Malis Sazdovska discusses organized crime and environmental crime connections. One of the most profitable activities of organised crime is the illegal holding of waste which is near the top of criminal activities at present in the Former Yugoslav Republic of Macedonia. A discussion on understanding, detecting, solving environmental crimes and the police methods utilized is enriched by special investigation measures and activities undertaken by the special police units which deal with crimes against the environment.

Bojan Dobovšek and Robert Praček, in Chap. 8, present the analysis of Slovenian state social control institutions in response to environmental crime in crime scene investigations and propose solutions for better working relations between police and prosecutors in the future. In-depth target interviews with police officers, prosecutors and judges were conducted in order to define appropriate crime scene investigation procedures for recovering the evidence supporting the ensuing procedures in courts were conducted. For this reason, these in-depth interviews consist of questions regarding different views on the problems of investigating environmental crime and, in particular, the collection of evidence gathered form a crime scene. In addition, crime scene security and safety issues are defined. The primary problem in investigating environmental crimes is that, according to our latest experience, it is generally difficult to collect proper evidence on which to substantiate relevant indictment charges. More specifically, though the circumstances, attributes and consequences of such crimes may be readily apparent the major remaining unresolved issue is usually the identity of the offender(s). In such cases, good cooperation between the correspondent institutions is of crucial importance. The dividing line between the competences or jurisdiction of particular institutions (e.g. the police, forensics, and the inspectorate, among others) is very thin, and it often happens that these have to solve an issue falling under their mutual jurisdiction. Finally, solutions for future crime scene procedures by the police and harmonisation of the procedures regarding environmental crime are proposed.

Damir Kulišić, in Chap. 9, discusses the possibility for a preliminary professional analysis of possible scenarios of complex accidents with hazardous materials, where a crime scene investigation easily becomes "a mission impossible". Such a complex situation can lead to irreparable and fatal failures of securing the crime scene in all phases of the crime scene investigation. The author analyses some of the (possible) hazardous materials major accidents (citing examples from Croatia) and explains the importance for the investigation of easily accessible professional documents of health, safety and security (HSS) hazards analysis and risk assessment, plans and programs for prevention and protection. Plans for emergency preparedness and response, for all those companies that manufacture, store or transport hazardous materials are also examined by the author. The possibility of police locating those documents quickly and professional insight into them enables engaging adequate forensic experts as members of crime scene investigation teams. That includes quality planning of all phases and procedures of crime scene investigation; at least according to available knowledge about relevant features of the accident under investigation. In this regard, there will be fewer possibilities of losing key evidence, redundant investigation procedures/costs, unnecessary dealing with less important parts of procedural, technical and operative documentation. It is necessary to update current education programs for all accident investigators. Analytical methods, techniques and software tools used for making those professional documents can be of great help in searching, finding and testing logical

1 Introduction to Understanding and Managing Threats to the Environment

connections of relevant macro- and micro-traces at the scene of accident with other important clues.

In Chap. 10, Zoran Keković discusses the relationship between security and environmental issues and presents evidence of the complex interconnected military and environmental issues based on a large accident dataset from an explosion in the ammunition storage facility "Paraćinske Utrine" at the Karađorđe Hill in Serbia. The case study examined a complex situation of assessment the risks of storing hazardous materials, crisis management tools and methods that provide the basis for the application of methods, as well as ideas for policy-making purposes and future preventative policies. Additionally, the author's discussion emphasises the need for unification and harmonisation of the Serbian national regulation with the European Union legislation.

Nanotechnology: The Need for the Implementation of the Precautionary Approach beyond the EU by Dejana Dimitrijević is examined in Chap. 11. Nanomaterials are widely seen as having huge potential to provide numerous benefits to existing consumer and industrial products. However, nanomaterials might also bring new classes of problems for human health and the environment. The author analyses the development of nanoscience and nanotechnology through the study of previous research and patent applications regarding the environment, health and safety in ten South Eastern European countries. Although this is not a complete analysis of this emerging field, the author provides an outline of the general direction of efforts to date. The current state of knowledge concerning the environment, health and safety of nanomaterials emphases in general, the paucity of health, environmental and safety data, is outlined. It is difficult to anticipate with any certainty what will happen when nanomaterial is released to the general public or into the environment at large. Regulation of nanomaterials has been quite a controversial issue in recent years as well as the capacity of international governance to deal with complex regulatory demands. The author urges responsible governments of South Eastern European countries to adapt a precautionary approach, develop standards, and to be more involved in global regulatory cooperation.

Miroljub Milinčić and Tijana Đorđević, in Chap. 12, discuss the local and regional problems in South Eastern Europe (SEE) regarding of the lack of water resources and the tradition of constructing canals, accumulation and spatial and temporal redistribution of natural waters. The authors address many water resources related issues in South Eastern European countries and their recent experiences which show that dealing with the results of environmental risks of water deficits was the most expensive, inefficient, time-limited, and the overall environmental and socio-economic consequences were often harmful. This chapter points out, especially in the case of Serbia, that the accumulation of surface water resources is a relevant and important factor of various positive impacts on the management of quality and quantity of water resources and prevention of environmental risks.

In Chap. 13, Ivica Ristović analyses the influence of coal utilization, excavation, conveyance and combustion in the energy generation practices as potential environmental threats to air, water and soil in Serbia as well as SEE countries. The author addresses the adverse effects on the environment during all phases of coal producing and utilization operations. The major emphasis of this chapter reflects conditions regarding the coal mining industry in Serbia. According to Ristović, conditions that surround the energy-mining industry in the SEE region are similar to those in Serbia. Therefore, the environmental problems linked to the energy-mining industry should not only be regarded as a local, but perhaps more so a regional problem. The author argues that the implementation and utilization of advanced coal excavation and conveyance practices, which are based on the latest scientific knowledge, as well as implementation of clean coal technologies, are of high significance. Increased responsibility of all parties in a process of planning, production and utilization of energy resources in Serbia and the region are important factor of sustainable development of mining and energy sector.

The importance of solid municipal waste management on a global scale and in particular, in the city of Chernivtsi, Ukraine, is examined in Chap. 14. Igor Winkler and Grygoriy Zharykov show that the influence of the landfill in Chernivtsi on the nearby atmosphere and soils is moderate whereas serious declines in ground water quality are registered in some areas. The authors analyse and discuss current conditions of the ground and surface water pollutants patterns and concentrations in the municipal landfill of Chernivtsi. The pollution is caused mainly by general organic contamination from the filtrate, which soaks into the soils and spreads pollution around the dump area. The authors recommend possible approaches as interim alternatives to reduce the negative environmental effects on the municipal landfill in the context of a complex sorting/utilization of solid municipal waste with partial waste composting being the most favourable way for the problem mitigation. That would provide some reduction in amount of the toxic waste decomposition products within the landfill.

In Chap. 15, Sonja Šostar-Turk and Sabina Fijan discuss environmental risk factors in connection with disposing of hospital wastewater. Textile laundering processes in hospitals use significant amounts of water as well as chemicals such as detergents, bleaches and disinfectants which can pollute waste water. In the context of reducing the effluent and as a result achieving a sustainable laundering procedure, the authors determine before and after optimization; the disinfection effect, the laundering quality, and the waste water parameters of the program for hospital linens. Furthermore, the optimized program for hospital bed linens demonstrates that a transformation from thermal to chemo-thermal laundering procedures can be environmentally friendly while at the same time, ensure disinfection. Besides the optimization of washing agents, the authors make several recommendations for further improving laundering procedures to become more effective, sustainable and environmentally friendly. They emphasize the necessity for effluent treatment methods prior to emission into the public drains.

Elizabeta Mičović, Mario Gorenjak, Gorazd Meško and Avrelija Cencič, in Chap. 16, outline the risk assessment of chemicals in food, especially in the context of protecting vulnerable populations. They emphasize that the vast majority of toxicity studies and risk evaluations deal with single chemicals. However, humans are exposed to large numbers of chemicals and the possible effect of mixtures of chemicals and the interactions between them, have not been understood and

1 Introduction to Understanding and Managing Threats to the Environment

analysed fully. The authors examine the risk assessment of exposure to food additives, such as preservatives, sweeteners and colours, among preschool children in Slovenia. While this exposure does not present any harm or threat to the children observed, regarding estimated daily intake and acceptable daily intake of each food additive, nevertheless, exposure to preservatives is the highest among all observed food additives. The authors make strong recommendations for exposure assessments of the mixture of chemicals and interactive influence among them as more reliable and comprehensive part of the risk assessment process.

In Chap. 17, Midhat Jašić and Dejana Dimitrijević address the methods of assessing and quantifying risk in the chain of food production, from "farm to table", that can be used in the implementation of a food safety management system. Current knowledge and requirements of risk assessment in the food chain, which is comprised of hazard identification, hazard characterization, exposure assessment and risk characterization steps, is reviewed in this chapter. The authors provide some insight into the possible causes of contamination of food that can originate from air, water and soil pollution. The most common contaminants in food and approaches to the process of assessment and quantification of the risks in the save-for-health food production are also addressed in this chapter.

In Chap. 18, Teodora Ivanuša, Matjaž Mulej and Iztok Podbregar explain that success against a variety of environmental threats can be improved significantly if the participants and stakeholders overcome their narrow specializations and follow the principles of interdisciplinary and creative cooperation. In this interdisciplinary and creative cooperation, the natural, technological, and social sciences can and must be found equally important so that no essential oversights and flaws could result. The General Systems Theory was created during the Second World War in order to overcome over-specialization and to promote the worldview of holism of approaches and wholeness of outcomes. This concept had been partly supported by a related methodology until Mulej's Dialectical Systems Theory was created in 1974 and revised during the following decades. The chapter provides an opportunity for the readers to reflect upon these facts and put them in the context of environmental threats.

Chapter 19, by Katarína Kampová and Tomáš Loveček discusses the close interdependency between nature and society which can be perceived from examining the risks impacting human security. The purpose of this chapter is to present a systematic approach to the process of risk analysis associated with human security, focusing on the processing of uncertainty within it. The risks related to human security are generally referred to as social risks. Various types of uncertainty present within the process of quantitative analyses of social risks are identified in this chapter, and the authors deal with the sources of uncertainty and perception of uncertainty with regard to the nature of social risks related to environmental threats. This approach allows the creation of a fundamental theoretical framework within a quantitative model of social risk, which is a precondition of the social risks analysis describing and processing the existing uncertainties.

In the final chapter, Katarína Jelšovská analyses environmental conflicts. It is pointed out that environmental conflicts among stakeholders and long-term impacts may result in serious crises. A precondition of effective prevention and conflict resolution is an examination of the nature and causes, as well as possible consequences of every single environmental conflict.

We believe that the unique contributions and perspectives of the authors are most important to the development of research and reflections on understanding threats to the environment and management of such threats in contemporary South Eastern Europe.

First, we would like to express enormous gratitude to the NATO for financial support (grant number 983909) for the ARW *Managing Global Environmental Threats to Air, Water and Soil - Examples from South Eastern Europe* Workshop held in Ljubljana on June 28–30, 2010 as this publication is a selection of the best contributions from this Advanced Research Workshop. In addition, thanks also go to the Slovenian Research Agency for financing a national basic research project on crimes against the environment entitled *Environmental Crime – Criminological, Victimological, Crime-preventative, Psychological and Legal Aspects (orig. Ekološka kriminaliteta – kriminološki, viktimološki, kriminalno preventivni, psihološki in pravni vidiki,* contract number 1000-09-212171, J5-2171, 2009–2011), headed by Gorazd Meško. International partners in this research project are Charles B. Fields of the Eastern Kentucky University, USA (crimes against the environment) (Edwards et al. 1996; Eman et al. 2009) and Tim Hope of Salford University of Greater Manchester (previously Keele University), UK (crime prevention) (Hope and Sparks 2000; Hope 2009). One third of chapters in this monograph come from the national research project on crimes against the environment.

Acknowledgements are due to the authors who contributed nineteen reflections on contemporary environmental issues in SEE. Special thanks go to Professor Rob White of the University of Tasmania, Australia, for a foreword to this publication and peer reviewers Professor Terry Edwards, the University of Louisville, Professor Alexis Miller, Northern Kentucky University, Kentucky, USA, Lecturers Edward Green and Carl Root, Eastern Kentucky University, Kentucky, USA, dr. Stanko Blatnik, Institute for district energetic Velenje, Slovenia, dr. Shmuel Brenner, The Arava Institute for Environmental Studies, Israel, Professor Pavel Danihelka, Faculty of Safety Engineering, Technical University of Ostrava, Czech Republic, dr. Mensur Dlakić, Montana State University, Montana, USA, Professor Nils Petter Gleditsch, The Peace Research Institute, Norway, Professor Zlatko Ivančić, Karlovac University of Applied Sciences, Republic of Croatia, dr. Zoran Jeftić, Ministry of Defence, Republic of Serbia, Professor Vladislav Kecojević, West Virginia University, West Virginia, USA, Professor Igor Linkov, Carnegie Mellon University, Pittsburgh, USA, Professor Vojko Potočan, Faculty of Economics and Business, University of Maribor, Slovenia, Professor Tanja Pušić, Faculty of Textile Technology, University of Zagreb, Republic of Croatia, Professor Steve A. Quarrie, Newcastle University, United Kingdom, Professor Edward W. Randall, Queen Mary University of London, United Kingdom, dr. Peter Slaveikov, Sofia University St. Kliment Ohridski, Bulgaria, Professor Ladislav Šimák, Faculty of special engineering, University of Žilina, Slovakia, Professor Drago Šubarić, J.J. Strossmayer Faculty of Food Technology University of Osijek, Republic of

1 Introduction to Understanding and Managing Threats to the Environment 9

Croatia, Professor Trahel Vardanyan, Faculty of Geography and Geology, Yerevan State University, Armenia, dr. Katarína Zánická Hollá, Faculty of special engineering University of Žilina, Slovakia, and dr. Zdenka Ženko, Faculty of Business and Economics, University of Maribor, Slovenia. We would like to thank Tanya Clark, Ashley Farmer, Brian Ink and Lindsey Upton, postgraduate students in the College of Justice and Safety, Eastern Kentucky University, Kentucky, USA, who visited the Faculty of Criminal Justice and Security, University of Maribor, in Ljubljana in September 2010 and helped us to edit the book chapters. Thanks also go to Maja Jere, a Ph.D. student and junior research fellow at the Faculty of Criminal Justice and Security, University of Maribor, for technical support. We are greatly indebted to Professor Terry Edwards of the Department of Justice Administration, University of Louisville, USA, and Katja Eman, a Ph.D. student and junior research fellow at the Faculty of Criminal Justice and Security, University of Maribor, for their significant contribution throughout the entire process of editing this volume. However, the authors and editors alone are responsible for any faults the reader may discover.

References

Edwards SM, Edwards TD, Fields CB (eds) (1996) Environmental crime and criminality. Garland Publishing, New York

Eman K, Meško G, Fields CB (2009) Crimes against the environment: green criminology and research challenges in Slovenia. Varstvoslovje (J Crim Justice Secur) 11(4):574–592

Hope T (2009) Evaluation of safety and crime prevention policies in England and Wales. In: Robert P (ed) Evaluating safety and crime prevention policies in Europe. VUBPRESS, Brussels, pp 91–122

Hope T, Sparks R (eds) (2000) Crime, risk and insecurities. Routledge, London

Chapter 2
Slovenian Environmental Policy Analysis: From Institutional Declarations to Instrumental Legal Regulation

Andrej Sotlar, Bojan Tičar, and Bernarda Tominc

Abstract This chapter identifies and analyses the existence of a comprehensive environmental and security policy (or its fragments) and a security system in the fields of environmental crime and protection of the environment. The chapter also analyses the administrative- and criminal-legal regime of environment protection in the Republic of Slovenia (RS) as a member of the European Union (EU). The study is based on the analysis of the existence and content of strategic documents (declarations, programs, strategies) and the implementing documents (regulations and codes) that define the organisational and functional aspects of state bodies, including those of the national security system, and which should tackle the intentional and unintentional threats to the environment. The study is also based on comparative and descriptive analysis of the *de lege lata* legal system of environment protection in the RS as a member of the EU. The study shows that there have been many sectoral (line) and cross (interdepartmental) approaches and policies (e.g. policy of protecting and conserving the natural environment and space; policy of protection against natural and other disasters; national environmental action programme, etc.) dealing with the question of environment protection (and thus society) in the widest sense. Nevertheless, it is still difficult to identify a single comprehensive and consistent "environmental security policy" that would address how a wide variety of sources and types of threats to the environment are going to be tackled. On the other hand, instrumental legal regulation is divided into two major areas: (1) pollution control and remediation and (2) resource conservation and management. Instrumental legal regulation is often media limited (i.e., it pertains only to a single environmental medium, such as air, water, soil, etc.) and it controls both emissions of pollutants into the medium, as well as liability for exceeding permitted emissions and responsibility for clean-up.

A. Sotlar(✉), B. Tičar, and B. Tominc
Faculty of Criminal Justice and Security, University of Maribor, Kotnikova 8,
1000, Ljubljana, Slovenia
e-mail: andrej.sotlar@fvv.uni-mb.si

G. Meško et al. (eds.), *Understanding and Managing Threats to the Environment in South Eastern Europe*, NATO Science for Peace and Security Series C: Environmental Security, DOI 10.1007/978-94-007-0611-8_2, © Springer Science+Business Media B.V. 2011

2.1 Introduction

The great Roman philosopher and writer Seneca once stated: "*If one does not know to which port one is sailing, no wind is favourable.*" (Seneca 2010) and it seems that two millenniums later contemporary society, including countries, nations, governments and individuals are still endlessly trying to solve this problem when making important, long-term decisions. If the meaning of consistent decision-making on the level of an individual is in this case set aside, the process in countries and international political-security organisations cannot be ignored. The output of the policy making process always produces a crystallized form of a certain sectoral policy, e.g. foreign, defence, security, economic, etc. Simply stated, there are as many public policies as there are (public) divisions or sectors and problems connected to them, or as Dunn (1994: 70) states: "*Public policies, a long series of more or less related choices (including decisions not to act) designed by governmental bodies and officials, are formulated in issue areas, ranging from defence, energy and health, to education, welfare, and crime*". In consideration to Seneca's statement, such policies could also be understood as a form of "rational plan of action" that should lead (in this case – countries, nations, organisations and mankind) to the final goal. This should be eligible even for sample cases when determining national and national-security goals as well as strategies for reaching these goals, which is not always an easy task because the realizations of goals also means a certain sacrifice of the entire society. This is why the decision-makers on the country level (government, parliament, president) must always consider whether the public accepts and supports the goals defined by the security policy and if they don't, there are two possibilities. Firstly, it should be considered whether the goals can be adjusted in a manner that would still enable political implementation and at the same time be acceptable to the public. If the first is not an option, secondly, the decision-makers are forced to adjust the policy itself (International Military and Defense Encyclopedia 1993).

But what should be done when dealing with a special security unit. The so-called "environmental security" that assumes even the existence of a suitable sectoral policy that also needs to be adequately inserted into the (national)security policy? Indeed, the environment (and consequently the threats that derive from it) is in its relation with decision makers an independent variable, therefore correcting target goals that refer to its security and protection (e.g. stopping environmental destruction and influencing climate change) is not simple and cannot be taken for granted, as it is often the case in the field of "classical" security. When it comes to the environment, the goals are mainly determined from the "outside" (they are solar) and are therefore not strongly influenced by particular states (e.g. governments), which is why it leaves decision makers to deal mainly with policies and policy strategies.

This chapter therefore has a dual intention – to analyse strategic documents (strategies) that the Republic of Slovenia has formulated in the field of environmental security/protection and national security in the past decade and from which it is as well possible to recognize (at least on the declaratory level) an outline of suitable sectoral policies in this domain (e.g. environmental policy, security policy); and it is intended to be used in the analysis of an actual legal frame of environmental protection that should at least theoretically draw its legitimacy from strategic documents. In this chapter, a method of policy (and legislation) content analysis is used, perhaps the most frequent research approach, despite that the borders among various approaches are not precisely defined and the researchers certainly (even unconsciously) vary among them.[1]

2.2 The Environment and Security

2.2.1 Environmental Protection as Part of (National) Security: A Political Construct or Natural Fact?

The dilemma exposed in the title will probably seem as a waste or even false to many readers because the environment keeps exposing its "security" component. After all, numerous domestic and foreign authors (see e.g. Homer-Dixon 1994; Homer-Dixon 1998; Levy 1994; Kajfež-Bogataj 2006; Prezelj 2007; Grošelj 2007) have been reminding us about this issue for many years and they have all exposed the meaning of environmental factors in connection to national and international security. Even more – environmental protection is and will remain an element of national security because national security "needs" a natural environment as a context (and relation) where it establishes and ensures the basic functions of social environment. Besides, environmental problems and threats are more easily solved on a group level, even through co-(operation) of national states who still remain the main actors and creators of (national and international) security.

If this is true, then it would be expected that modern states would adjust and harmonise all sectoral policies that deal with different sources of threats. As Malešič (2009) has actually discovered, national security policies and systems based on these policies are only slowly adjusting to the new security-political environment that is (directly or indirectly) defined also by environmental threats. The changes that may be noticed usually do not concern the total national security system but mainly its individual elements. This all leads to a conclusion that the dilemma mentioned earlier is not completely exceeded and that only an accurate analysis of sectoral policies in individual states can answer what may be a rhetorical question at first glance.

---

[1] Hill (1997: 2) for an example, roughly divided policy analysts into those who are interested only in understanding (the content of) policies (1); those who dedicate themselves to finding answers on how to improve the process of its formation (2) and those who believe that the process of forming and the content are so closely connected that it is not possible or logical to study them separately (3). Hogwood and Gunn (1984: 26–31) identified the following seven different approaches in policy studies within policy analysis: studies of policy content, studies of policy process, studies of policy outputs, evaluation studies, information for policy process, process advocacy and policy advocacy.

2.2.2 National Security Interests and Goals of the Republic of Slovenia

While the physical dimension of human security is fortunately and progressively losing its original meaning in the contemporary world (nowadays, being "secure" means much more than just not being physically threatened), the number and complex sources of threats have on the other hand risen to such significance that it is in most cases impossible to successfully discern, treat, canalise, invigilate or manage them individually, selectively or partially. Accordingly, various security concepts have been developed and are being developed in the past years and decades. All of them contain security as their common denominator, but their operationalization is at its largest extent dependant on the entity or entity ownership that is being asserted.

The concepts accepted, particularly on the state/nation level, are "national security", "homeland security" and "human security"; the latter being used more and more in recent years. Even though the concepts should not exclude each other, human security seems acceptable for many different entities (states, NGOs, individuals) since it sets the individual, his needs and expectations in the centre of security approaches (Prezelj 2008). Through the prism of "international/global/ universal security", the most successfully established concepts in the international community were "system of collective security" (UN-like) and "system of collective defence" (NATO-like). The mentioned concepts may co-exist in practise; they are not exclusive and they may even function mutually. However, it should be noticed that there is some kind of transition phase occurring, where security concepts originating from security needs, expectations and demands of individuals will proceed towards (superior) group/state entities ("bottom-up" approach) and not vice versa. In connection to the subject of study, an interesting paradox may be recognized. Since environmental threats do not acknowledge national borders, it is only possible to systematically defeat them when collectively-managed, even conducted from the top down, i.e. from national and supranational entities to individuals, being positioned again at the familiar "top-down" approach. The thesis of "transition" must therefore be somewhat relativized!

Slovenia is a small and, in regard to resources (economical, human, financial, etc.), a limited country and is as such especially vulnerable to numerous sources of threats that it may be harmed by. On the other hand, Slovenia is so closely tied up in the international environment (political, economic, security etc.) that it cannot be isolated and immune to the trends and paradigms delivered by the emerging security concepts in the contemporary world. Despite all, the concept of "national security" launched around twenty years ago is very deeply rooted in the Slovenian society, so where security issues in society are concerned, not only are initiatives expected from "the state" (e.g. government) but also management and especially financing certain elements of the national security system, and this just makes a redirection back to the question of existence and the content of the security and environmental policies.

In the past two decades, Slovenia has slowly but reliably consolidated (theoretically) the meaning of national security where it at least in its content (if not also organisationally) comprehends the security of territorial land, including its air space and territorial waters, the safety of human lives and their property, the preservation and support of national sovereignty and the execution of basic society functions – social, economic, socio-political, cultural, ecological and others (Grizold 1992; Grizold 1999). This is the theory (on a technical or executive level, national security is defined significantly more indirectly) that is recognized in the definition of interests and goals that the Republic of Slovenia desires to exercise in the security field and has addressed these issues in both the current fundamental strategic document on security, *Resolution on the National Security Strategy of the Republic of Slovenia* (Resolucija o strategiji nacionalne varnosti Republike Slovenije [ReSNV-1] 2010) as well as in the previous strategy document issued in 2001. So what will Slovenia strive for and attempt to achieve? *The Resolution* states that the protection and defence of national interests and goals "present the essence in ensuring national security". National interests are therefore "*vital*" and "*strategic*". According to the *Resolution*, two *long-standing* and *vitally important interests* of the Republic of Slovenia are "to preserve independence, sovereignty and territorial integrity of the state, and to preserve the national identity, culture and originality of the Slovene people, both within Slovenia's internationally recognized borders and abroad". And *strategic interests* mean "acknowledging and respecting the inviolability of internationally recognized borders and state area, including access to open waters through territorial seas, the functioning of a democratic political system, the respect for human rights and basic freedoms, the strengthening of a legal and social state, the well-being of its population and integral social development, the protection of life and a high degree of security of the population in all forms, the preservation and development of Slovenian autochthonous national communities in neighbouring countries, international peace, security and stability and the *preservation of the environment and national sources* in the Republic of Slovenia" (Resolucija o strategiji nacionalne varnosti Republike Slovenije [ReSNV-1] 2010).

Interests should however be carried out by bringing the "*national security objectives*" into effect. They present an "efficient activity of a legal and social state, a high degree of security that is based on appropriate precautions, arrangement and preparedness of all capacities necessary for efficient and timely detection and response to contemporary sources of threats and security risks, *efficient environment protection and preservation of natural sources and assuring strategic sources*, invigorating good relations with neighbouring and other countries, a solid and stable international political-security position of the Republic of Slovenia, promoting peace and strengthening security and stability in the international community" (Resolucija o strategiji nacionalne varnosti Republike Slovenije [ReSNV-1] 2010).

2.2.3 Sources of Threats to the National Security of the Republic of Slovenia with Emphasis on Environmental Threats

Interests and goals are one matter and reality is frequently something completely different. In fact, sources of threats are what determine the security policy and consecutively, the interests and goals that the state appoints in the field of security (Sotlar 2008).[2] If anything has significantly changed in the past decade in Slovenia, then it certainly relates to the perception of sources of threats to the national security. Sources of threats may be divided according to several criteria. Very traditional classifications are those between military and non-military sources of threats (criterion of use of armed forces), external and internal (criterion of area), natural and anthropogenic (criterion of threat carriers) or those accordant with either the criterion of emerging form or the content of threat: military, political, societal, economic, environmental, and others. However, analysing and realistically estimating sources of threats is not so simple because at least three levels of perception are being dealt with, *declarative* (those stated in strategic security documents), *perceived* (among the public) and *real* (that frequently occur and are statistically recognized) threats.

2.2.3.1 Declarative (Political) Threats

It's been only ten years since the *Resolution on the National Security Strategy of the Republic of Slovenia* issued in 2001[3] (Resolucija o strategiji nacionalne varnosti Republike Slovenije [ReSNV] 2001), which expressed the following sources of threats to be of greatest importance; subversive activity, threats of aggression, military attack, mass migration, terrorism, organised crime, *ravaging of the environment*, economic blockades-including an energy crisis, information of cyber blockades or actions, health and epidemiological threats, natural and other disasters, transnational crime, *ecological risks (chemical, radioactive and other types of pollution and uncontrolled interventions in nature)*[4] etc.

However, just a decade later, an important shift in the perception of *threats and risks* to the Slovenian national security may be noticed. The perception in this case is limited to a few governmental experts and to the governmental and parliamentary political

---

[2] Sources of threats are in a dual relationship when it comes to the state's security policy making. Such a policy on one hand does and must contain them (the policy must answer the question on how and with which resources it will face specific sources of threats that the state is mostly threatened by) and on the other hand, sources of threats in the policy making process function as »a factor of pressure« (the decision makers are influenced by them and cannot ignore them, especially the most realistic and dangerous ones) (Sotlar 2008).

[3] Although this strategic document is formally and legally no longer valid, the sources from it are being used because the revised strategy on the national security of the Republic of Slovenia was adopted only in spring 2010 and it probably does not yet have proper implications. Comparing strategic documents of later dates makes declarative sources from 2001 interesting since they expose the transformation of sources of threats from "classical" to "postmodern".

[4] These ecological risks in a large-scale reflect planetary climatic changes, harmful genetic consequences, animal and plant diseases, food and water of dubious quality standards, and cross-border effects of some ecological disasters (ReSNV 2001).

elite[5] who worked together in the process of preparing, forming and adopting the new *Resolution on the National Security Strategy of the Republic of Slovenia* (ReSNV-1 2010). The sources of threats and risks are presented and classified in the *Resolution* by source of threat and concerning global, supranational and national levels.

Global sources of threats and risks to the national security are identified as *climate changes* (consequences may already be recognized in a higher number of natural disasters), financial, economical and social risks and critical focal points. *Supranational sources of threats* mainly include terrorism (although currently relatively low), illegal activities connected to conventional weapons, weapons of mass destruction and nuclear technology, organised crime, illegal migration, cyber threats, abuse of informational technologies and systems, foreign intelligence service activities and military threats. *National sources of threats* and risks to the national security mainly concern threats towards public security, *natural and other disasters* (especially earthquakes, flooding, thunderstorms, droughts, extensive fire outbreaks, emerging infectious diseases in humans, animals and plants, road traffic accidents, industrial disasters, as well as disasters that are a consequence of various forms of intentional threatening), *limited natural sources and the degradation of the living environment* (Slovenia depends on the import of power supply and raw materials; because of global warming and environmental pollution it is possible that in the future it will have to deal with problems in supplying qualitative drinking water and providing natural food sources), epidemiological threats and certain factors of uncertainty (mostly poverty, negative demographic movement, negative consequences of globalisation and commercialism, dangers to critical infrastructures).

When it comes to highest strategic documents adopted by the National Assembly of the Republic of Slovenia which have a direct or indirect impact on protecting the environment, at least three more documents need to be mentioned. The *Resolution on the Prevention and Suppression of Crime* adopted in 2006 (Resolucija o preprečevanju in zatiranju kriminalitetne [RePZK] 2006) points out that "the number of crimes connected to the protection of the environment and natural resources is rising and it's not just a consequence of a higher conscious related to the necessity of environmental protection. Especially problematic are unregulated construction and asbestos-cement waste disposals and an insufficient number of sanitary purifying plants. The environment is burdened by numerous interventions such as illegal excavation where the damage is both ecological and material". A similar document was adopted a year later, *Resolution on the National Programme of Prevention and Suppression of Crime for the Period of 2007–2011* (Resolucija o nacionalnem programu preprečevanja in zatiranja kriminalitetne za obdobje 2007–2011 [ReNPPZK0711] 2007), dedicating a whole subchapter to environmental protection. The key problems regarding *environmental protection* in the Republic of Slovenia are according to this *Resolution* "extensive dark field in discovering violators,

---

[5] It is necessary to point out that elites (politicians, high state officials, influential intellectuals and reporters) have an important impact on forming public opinion because their opinions and standpoints are broadcasted by mass media. Even the field of security and consequently the perception of threat is not an exception (Burk; in Malešič and Vegič 2009: 101).

realistic power (or powerlessness) of punishment, personnel crisis in institutions, general level of environmental consciousness, and mainly uncoordinated or unsuitable functioning in authoritative institutions in different cases of pollution". The *Resolution* recognizes the causes for these conditions in the following:

– Managing hazardous waste recycling requires proper equipment, qualifications and financial-material funds that are not sufficiently provided by manufacturers of hazardous wastes and other dangerous substances due to a relatively low risk factor.[6]
– Absence of systematic and proactive detection and prevention of crimes and other acts that are dangerous for the environment, space and natural resources.

Finally, there is the *Resolution on the National Programme of Protection against Natural and Other Disasters for the Period of 2009–2015*[7] (Resolucija o nacionalnem programu varstva pred naravnimi in drugimi nesrečami v letih 2009 do 2015 [ReNPVNDN] 2009). It defines natural and other disasters that present a threat to the Republic of Slovenia, the people, property, cultural inheritance, environment and other goods. The most important disasters influenced by climate changes include thunderstorms with hail and strong winds, floods and avalanches, drought, sleet and other disasters. The security and welfare of the inhabitants of the Republic of Slovenia and of the environment is threatened by environmental disasters and excessive exploitation and destruction of natural resources, health related epidemiological threats and other harmful occurrences. Such occurrences influencing the Republic of Slovenia from overseas are also not excluded. Meanwhile, according to the *Resolution*, terrorism, the use of means and weapons of mass destruction and repeated intensified military sources of threat in the region are less or hardly probable, respectively.

2.2.3.2 Perceived Threats

When it comes to issues regarding the security of the state and its inhabitants, no political decision can be legitimate if it does not (at least to some extent) consider public opinion. The Faculty of Social Sciences (University of Ljubljana) has been performing public opinion research in the field of security (Defence Research Centre). Table 2.1 presents threats to security as perceived by Slovenian residents. Threats/factors that are indirectly or directly connected to the environment are in bold. The results show that the Slovene public believes that some classical sources of threats (e.g. terrorism, military threats of other countries, extreme nationalism,

---

[6] In a sense that they will be discovered and prosecuted by the state agencies.

[7] The authors are aware that the *Resolution on National Environmental Action Programme 2005– 2012* (Resolucija o Nacionalnem programu varstva okolja 2005–2012 [ReNPVO] 2006) is the guiding declarative-political document when it comes to environmental protection but in a sense of sources of threats, therefore the security component, it is not comparable to other classical "security" strategic documents.

2 Slovenian Environmental Policy Analysis: From Institutional Declarations

Table 2.1 Factors endangering Slovenia's security ($N = 1000$)

Threats/Year	1994	1999	2001	2003	2005	2007
Traffic accidents[a]	–	3,21	3,24	3,16	3,12	3,34
Crime	3,14	3,46	3,28	3,28	3,20	3,20
Drugs, narcotics	2,95	3,45	3,41	3,28	3,21	3,17
Degradation of environment	**3,17**	**3,35**	**3,07**	**2,91**	**3,06**	**3,04**
Sell-out of social property	3,01	3,14	2,87	3,06	2,96	3,03
Poverty	–	3,13	3,05	3,08	3,05	2,99
Low birth rate	2,25	3,29	3,00	3,09	3,14	2,98
Unemployment[a]	–	3,35	3,14	3,26	3,24	2,97
Natural and technological disasters	**2,80/2,76**	**3,19**	**2,76**	**2,62**	**2,73**	**2,85**
Suicides	–	3,08	2,88	2,82	2,72	2,74
Economic problems	3,08	3,22	2,99	2,92	2,85	2,69
Refugees, illegal immigrants	2,68	2,98	2,74	2,59	2,49	2,52
Internal political instability	2,89	2,94	2,53	2,59	2,45	2,51
Lagging behind in the field of science and technology	2,66	2,83	2,33	2,47	2,55	2,41
Contagious diseases, AIDS, etc.[a]	–	2,77	2,43	2,21	2,28	2,22
Conflicts on the territory of former Yugoslavia	2,72	2,74	2,09	2,31	2,22	2,15
Extreme nationalism	2,48	2,53	2,20	2,14	2,15	2,07
Terrorism	2,45	2,64	2,09	1,87	1,90	1,91
Military threats of other countries	2,36	2,21	1,79	1,76	1,68	1,70

Note: Table shows average values on the scale 1–4, where 1 means "not at all a threat", and 4 means "a strong threat"

[a] Question was not posed (*Source*: Malešič and Vegič 2009: 103, 105)

conflicts on the territory of former Yugoslavia, etc. – all perceived as "weak threats") are in decline and that much stronger sources of threats that the respondents are dealing with on a daily basis and are presented in the media are traffic accidents, crime, illegal drugs, *degradation of the environment*, poverty, unemployment, and also *natural and technological disasters*. These threats are mostly considered as "medium threats", some even go as far as estimating them as "strong threats".

2.2.3.3 Real (Statistical) Threats

Despite the elite's beliefs about which are the highest threats or what the public believes to be true,[8] the most important sources of threats are those that have

---

[8] There is a saying too often true in life: "it doesn't matter what is true, what matters is what others believe to be true", and this is risky when it comes to security issues. In order to achieve a suitable response to a certain threat where the response will prevent unnecessary victims, the state may through its institutions on the other hand strengthen or mitigate and especially guide certain perceptions that people have. Burk's theory on the elite's influence on public opinion mentioned above is in this sense correct and in such cases also utterly legitimate.

Table 2.2 Environmental crime – police statistics

Year	1998	1999	2000	2001	2002	2003	2004	2005	2006	2007	2008	2009
No. of criminal offences	100	133	128	131	155	152	136	137	133	116	145	201

Sources: Resolution on the prevention and suppression of crime (RePZK 2006) and crimes – statistics through 2005–2009 (http://www.policija.si/index.php/statistika/kriminaliteta)

actually occurred, therefore actual threats recorded and statistically verified by agencies and services of the national security system. Police statistics may be used when providing a microscopic view of only those threats that illustrate *environmental crime* (officially, these are criminal offences against the environment, space and natural resources), therefore acts of individuals and companies that have intentionally or in negligence caused serious threats to the natural, and consequently to the living environment. Table 2.2 shows the movement of environmental crime activities as recorded by the Slovenian police, in accordance to the Penal Code[9] of the Republic of Slovenia.

At a glance, a relatively low number of recorded criminal activities against the environment and space (up to 100 in 1998 and up to 201 in 2009) may be noticed, which leads to the conclusion that there is probably a considerable grey field of environmental crime, especially since the environment does not have an "individual owner" that would report each criminal activity to the police. Therefore, including a certain number of pollutions and endangerment cases to the environment that do not contain all signs of criminal activity makes the rose-coloured image of environmental security significantly relativized.

2.3 Towards a Common Environmental and Security Policy?

Now that the national security interests and goals have been recorded and the declarative, perceived and real sources of threats, including environmental threats, described, there will be an attempt to identify the state's intentions, strategy-ies, policy-ies that should respond to the challenges of security threats and of achieving national security and environmental goals. Keeping this in mind, the third chapter will focus mainly on the following:

– The identification of sectoral policies (environmental, security).
– The identification and (partial) analysis of the basic strategic and directing documents that present the foundation and/or the results of a suitable sectoral

---

[9] The old Penal Code of the Republic of Slovenia was valid between January 1, 1995 and October 31, 2008 (Kazenski zakonik Republike Slovenije [KZ] 1994), and the new Penal Code has been valid since November 1, 2008 (Kazenski zakonik Republike Slovenije [KZ-1] 2008). Regarding Chap. 32 – Criminal offences against the environment, space and natural resources, the new Penal Code introduces two new criminal offences, further, one is renamed and another one deleted (actually, it is incorporated in another criminal offence).

policy in the field of environmental security, including prohibited/criminal acts ("environmental crime") and protection from natural and anthropogenic disasters.
- The nature of connection between the environmental and security policy and possibility/reasonability/necessity of forming of a common "environmental-security policy".

An example case of work realization within the framework of the environmental policy, manifested through the instrumental legal regulation of environment protection, will be presented in the fourth chapter.

2.3.1 (Fragments of) the National Security Policy of the Republic of Slovenia: Environmental Components

According to the *Resolution on the National Security Strategy of the Republic of Slovenia* (ReSNV-1 2010), the national security policy of the Republic of Slovenia is a "balanced integrity of visions, strategies, programs, plans and activities of the state, required to respond to sources of threats and risks to its national security and thus realizing its national security goals aimed at protecting Slovene's national interests". It consists of the foreign policy, the defence policy, the policy of ensuring internal security and the policy of protection against natural and other disasters. "In accordance to the modern multidimensional comprehension of the concept of national security, it also considers the security aspect of the state's policy in the economical, social, *environmental*, health, demographic, educational, scientific-technological, informational and other areas". This is an important declaration of the (latest) resolution because the *Resolution on the National Security Strategy of the Republic of Slovenia* issued in 2001 stated that two other public policies – the economic policy and the *policy of protecting and preserving the natural environment and space*[10] are part of the national security policy. In terms of this discussion, the absence of particularly the latter is significantly noticeable.

How can Slovenia then conceive a planned response to environmental sources of threats? In terms of *climate changes*, the Republic of Slovenia will join determined global and universal measures and will by itself "prepare and implement policies to

---

[10] From the *Policy of protecting and preserving the natural environment and space*: "In order to protect the natural environment as much as possible, the Republic of Slovenia strives for a technologically advanced economy, which will have the least possible negative effects on the environment. All-embracing, timely and coordinated activities of the state agencies, public administration and local self-government will prevent disasters or ensure their rapid and effective consequence management. The measures for environment protection and use of natural goods are aimed at handling the environment problems in the country, with the emphasis on giving the priority to major environment problems. The stress is laid on the enforcement of all environment protection principles defined by law and the inclusion of the environment issues and the principles of development in this area into the programmes of various ministries" (ReSNV 2001).

adjust to climate changes where it will restrict and reduce hazardous consequences and ensure risk management in variable climate conditions. These policies will be based on UN and EU directions and activities and will take into account specific national conditions of the Republic of Slovenia. Within the mentioned organisations, the Republic of Slovenia will strive to conclude an international agreement to mitigate the consequences and accommodate to the climate changes, with a goal to stop the average temperature on Earth from rising. In connection to the development of the high growing technological economy it will strive to attain international standards to reduce hazardous emission in the environment. The Republic of Slovenia will in particular strive to efficiently adapt the whole society to climate changes, with emphasis on interdisciplinary and intersectoral coordinated measures. In the short term, the state will provide priority attention to upgrading early warning systems for extreme weather conditions." (ReSNV-1 2010).

Regarding the response to *limited natural resources and the degradation of the living environment*, the Republic of Slovenia will consider goals and standards, valid in both Slovenia and EU. Response will be aimed primarily at reducing greenhouse gas emissions, preserving biodiversity, reducing soil pollution, effective water management, improving air quality, reducing the quantity of waste and promoting its recycling and reducing both noise and electromagnetic emissions. In concern for social development, all legitimately declared principles of environment security will be considered, integration of environmental issues and sustainable development principles into programs of other departments will be ensured and the awareness of citizens on the importance of environmental security will be strengthened.

The Republic of Slovenia will strive for the conservation, protection and expansion of existing agricultural land and other resources for food and animal feed production, and it will focus on adapting agriculture to the consequences of climate change. The Republic of Slovenia will provide sufficient supply of key energy resources with proper diversification of supplies. Special attention will also be devoted to greater use of local renewable energy resources and the transition to alternative ones. It will also support peaceful uses of nuclear energy, based on the highest standards of technical and physical security (ReSNV-1 2010).

In managing natural and other disasters, the Republic of Slovenia will strive for a more conformable, efficient and rational development of a disaster management system. More attention will be given to preventive activities, such as earthquake-safe construction, rational land intervention, measures to reduce the effects of hail and drought and fire prevention measures in the most deprived areas in the west of the country. Improved preventive activities will be related to flood protection and protection against avalanches. The Republic of Slovenia will strive to further develop its own capacities to respond to natural and other disasters, taking into account the multiplicative effect of individual sources of security threats and risks. Emphasis will be given on upgrading infrastructure systems, particularly information-communication systems and improving conditions for the functioning of services, units, associations and other non-governmental organisations. The development of forces for protection, rescue and assistance will be based on a modular structure and be able to adapt in specific circumstances. The Republic of Slovenia will continue to closely cooperate with neighbouring countries, especially in border areas, but will seek to strengthen the civil protection mechanism of the European Union and to jointly provide assistance to EU Member States and other countries. On the basis of international treaties, the Republic of Slovenia will be active in international organizations, mainly in the form of mutual information exchange about dangers and consequences of natural and other disasters, as well as mutual aid in emergencies (ReSNV-1 2010).

The *Resolution on the National Programme of Prevention and Suppression of Crime for the Period of 2007–2011* (ReNPPZK0711 2007) has already been mentioned in the chapter on declarative sources of threats. It's an interesting document because besides stating the causes for conditions in the field of environmental security, it also offers/predicts corresponding solutions. In this sense, there should be an establishment of "planned, coordinated and consistent acting and functioning of authoritative institutions, sufficient information flow between inspecting and persecuting bodies and an elaboration plan for preventing and detecting incriminating acts endangering the environment, space and natural sources". There should be one single goal to all this – "to improve the efficiency of state bodies and institutions serving the field of environmental security" (e.g. Ministry of the Interior, the Police, Ministry of Defence, Ministry of the Environment and Spatial Planning, Ministry of Justice, autonomous local communities and research institutions).

Of course, these are mere supplements of the security policy, which can be realized only after the adoption and implementation of appropriate legislation – the most important will be mentioned in section 1.3.3.

2.3.2 (Fragments of) the Environmental Policy of the Republic of Slovenia

Although every single individual must (as a "citizen of the world", if somewhat pathetically expressed) take responsibility of the security and preservation of the environment, the case with sources of threats (which do not recognize state boundaries) is that the most efficient and best way to deal with them is on a group level, a state level in this matter. Therefore, states establish specialized agencies and services that (when provided by law) use repressive approaches to sanction those whose actions intentionally or unintentionally endanger the environment, natural sources and consequently, the living environment of humans and animals. States may protect themselves from sources of threats by either abolishing and preventing them or by reducing their own vulnerability. This is exactly why a "rational plan of action" as mentioned in the introduction is required. Environmental threats cannot be fought only in a repressive-curative manner, which is still typical for classical state security policies, but both, a specific and integral sectoral policy – *an environmental policy*, is required.

Environmental policy may be understood as "a set of principles and intentions, used to guide the decision making process on human management of environmental capital and environmental services" (Roberts 2004: 2), whereas in accordance to this article, the main interest is in managing concealed threats produced (often assisted by a person) by the natural environment that the social environment may only anticipate, plan, be prepared for or fight against. The *Resolution on National Environmental Action Programme 2005–2012* (ReNPVO 2006), the basic strategic document[11] in the field of environmental protection "where the goal is a general improvement of the environment and the quality of life and the protection of natural sources", is certainly in context with Roberts's definition.

Basic principles and strategic guidelines of the National Action Programme are related to drafting and adopting new legislation and consistent implementation of the existing one; sustainable use of natural resources; inclusion of environmental protection requirements in spatial development planning; integration and consideration of environmental contents in department policies; environmental technologies; promotion of sustainable production and consumption; economic policy for environmental protection; raising environmental awareness, maintaining dialogue with interested parties and public participation; rehabilitation of degraded areas (ReNPVO 2006).

To make this possible, the third chapter of the National Action Programme provides an assessment of the environmental situation in Slovenia in terms of climate changes, nature and biodiversity, quality of life and waste management. Chapter four is the key chapter, it includes objectives and programmes of measures for environmental protection. It actually touches all aspects of the environment and the life in it, including climate changes, preserving biodiversity, monitoring the conditions of forest ecosystems, soil, genetically modified organisms, quality of life, the use, protection and management of water, air and chemicals, noise, electromagnetic radiation, urban development, waste and industrial pollution (ReNPVO 2006).

Who is responsible for implementing the national programme of environmental security? The *Resolution* defines the main actors, namely the state administration, local authorities, public environmental protection services, non-governmental organisations and business and economic entities (Chamber of Commerce and Industry of Slovenia – Environmental Protection Department; Chamber of Crafts of Slovenia). In addition, the state is logically embedded in the international efforts to protect and preserve the environment and the *Resolution* in particular emphasises

---

[11] The National Assembly of the Republic of Slovenia adopted this document according to Article 35 of the *Environment Protection Act* (Zakon o varstvu okolja [ZVO-1] 2004), Article 94 of the *Nature Conservation Act* (Zakon o ohranjanju narave [ZON] 2004), Article 54 of the *Waters Act* (Zakon o vodah [ZV-1] 2002) and in accordance to *Decision No 1600/2002/EC of the European Parliament and of the Council of 22 July 2002 laying down the Sixth Community Environment Action Programme*). It's interesting because the strategic document is being formulated on the basis of Slovenian's partial, sector legislation and not the other way around! Such practice was realised even in the first *National Environment Protection Action Programme* issued in 1999 (Nacionalni program varstva okolja [NPVO] 1999).

educating, tutoring and providing public awareness as well as supporting public participation in the environmental policy making process (ReNPVO 2006).

At this point, two basic laws on environmental security that co-mark the Slovene environmental policy will not specifically be dealt with, these are the *Environment Protection Act* (ZVO-1 2004) (hereinafter referred to as the ZVO-1) and the *Nature Conservation Act* (ZON 2004). Among other reasons, these acts are also important because they impose certain powers and duties in the field of the security of the environment and nature to individual bodies and services in Slovenia. Together, they shape something that could be called "an environmental security system", which will be discussed in the next subsection.

2.3.3 A Common Environmental and Security Policy Provides a Common "Environmental Security System"?

Each policy is of course not an end in itself, nor can it be all that successful if it does not establish a proper security system that will possess the appropriate mechanisms and instruments to respond to/prevent threats. If this discussion is transmitted to the field of environment and environment security or even an "environmental security system", then the political-legal framework of a joint environmental-security policy must first be analysed in order to further identify the key bodies and services engaged in environmental protection in Slovenia.

A few strategic documents concerning the field of national security have already been mentioned so far, crime prevention, protection against natural and other disasters and environmental security. Combining these documents, a rough image of the following is provided:

- Resolution on the National Security Strategy of the Republic of Slovenia (ReSNV 2001)[12]
- Resolution on the National Security Strategy of the Republic of Slovenia (ReSNV-1 2010)
- Resolution on the National Environmental Action Programme 2005–2012 (ReNPVO 2006)
- Resolution on the Prevention and Suppression of Crime (RePZK 2006)
- Resolution on the National Programme of Prevention and Suppression of Crime for the Period of 2007–2011 (2ReNPPZK0711 007)
- Resolution on the Programme for Protection against Natural and Other Disasters for the Period of 2009–2015 (ReNPVNDN 2009)

---

[12] Although formally no longer valid, the Resolution on the National Security Strategy of the Republic of Slovenia issued in 2001 remains significant since systemic changes are possible only over a longer time period.

The legislation that reflects the "political environmental output" in principle originates from strategic documents. The following are among the most important legal documents:

- Constitution of the Republic of Slovenia (Ustava Republike Slovenije [URS] 1999)
- Environment Protection Act (ZVO-1 2004)
- Nature Conservation Act (ZON 2004)
- Waters Act (ZV-1 2002)
- Management of Genetically Modified Organisms Act (Zakon o ravnanju z gensko spremenjenimi organizmi [ZRGSO] 2002)
- Act on Protection against Ionising Radiation and Nuclear Safety (Zakon o varstvu pred ionizirajočimi sevanji in jedrski varnosti [ZVISJV] 2002)
- Act on the Protection against Natural and Other Disasters (Zakon o varstvu pred naravnimi in drugimi nesrečami [ZVNDN] 2006)
- Fire Protection Act (Zakon o varstvu pred požarom [ZVPoz] 2007)
- Fire Service Act (Zakon o gasilstvu [ZGas] 2005)
- Penal Code (KZ-1 2008)
- Minor Offences Act (Zakon o prekrških [ZP-1-UPB4] 2007)
- Police Act (Zakon o policiji [ZPol-UPB6] 2006)
- Local Police Act (Zakon o občinskem redarstvu [ZORed] 2006)
- Private Security Act (Zakon o zasebnem varovanju [ZZasV] 2003)
- Decree on Obligatory Setting-up of Security Service (Uredba o obveznem organiziranju službe varovanja 2008)
- Defence Act (Zakon o obrambi [ZObr] 1994)
- Service in the Slovenian Armed Forces Act (Zakon o službi v Slovenski vojski [ZSSloV] 2007)

When it comes to protecting the environment and punishing illegal interventions and acts in its regard, the substantive analysis of the mentioned documents exposes a whole variety of terminology concepts. This brings up a logical question on consistency and effectiveness of such (e.g. environmental, security) policy, which is even not consistent in the terminology?

However, on the other hand, political and legal acts indirectly or directly define the subjects within the environmental security system. Here, they can also be divided into either political-decision making ones or executing-operational ones. The most important bodies in the field of environmental security policy are: National Assembly of the RS, Government of the RS, Ministry of Environment and Spatial Planning, Environment Protection Council of the RS, National Security Council, Environmental Agency of the RS, Government Office of the Republic of Slovenia of Climate Changes, Slovenian Nuclear Safety Administration etc.

But, which are the bodies and services who by law deal with the prevention, detection and prosecution of environmental crime and illegal acts and who deals with the damage caused by such acts? In fact, they are found in all three subsystems of the Slovenian national security – the internal security system, the defence system and the disaster management system.

2 Slovenian Environmental Policy Analysis: From Institutional Declarations

The internal security system is the largest and the most diversified. Practically, there is no entity within this system that would not also have certain duties in the field of environmental security. The most important departments are the Police, the State Prosecutor's Office, various inspection services, municipal warden services, private security services etc.

The defence system consists of the Slovenian Armed Forces and civil defence, while the disaster management system or the system of protection against natural and other disasters includes a large number of national, public and non-governmental bodies, services, organizations and associations:

- *Civil protection* (mainly the Administration of the Republic of Slovenia for Civil Protection and Disaster Relief, units and staffs of Civil Protection).
- *Units and services of associations and other nongovernmental organisations* (voluntary firefighter's associations, Slovenian mountain rescue service, speleological rescue service, underwater rescue service, search and rescue dog units, Scouts Association Slovenia and the Association of Slovenian Catholic Scouts, Red Cross Organisation, Association of Radio Amateurs of Slovenia).
- *Companies, public institutions and other organizations* (professional fire fighting brigades, emergency medical services (medical institutions as part of the public health service); the unit for identifying the deceased (Faculty of Medicine, Institute of Forensic Medicine); the unit for hygienic-epidemiological services (Institute of Public Health); the mobile unit for meteorology and hydrology (Ministry of environment and spatial planning, the Environmental Agency of the Republic of Slovenia); the ecological laboratory with a mobile unit – ELME (the Jožef Štefan Institute and MEEL, organized by the Public Health Association of Maribor); the unit for search and rescue in accidents with chlorine and other corrosive substances (TKI Hrastnik); the rescuing unit in mine disasters (Velenje mine); the Mobile Information Centre, the service for search and rescue in ecological and other disasters at sea).
- *The Police* (especially in providing security, public order and peace and participating in (mountain, traffic) rescue operations by helicopter and other forces).
- *The Slovenian Armed Forces* (in accordance with the law, their structure, fitting and training; particularly its air units, nuclear, chemical and biological defence units, engineering units, medical services and other units).

And finally, no preventive measure of the state and its institutions can be successful in the long term if it is not acceptable to the people, to the civil society. Therefore, raising the security culture and environmental awareness is a prerequisite for successful coping with environmental threats. When it comes to intentional or unintentional nuisance/pollution/threat to the environment, the key in this context is developing a self-restricting attitude in individuals and in the society. The strategic decision of the state, originating in the *Resolution on the National Environmental Action Programme 2005–2012* (ReNPVO 2006) and according to which the public or non-governmental organizations will appropriately cooperate in the environmental policy-making process, therefore seems even more important.

2.4 Instrumental Legal Regulation

2.4.1 Introductory Aspects of the Instrumental Legal Regulation of Environment Protection

The instrumental-legal regime of environmental protection under the law of the Republic of Slovenia will be presented below. First of all, a general European and Slovenian constitutional legal platform in environment protection will be defined, followed by an introduction of the fundamental, valid administrative and criminal law regulations in the Republic of Slovenia.

The substantive foundation of contemporary transnational and national regulation of environmental protection in administrative law in democratic societies in the last decades is no longer an anthropocentric perception of nature, i.e. that nature is exclusively subordinate to humans (Greek: *anthropeioi nomos*), but rather a new, ecocentric perception of nature (Greek: *nomos theios*). According to ecocentrism, the centre of legal protection is nature and not the exploitation of nature by humans. In light of the fact that anthropocentrism (i.e. in the sense of exploitation by humans) as a basis for human interaction with nature has led to the damaging and in some places even to the complete destruction of nature, there is no doubt that the ecological reasoning which lies within the anthropocentric perception of nature cannot lead to its protection. Legal protection of nature can only be achieved by establishing an ecocentric foundation in positive law regulation. This direction must also be followed by contemporary national and transnational legislation – Greek *nomos* (Pličanič 1999: 16–30; Kirn 1992: 5–22).

The ecocentric foundation of the regulation of nature has been followed in the last two decades by cogent (Latin: *ius cogens* – compelling law) regulation of European and Slovenian administrative law through the system of positive law (Latin: *ius positivum* – applicable law) institutional and instrumental regulations. The term *institutional regulations* primarily refers to the highest legal acts which establish a value framework for the legal system (Parsons 1978). These are especially the primary European legislation (i.e. the Treaty on European Union – hereinafter referred to as the TEU) and the Constitution of the Republic of Slovenia (hereinafter referred to as the Constitution). The term *instrumental regulations*, on the other hand, refer to national laws and regulations, which implement the value-based aims established at the institutional level in positive law.

The national regulation of environmental protection in administrative law in the Republic of Slovenia is a fairly new legal discipline that has existed and been developing only for the last two decades within the framework of graduate and postgraduate studies at Slovenian universities. Conceptually, this entails studying environmental law, which at the transnational and national levels enacts moderation in activities affecting nature. From the perspective of the primary law of the EU, environmental law is based on the four fundamental principles of environmental protection contained in the TEU. These are the following (the second paragraph Article 174 of the TEU):

2 Slovenian Environmental Policy Analysis: From Institutional Declarations

1. The precautionary principle (*le principe de précaution, der Grundsatz der Vorsorge*)
2. The principle of preventive action (*le principe d'action préventive, der Grundsatz der Vorbeugung*)
3. The principle of correction at source (*le principe de correction à la source, der Grundsatz, Umweltbeeinträchtigungen mit Vorrang an ihrem Ursprung zu bekämpfen*), and
4. The polluter pays principle (*le principe de pollueur-payeur, das Verursacherprinzip*)

With reference to such, it can be established that all the above-mentioned principles are also in conformity with the Slovenian Constitution, which in Section III, regulating economic and social relations, ensures everyone the protection of a healthy living environment (Article 72 of the Constitution) and the protection of the natural heritage (Article 73 of the Constitution). In this connection, the Constitution also addresses the concepts of national asset and natural resource (Article 70 of the Constitution). With reference to all the above-mentioned constitutionally protected values, the Constitution instructs the legislature to adopt the appropriate national legislation there on. In addition, Article 3a of the Constitution, which was added in 2003 as the basis, and one of the requirements, for Slovenia to join the EU, cannot be neglected. On the basis of this provision the Republic of Slovenia recognizes the supremacy of the primary EU legislation over national legislation in the creation of the national legal order. The above-mentioned principles and constitutionally protected rights are implemented in the Slovenian legal order through national regulations, first of all through the *Environmental Protection Act* (ZVO-1 2004).

However, the ZVO-1 is not the only Slovenian regulation that regulates the environment. There is a whole range, depending on the special regulation which each individual aspect of the environment requires. Slovenian environmental law can, in the broader sense, be divided into five parts related in terms of content (Scripta UL PF No. 35 2008, pp. 166–168), namely: (1) the legal regulation of natural assets and the conservation of biotic diversity, (2) the legal regulation of water management, (3) the legal regulation of mineral matters management (non-renewable natural resources), (4) the legal regulation of wild animal management (hunting and fishing), and (5) the legal regulation of forests and barren land management.

This chapter does not deal with specific and special regulations. It can be mentioned, however, that for a profound understanding of the structure of the national legal order in the field of environmental law, the rule *lex specialis derogat lex generalis* must be applied. This entails that in the event of a collision of statutory rights and obligations, when two laws regulate the same legal relation, the regulation of the law which is more specialised – in the sense of Latin *lex specialis* – prevails. The ZVO-1 is a general regulation. In individual specific environmental law fields, the regulation determined in special regulations will thus prevail, in as much as they regulate specific environmental relations in a specific and special manner.

The central term of modern environmental law is the subjective law of persons subject to environmental law – their *legal rights and duties*. They are based on the general constitutional right of individuals to a healthy living environment. This is a general right which is followed by special rights and duties regulated by special laws in accordance with the principle of speciality. The right is in general a legally protected entitlement (Latin: *facultas agendi*) that a legal subject can act in a certain manner. The right is composed of two entitlements, namely of a fundamental entitlement and a claim. A fundamental entitlement enables subjects to fulfil their own interests if in compliance with the legal aim of the entitlement. A claim, on the other hand, contains the possibility that the state will impose a coercive sanction in the interest of the subject if another subject does not act in compliance with the obligation. The right thus contains, on one hand, an entitlement of one subject, and, on the other hand, the duty or obligation of another subject. Therefore, from a broader perspective (e.g. in interest-volition theory), a legal right is both – a right and a duty entitlement of the legal subject (Leksikon Cankarjeve Založbe: Pravo 2003: 264).

In what manner administrative law rights and the duties of subjects to environmental law are implemented in accordance with the Slovenian ZVO-1 will be discussed below. At the next point, focus will be drawn to the question of environmental protection consent as an individual administrative act and through this, the regulation of the supervision of state authorities regarding violations of administrative law statutory provisions from the field of environmental protection.

2.4.2 Environmental Protection Consent in Administrative Law in Slovenian Legislation

The environmental protection consent is a central concrete and individual administrative act that every entity responsible for a planned activity must obtain in accordance with Slovenian legislation (Article 57 of the ZVO-1). In order to operate an installation or plant where any activity that might cause large-scale environmental pollution will be carried out, the operator must obtain an environmental protection consent. However, the ZVO-1 introduces various types of environmental protection permits which differ in the scope of the requirements that must be fulfilled in order to be obtained, in the time limits for issuing the permit, and in the participation of the public; all types of permits are, however, issued by the Ministry of the Environment and Spatial Planning or its affiliated body, i.e. the Environmental Agency of the Republic of Slovenia (hereinafter referred to as the EARS). An environmental protection permit must be obtained prior to the beginning of the operation of the installation or plant and for every substantial change in the operation of an installation or plant. A substantial change in the operation is: (1) any change in the installation or its extension that modifies any principal technical characteristic of the installation or its capacity, which as a consequence causes a change in the quantity or type of emissions into the environment or the waste for which the environmental protection permit was obtained, (2) a considerable increase in the quantity of a hazardous substance, (3) a significant change in the chemical or physical properties of a hazardous substance, or (4) any other change in the technological

2 Slovenian Environmental Policy Analysis: From Institutional Declarations

process in a plant where a hazardous substance is used. An environmental protection permit must be obtained in three instances (Viler-Kovačič 2006: 6–7), namely:

1. For installations that may cause *large-scale* pollution, i.e. installations subject to the IPPC Directive. The abbreviation derives from the title of the EU directive, namely Council Directive 96/61/EC Concerning Integrated Pollution Prevention and Control or abbreviated as the IPPC Directive. This directive regulates IPPC installations.
2. For other installations that do not cause large-scale pollution (installations determined in Article 82 of the ZVO-1):

   (a) If an activity is pursued in these installations that causes emissions into the air, water, or soil for which limit values are prescribed.
   (b) If an activity is pursued in these installations for which the obligation to obtain an environmental protection permit is prescribed in other regulations.
   (c) If in these installations waste is treated or disposed of.

3. For installations that pose a major environmental risk.

Pursuant to the ZVO-1, an installation is any stationary or mobile technical unit for which it has been established that it may cause environmental burden as a consequence of one or more specified technological processes taking place in that installation, and of any other technology-related processes taking place at the same site (item 8 of Article 3 of the ZVO-1). In accordance with the same Act, a plant is defined as the entire area managed by one operator where there are one or more installations, including the accompanying or associated infrastructure and technological processes related thereto, in which hazardous substances are produced, stored, or used in any other way (item 9 of Article 3 of the ZVO-1).

In what manner European legislation is applied at the national level will be demonstrated through the case study of the application of the Seveso Directive into the Slovenian ZVO-1 (Viler-Kovačič 2006: 6–7):

CASE STUDY: A particularity of the environmental protection permit in the Slovenian legal system is also the transposition of the Seveso European Directive for plants into the ZVO-1 (see Article 86 of the ZVO-1). The Council of the European Union adopted the Seveso Directive (i.e. Council Directive 96/82/EC of 9 December 1996 on the control of major-accident hazards involving dangerous substances) following a major accident in Seveso in Italy in 1976, when several kilograms of dangerous dioxin was suddenly and uncontrollably released from the chemical plant and contaminated more than 15,000 m^2 of land and vegetation. As many as 2000 people were treated for dioxin poisoning and more than 600 had to be evacuated from their homes (EC 2010).

The above-mentioned directive was transposed into the ZVO-1 through the requirement that when operating a plant, the person posing a risk to the environment must implement the prescribed measures for the prevention of any major accident and for the mitigation of consequences there from for humans and the environment, and in particular draw up a scheme for the reduction of environmental risk and a safety report. A person posing the risk to the environment must obtain an environmental protection permit for the plant, which is named an environmental protection permit issued in accordance with Article 86 of the ZVO-1 or a Seveso permit, in order to distinguish it from other environmental protection permits.

Operators of such plants are termed persons posing a risk to the environment, whereas a risk to the environment in accordance with the ZVO-1 entails the probability that in certain circumstances or within a certain time interval an activity would harm, directly or indirectly, the environment, or human life or health. The measures that the plant operators must undertake for the prevention of major accidents and the mitigation of their consequences for humans and the environment as well as safety measures connected therewith which plant operators must meet, are regulated by the Government Decree on the Prevention of Major Accidents and the Mitigation of their Consequences (Uredba o preprečevanju večjih nesreč in zmanjševanju njihovih posledic 2005). Pursuant to this Decree, plants are divided into plants posing a lower risk to the environment and plants posing a greater risk to the environment, regardless of the type and quantity of hazardous substances that are in the plant. A formula for the assessment of a plant posing a lower risk to the environment and posing a greater risk to the environment is provided for in the annex to the Decree.

In the procedure for issuing the environmental protection permit, the EARS must make the permit application and the draft decision on the environmental protection permit available to the public. The provisions that apply to entities subject to the IPPC Directive apply, mutatis mutandis, regarding the participation of the public. The time limit for issuing a permit is three months from receiving a complete application.

The environmental protection permit stipulates in detail measures for the prevention of major accidents and the mitigation of their consequences for humans and the environment; the environmental protection permit is issued for a period of five years from the day of the beginning of the operation of the plant (and not from the day this administrative act takes effect). In the event that it is assessed that the consequences of a major accident in a plant could have an impact on the environment of another state, or when the latter so requests, the competent authority of that state is to be informed of the environmental protection permit issued. If on the basis of such information another state so requests, it must be forwarded the safety report of the plant in question.

Such environmental protection permits may also be extended, updated, or withdrawn under the conditions and in the manner determined by the ZVO-1 (2004).

2.4.3 Administrative Supervision of Environment Protection in Slovenian Legislation and Practice

Administrative, particularly inspection, supervision is one of the administrative functions of the state that ensures supervision of the implementation of the adopted legal order and provides feedback so that administrative bodies can learn of the effects of the adopted regulations and introduce appropriate modifications and measures. The coercive nature of inspection tasks ensures that addressees of legal norms respect regulations, whereas at the same time these norms protect the rights and legal benefits of individuals that were recognized to them by laws and other regulations. Inspection supervision is thus supervision of the implementation of laws and other regulations. The tasks of inspection are performed by inspectors as officials with special authorisations and responsibilities (Likozar-Rogelj 2006: 16–17).

If an environmental inspector establishes that a person liable for the operation of a plant or installation does not have the necessary environmental protection permit, he or she does not instruct the person liable to obtain such a permit, as the author of the above cited article erroneously thinks, but prohibits the operation of the installation or plant if it is operating without an environmental protection permit (see item 6 of the first paragraph of Article 157 of the ZVO-1). As the legislature determined that the operation of a plant or installation without obtaining a prior environmental protection permit is prohibited, it provided for the imposition of sanctions; a legal entity may impose a fine in the amount of €1,200 to €360,000. Due to the fact that the amount of the prescribed sanctions that can be imposed is high, accelerated minor offence proceedings are not allowed (Likozar-Rogelj 2006: 16–17).

It proceeds from the above-mentioned that inspectors must respect the principle of legality when performing their tasks. They supervise whether the adopted legal regulations are respected and take appropriate measures in cases of violations of the regulations. The repressive nature of inspection tasks is demonstrated solely through measures which inspectors are obliged to impose *(ex-offo)* if they establish that the legal order is not respected.

In 2006 the Inspection Act (Zakon o inšpekcijskem nadzoru [ZIN] 2007) (hereinafter referred to as the ZIN) introduced an amendment that authorises inspectors to apply preventive measures (Article 33 of the ZIN) which concretise the principle of the protection of the public and the private interests in order to ensure that the regulations are respected. Inspectors can only issue a warning if they find irregularities when performing their inspection tasks and assess that such is an appropriate measure regarding the seriousness of the violation. They also determine a time limit in which the irregularities must be remedied. If the irregularities are not remedied, inspectors impose other measures in accordance with the law. The above-mentioned discretionary right of inspectors importantly influences the preventive action; in light of the seriousness of the violation, inspectors assess that a warning will suffice and that the person subject to inspection will respect the norms of the regulation and perform the omitted prescribed obligation or remedy the violation which they have caused with their operation. If, however, inspectors establish that after a warning has been issued, a person subject to inspection does not respect the imposed measure, they are to impose other, graver measures prescribed by law. A person subject to inspection is obliged to remedy the irregularities or deficiencies within the determined time limit. If they do not do so themselves, the inspectors force them to remedy such proceedings with the execution carried out by a third person or by a fine. Inspectors thus do not only impose measures, but also ensure that such measures are implemented (Likozar-Rogelj 2006: 17).

2.4.4 Environment Protection in Slovenian Criminal Law

Slovenian criminal law is Slovenia's national law and is specific for the Republic of Slovenia. The primary function of the Slovenian criminal law is the protection of the Slovenian society from the most serious forms of deviant behaviour in all spheres of social life. The latter is realized through substantive and procedural criminal law through the specific and general precautionary principle. This is how the Slovenian criminal law contributes to the strengthening and development of the social consciousness, providing social harmony in the Republic of Slovenia as a whole and as part of the European Union and an integral part of the global human community.

In the most successful and acceptable manner, the Slovenian criminal law reinforces the social consciousness of the moral unacceptability and danger of those social actions that apply to *lat: mala in se* (evil in itself). Socially less acceptable are offenses that are presented in the national criminal law as *lat: mala prohibita* (positive-legal prohibited acts). The moral and ethical significance for these offenses has not yet solidified in the social consciousness, and such acts are not yet evil in themselves (Dežman 1993: 22).

Regarding the phenomenon of *ecocentrism* – in contrast to the previous principle of *anthropocentrism* – in law and the actual crisis of the environment, it may be concluded that the illusions of nature's inexhaustibility and indestructibility are lost. Imbalance between man and his natural environment raises the categorical request of respecting ecological rules. This means that threatening or destroying natural assets present proceedings that completely reveal their negative moral and ethical foundation.

Therefore, a criminal justice intervention in the area of limiting ecological crisis has all its moral and ethical justification. Acts against natural resources today are no longer just *lat: mala prohibita*, but increasingly subject to *lat: mala per se*. In the Penal Code (KZ-1 2008), the Slovenian criminal law has developed a precise and clear legal description of criminal acts which ensure an adequate interpretive scheme for judicial review.

The offenses against the environment are classified in the criminal code predominantly under *Criminal Offences against Environment, Space and Natural Resources* in Chap. 32. Given the positive legal system, the act of crimes against the environment in the Slovenian criminal law may be classified in the manner shown in Table 2.3.

A quick overview of the current criminal justice protection of the environment in the Republic of Slovenia suggests that criminal law may also be a means of the society to overcome environmental difficulties. Its application requires such a preventive system that it can distinguish between environmental-harmful events and coincidences. It must be taken into account that most of the acts at debate are committed in difficult circumstances; therefore, it is reasonable to raise the question of the possibility of a precise criminal justice evaluation. It is also in the domain of competent national authorities for them to, in joint efforts, ensure higher effective security of the environment in the Republic of Slovenia. Legally, they have sufficient instruments of repression available.

2 Slovenian Environmental Policy Analysis: From Institutional Declarations   35

Table 2.3 Criminal offences against environment, space and natural resources in penal code of the Republic of Slovenia

Article No.	Criminal offence	Punishment
332	Burdening and destruction of environment and space	…. shall be punished by imprisonment of up to five years
333	Sea and water pollution by ships	…. shall be punished by imprisonment of up to five years
334	Import and export of dangerous substances into the country	…. shall be punished by imprisonment of up to five years
335	Unlawful acquisition or use of radioactive or other dangerous substances	…. shall be punished by imprisonment of up to five years
336	Pollution of drinking water	…. shall be punished by imprisonment of up to three years
337	Tainting of foodstuffs or fodder	…. shall be punished by imprisonment of up to three years
338	Unlawful occupation of real property	…. shall be punished by a fine or by imprisonment of up to one year
339	Destruction of plantations by a noxious agent	…. shall be punished by a fine or by imprisonment of up to two years
340	Destroying of forests	…. shall be punished by imprisonment of up to one year
341	Torture of animals	…. shall be punished by a fine or by imprisonment of up to six months
342	Game poaching	…. shall be punished by a fine or by imprisonment of up to six months
343	Fish poaching	…. shall be punished by a fine or by imprisonment of up to one year
344	Illegal handling of protected animals and plants	…. shall be punished by imprisonment of up to five years
345	Transmission of contagious diseases in animals or plants	…. shall be punished by a fine or by imprisonment of up to one year
346	Production of injurious medicines for treatment of animals	…. shall be punished by a fine or by imprisonment of up to one year
347	Unconscientious veterinary aid	…. shall be punished by a fine or by imprisonment of up to one year

2.5 Conclusions

In the last decade, many sectoral (line) and cross (interdepartmental) approaches and policies have been formed (e.g. the policy of protecting and conserving the natural environment and space; the policy of protection against natural and other disasters; the national environmental action programme, etc.) and they all deal with the question of environmental protection (and thus society) in the widest sense. Nevertheless, it is still difficult to identify a single, comprehensive and consistent security policy that would address how a wide variety of sources and types of threats to the environment will be tackled. In this context, the strategic and operational documents describe environmental crime as a specific (intentional) source of threat that receives relatively little attention, although many environmental threats (e.g. anthropogenic disasters) may be a result/outcome of intentional human acts or omissions. In this context, it is interesting (and concerning) that the Resolution on National Security of the Republic of Slovenia adopted in 2010 no longer classifies the *Policy of protecting and preserving the natural environment and space* as part of the national security policy framework, as it did since 2001. If nothing else, it's a symbolic significance, since strategic documents that deal with environmental security dedicate a great deal of attention to environmental threats (e.g. to those that are derived from climate changes) but while facing these and similar threats they don't clearly define a sectoral policy or a kind of an "environmental security policy"!

On the other hand, instrumental legal regulation is divided into two major areas: pollution control and remediation; and resource conservation and management. Instrumental legal regulation is often media-limited – i.e., it pertains only to a single environmental medium, such as air, water, soil, etc. – and it controls both emissions of pollutants into the medium, as well as liability for exceeding permitted emissions and responsibility for clean-up.

Finally, here are a few more or less rhetorical questions which may also require scientific verification, namely:

- Can climate changes, natural disasters, environmental threats and risks truly even be a matter of nations, regions, and the world if they, in their final consequence, mainly concern the individual?
- What are (on the other hand) atomized individuals of the society able to do on their own in preventative and curative purposes without activity coordination on a group – i.e. state level?
- Are existing security systems and concepts even capable of transformation which will allow the institutionalization of the concept of human security?
- Should the matter of a national security policy in particular stand for penalizing violators of environmental laws (as a "general prevention") and for sanitizing natural and other, even ecological disasters etc. or should it also, and above all, be a preventive activity in terms of promoting individual and corporative "environmental awareness" and providing successful environmental security practices?
- And last but not least, as a country, will Slovenia only "respond" to (threats and risks, policies, requirements and expectations of the EU and UN, etc.) or will its policies and strategies be proactive, initiative and anticipative?

References

Council Directive 96/82/EC of 9 December 1996 on the control of major-accident hazards involving dangerous substances (1996) Official Journal L 010, 14/01/1997, 13–33. http://eur-lex.europa.eu/LexUriServ/LexUriServ.do?uri=CELEX:31996L0082: EN:HTML. Accessed 25 June 2010

Dežman Z (1993) Kazenskopravno varstvo človekovega naravnega okolja (Eng. Protection of Human Natural Environment in Criminal Law). Pravna praksa 1993(12):22–24

2 Slovenian Environmental Policy Analysis: From Institutional Declarations 37

Dunn WN (1994) Public policy analysis: an introduction, 2nd edn. Prentice Hall, Englewood Cliffs

EC (2010) Environment: Seveso. Resource document. http://europa.eu.int/comm/environment/seveso/#1. Accessed 15 May 2010

Grizold A (1992) Oblikovanje slovenske nacionalne varnosti (Eng. Creation of the Slovenian national security). In: Grizold A (ed) Razpotja nacionalne varnosti: obramboslovne raziskave v Sloveniji (Eng. Crossroads of national security: the defence research in Slovenia). Faculty of Social Sciences, Ljubljana, pp 59–93

Grizold A (1999) Obrambni sistem Republike Slovenije (Eng. Defence system of the Republic of Slovenia). College of Police and Security Studies, Ljubljana

Grošelj K (2007) Okoljsko ogrožanje nacionalne varnosti (Eng. Environmental threats to national security). In: Prezelj I (ed) Model celovitega ocenjevanja ogrožanja nacionalne varnosti Republike Slovenije (Eng. Comprehensive assessment model of the endangering national security of the Republic of Slovenia). Ministry of Defence, Ljubljana, pp 147–165

Hill M (1997) The policy process in the modern state, 3rd edn. Prentice Hall, Harvester Wheatsheaf, London, New York, Toronto, Sydney, Tokyo, Singapore, Madrid, Mexico City, Münich, Paris

Hogwood BW, Gunn LA (1984) Policy analysis for the real world. Oxford University Press, Oxford

Homer-Dixon TF (1994) Environmental scarcities and violent conflict. Int Secur 19(1):5–40

Homer-Dixon TF (1998) Environmental scarcities and intergroup conflicts. In: Klare MT, Chandrani Y (eds) World security – challenges for new century, 3rd edn. St. Martins, New York, pp 342–364

International Military and Defense Encyclopedia (1993) Brassey's (US), Inc, Washington, New York

Kajfež-Bogataj L (2006) Podnebne spremembe in nacionalna varnost (Eng. Climate change and national security). UJMA 20:170–176

Kazenski zakonik Republike Slovenije [KZ] (1994) (Eng. Penal Code of the Republic of Slovenia). Uradni list Republike Slovenije, št. 63/1994, 70/1994, 23/1999, 40/2004, 95/2004

Kazenski zakonik Republike Slovenije [KZ-1] (2008) (Eng. Penal Code of the Republic of Slovenia). Uradni list Republike Slovenije, št. 55/2008, 66/2008, 39/2009, 55/2009

Kirn A (1992) Ekološka (okoljska) etika (Eng. Ecological (environmental) ethics. Aram, Maribor

Leksikon Cankarjeve založbe: Pravo (2003) Cankarjeva založba, Ljubljana

Levy MA (1994) Is the environment a national security issue. Int Secur 20(2):35–62

Likozar-Rogelj M (2006) Okoljevarstveno dovoljenje – inšpekcijski nadzor (Eng. environmental permit – inspection control). Pravna praksa 25(21):14

Malešič M (2009) Prenova nacionalnovarnostnih politik in sistemov (Eng. The renewal of national security policies and systems). Teorija in praksa 46(2):85–104

Malešič M, Vegič V (2009) Slovene public opinion about security issues: a coincidence or a consistent pattern? In: Caforio G (ed) Advances in military sociology: essays in honor of Charles C. Moskos. Contributions to conflict management, peace economics and development. Emerald, Bingley, pp 99–119

Nacionalni program varstva okolja [NPVO] (1999) (Eng. National Environment Protection Action Programme). Uradni list Republike Slovenije, št. 83/1999

Parsons T (1978) Action theory and the human condition. Free Press, New York

Pličanič S (1999) Človek ali narava? Novoveško pravo narave: "anthropeioi nomos" ali "nomos theios": (razprava o razlogih "uničenja Narave" in o novoveškem pravu narave kot pravu uničevalcev Narave) (eng. Human or nature? Modern law of nature: "anthropeioi nomos" or "nomos theios": (discussion on the reasons for "the destruction of the Nature" and on modern law of nature as a law of Nature destroyers). Zbornik znanstvenih razprav Pravne fakultete v Ljubljani (Eng Proceedings of Scientific Papers of Faculty of Law Ljubljana) 59(1999):273–319

Prezelj I (2007) Uvod v ocenjevanje ogrožanja nacionalne varnosti (Eng. Introduction to the assessing of the threats to the national security). In: Prezelj I (ed) Model celovitega ocenjevanja ogrožanja nacionalne varnosti Republike Slovenije (Eng. Comprehensive assessment model of the endangering the national security of the Republic of Slovenia). Ministry of Defence, Ljubljana, pp 7–25

Prezelj I (2008) Challenges in conceptualizing and providing human security. HUMSEC Journal. Issue 2. Graz. ETC. http://www.humsec.eu/cms/fileadmin/user_upload/humsec/Journal/Prezelj.pdf. Accessed 25 June 2010

Resolucija o Nacionalnem programu varstva okolja 2005–2012 [ReNPVO] (2006) (Eng. Resolution on the National Environmental Action Programme 2005–2012). Uradni list Republike Slovenije, št. 2/2006

Resolucija o nacionalnem programu preprečevanja in zatiranja kriminalitetne za obdobje 2007–2011 [ReNPPZK0711] (2007) (Eng. Resolution on the National Programme of Prevention and Suppression of Crime for the Period of 2007–2011). Uradni list Republike Slovenije, št. 40/2007

Resolucija o strategiji nacionalne varnosti Republike Slovenije [ReSNV] (2001) (Eng. Resolution on the national security strategy of the Republic of Slovenia). Uradni list Republike Slovenije, št. 56/2001

Resolucija o strategiji nacionalne varnosti Republike Slovenije [ReSNV-1] (2010) (Eng. Resolution on the national security strategy of the Republic of Slovenia). Uradni list Republike Slovenije, št. 27/2010

Resolucija o preprečevanju in zatiranju kriminalitetne [RePZK] (2006) (Eng. Resolution on the prevention and suppression of crime). Uradni list Republike Slovenije, št. 43/2006

Resolucija o nacionalnem programu varstva pred naravnimi in drugimi nesrečami v letih 2009 do 2015 [ReNPVNDN] (2009) (Eng. Resolution on the programme for protection against natural and other disasters for the period of 2009–2015). Uradni list Republike Slovenije, št. 57/2009

Roberts J (2004) Environmental Policy. Routledge, Taylor & Francis Group, New York

Seneca (2010) Quotations page. Resource document. http://www.quotationspage.com/quote/36804.html. Accessed 13 May 2010

Sotlar A (2008) Od globalne varnosti do individualne (ne)varnosti (Eng. From the global security to the individual (In)security). Delo in varnost 53(6):8–16

Uredba o obveznem organiziranju službe varovanja (2008) (Eng. Decree on obligatory setting-up of security service). Uradni list Republike Slovenije, št. 43/2008, 16/2009, 22/2010

Uredba o preprečevanju večjih nesreč in zmanjševanju njihovih posledic (2005) (Eng. Decree on the prevention of major accidents and the mitigation of their consequences). Uradni list Republike Slovenije, št. 88/2005

Ustava Republike Slovenije [URS] (1999) (Eng. Constitution of the Republic of Slovenia (1991). Uradni list Republike Slovenije, št. 33 -1/1991, 42/1997, 66/2000, 24/2003, 69/2004, 68/2006

Viler-Kovačič A (2006) Okoljevarstveno dovoljenje (Eng. Environmental permit). Pravna praksa 25(19):6

Zakon o gasilstvu [ZGas] (2005) (Eng. Fire Service Act (2005). Uradni list Republike Slovenije, št. 113/2005

Zakon o ravnanju z gensko spremenjenimi organizmi [ZRGSO] (2002) (Eng. Management of Genetically Modified Organisms Act). Uradni list Republike Slovenije, št. 67/2002

Zakon o inšpekcijskem nadzoru [ZIN] (2007) (Eng. Inspection Act). Uradni list Republike Slovenije, št. 43/2007

Zakon o občinskem redarstvu [ZORed] (2006) (Eng. Local Police Act). Uradni list Republike Slovenije, št. 139/2006

Zakon o obrambi [ZObr] (1994) (Eng. Defence Act). Uradni list Republike Slovenije, št. 82/1994, 44/1997, 87/1997, 13/1998, 33/2000, 87/2001, 47/2002, 67/2002, 110/2002, 97/2003, 40/2004, 103/2004, 138/2004, 53/2005, 117/2007, 46/2010

Zakon o ohranjanju narave [ZON] (2004) (Eng. Nature Conservation Act). Uradni list Republike Slovenije, št. 96/2004

2 Slovenian Environmental Policy Analysis: From Institutional Declarations

Zakon o policiji [ZPol-UPB6] (2006) (Eng. Police Act). Uradni list Republike Slovenije, št. 107/2006, 14/2007, 47/2009, 48/2009)

Zakon o prekrških [ZP-1-UPB4] (2007) (Eng. Minor Offences Act). Uradni list Republike Slovenije, št. 3/2007, 21/2008, 41/2010

Zakon o službi v Slovenski vojski [ZSSloV] (2007) (Eng. Service in the Slovenian Armed Forces Act). Uradni list Republike Slovenije, št. 68/2007, 58/2008

Zakon o varstvu okolja [ZVO-1] (2004) (Eng. Environment Protection Act). Uradni list Republike Slovenije, št. 41/2004, 17/2006, 20/2006, 28/2006 Skl.US: U-I-51/06-5, 39/2006-UPB1, 49/2006-ZMetD, 66/2006 Odl.US: U-I-51/06-10, 112/2006 Odl.US: U-I-40/06-10, 33/2007-ZPNačrt, 57/2008-ZFO-1A, 70/2008, 108/2009

Zakon o varstvu pred ionizirajočimi sevanji in jedrski varnosti [ZVISJV] (2002) (Eng. Act on protection against Ionising Radiation and Nuclear Safety). Uradni list Republike Slovenije, št. 67/2002

Zakon o varstvu pred naravnimi in drugimi nesrečami [ZVNDN] (2006) (Eng. Act on the Protection against Natural and Other Disasters). Uradni list Republike Slovenije, št. 51/2006

Zakon o varstvu pred požarom [ZVPoz]. (2007). (Eng. Fire Protection Act). Uradni list Republike Slovenije, št. 3/2007

Zakon o vodah [ZV-1] (2002) (Eng. Waters Act). Uradni list Republike Slovenije, št. 67/2002, 110/2002, 2/2004, 41/2004

Zakon o zasebnem varovanju [ZZasV] (2003) (Eng. Private Security Act). Uradni list Republike Slovenije, št. 126/2003, 7/2007, 16/2007, 102/2007, 96/2008, 41/2009

Chapter 3
Situational Crime-Prevention Measures to Environmental Threats

Gorazd Meško, Klemen Bančič, Katja Eman, and Charles B. Fields

Abstract Crimes against nature are far-reaching, dangerous and complex. Therefore, efforts to detect and deter environmental crimes should be effective. The purpose of this chapter is a presentation of possible forms of Clark's model of situational crime-prevention techniques in the field of environmental crime. The proposals are based on the analysis results of the review of detected environmental threats in Slovenia and the consequences that arise in this connection.

Situational crime-prevention has, in combination with rising of people's awareness on crime and other forms of threats, proved to be quite an effective form of prevention. These are four types of measures – aggravation of access, increasing the risk of the offender, reducing the potential for damage (reduction of perpetrator's profit) and prevention of excuses that someone did not know that a specific action is unacceptable, prohibited or illegal. These situational crime-preventive aspects of the response to environmental threats represent a novelty in the Slovenian practice of crime prevention methods in the environment protection field. The findings of the debate can be used in the planning of crime-preventive measures with regard to

G. Meško (✉)
Faculty of Criminal Justice and Security and Institute of Security Strategies,
University of Maribor, Kotnikova 8, 1000 Ljubljana, Slovenia
e-mail: gorazd.mesko@fvv.uni-mb.si

K. Bančič
Slovene Army, Ministry of Defence of the Republic of Slovenia, Ljubljana, Slovenia
e-mail: klemen.bancic@gmail.com

K. Eman
Faculty of Criminal Justice and Security, University of Maribor, Kotnikova 8,
1000 Ljubljana, Slovenia
e-mail: katja.eman@fvv.uni-mb.si

C.B. Fields
Department of Criminal Justice and Police Studies, College of Justice and Safety, Eastern
Kentucky University, Richmond, KY, USA
e-mail: chuck.fields@EKU.EDU

G. Meško et al. (eds.), *Understanding and Managing Threats to the Environment in South Eastern Europe*, NATO Science for Peace and Security Series C: Environmental Security, DOI 10.1007/978-94-007-0611-8_3, © Springer Science+Business Media B.V. 2011

environmental threats. Furthermore, these measures represent a very useful starting point for all further research and development of programs, resolutions and other documents about responding to environmental threats.

3.1 Introduction

Dealing with environmental harm will demand new ways of thinking about the world...

D. Heckenberg (2009)

Scientific and technological developments have become the obsession of man at the end of twentieth century, which continues in the twenty-first century. The industrial revolution was only the beginning of capitalist development and progress. Today, in an age of electronics, the internet and robots, people have to face with the other side of human development. Pollution and waste as a side or end-product of the production already influenced on the natural environment in which man lives so very negatively, that his life and survival are threatened. Environmental pollution and natural disasters, as part of the consequences, were almost rocketed to the top of the scale of security problems in modern society. Pollution and toxic waste, killing of plants and animals, destroying of soils and other different varieties of injuries and degradation linked to the use, misuse and poor management of the natural environment are known as environmental harms. As stressed by Heckenberg (2009), environmental harm can be conceptualized as involving acts and omissions that are both "legal" and "illegal", because defining harm is ultimately about values and priorities in the society and not just what the criminal laws state it is. These violations against the environment occur as a consequence of irresponsible human behaviour and endanger not only the environment but also human society. Furthermore, some of these illegal actions can affect national and global security. When talking about relationship environment – security, it is a question of the direct and/or indirect influence of the environment on the national or global security. For Homer-Dixon (1994) environmental degradation represents a central threat to the security, because of the political and partly the physical bond between the environment and security. These bonds cause the impact of the environmental degradation and its factors are a source for the formulation of the conflicts. Therefore, the question of environment and environmental politics is becoming more and more an important part of the public policies dealing with the question of the survival of the human race. The main need for research on these issues comes from the fact that this field is still unresearched, especially from the aspect of social sciences. The field of environmental criminality is also exceptionally wide and, on the whole, interdisciplinary. This presents a methodological problem in the access to knowledge, where cooperation between the natural and social sciences, especially criminology, is necessary.

Environmental criminality does not respect any boundaries – local, regional, national, international, not even physical; therefore it is important to note that environmental crimes are calling for the invocation of principles of (international) comparative criminology for its study and principles of international criminal legislation for environmental crime control. Furthermore, it is important to state that lack of theory in the field of environmental criminality is equally glaring in environment-specific crime research.

The role of prevention environmentally harmful activities is linked to criminology, because it was criminology that realized humans cannot conquer, nor control, nature. Furthermore, people are much better off constructing their lives within ecological constraints. Therefore, criminology is called "proactive green criminology". In this respect, it is emphasized that the solution of contemporary environmental problems does not lie in more restricted environmental protective legislation. It lies in redefining our relationship to nature in general, and to particular bioregions. Unfortunately our culture, values, and perceptions towards nature need to be re-examined. People have to pass over from reactive to proactive (preventive) behaviour towards the environment.

The purpose of this chapter is a presentation of possible forms of situational crime-prevention and the defensible space theory in the field of environmental crime. The proposals are based on the analysis results of the review of detected environmental threats in the Republic of Slovenia and the consequences that arise in this relation. As emphasized by Seis (1996: 123): "recognizing a problem is the first step in creating solutions" and in relation to this is stressed that it is necessary to change the way people perceive their relation to nature and to restructure their legislation in the sense of the environmental protection if they want to prevent the environmental crime. One of the possible solutions for environmental crime issues are situational crime-prevention methods. Situational crime-prevention offers solutions that more traditional methods did not manage to achieve. Furthermore, as stressed by Adler (1996), prevention planning needs to take into account new industrial development, economic trends and existence of financial problems in specific corporations and conform to them. Although situational crime-prevention does not necessarily involve more enforcement, it does involve more creative strategies.

Deriving from the above described situation in the field of environmental criminality, the objective of the chapter is the use of situational crime-prevention measures against the environmental threats. The chapter begins with the description of the basic terms and exposes the frames and problems of the environmental crimes as spreading threats against the human society, its health and life, and situational crime-prevention measures as one of the possible tools in responding to these crimes. In the third section, the analysis of the detected environmental crimes and their forms and particularities in the Republic of Slovenia, is presented. A review of present environmental and criminal legislation for the field of environmental threats represents the starting point for the implementation of the situational crime-prevention theory. The fourth section ends with the developed proposals of situational crime-prevention measures of environmental threats, made by using method of reflection and five theories: the theory of situational crime prevention (Clarke 1992, 1997) with the theories of rational choice theory (Keel 1997; Cornish and Clarke 1986), routine activity theory (Cohen and Felson 1979), control theory (Hirschi 1969) and lifestyle theory (Biderman and Reiss 1976). In the end, the presented aspects of the response to environmental threats are discussed, since the authors are aware that the proposed measures can have a limited effect.

Situational crime-preventive aspects of the response to the environmental threats represent useful ideas and methods to improve the current situation in the Republic of Slovenia and in the South Eastern European region. Both the rational choice theory and the routine activities theory are important for the development of crime-control policy and of situational crime prevention in the field of environmental criminality. Transferring situational crime prevention techniques to crimes against the environment involves the basic situational crime prevention principles (Clarke and Eck 2008): reduce the commission of crime by designing models that eliminate the crime opportunities (e.g. redesigning enforcement strategies to cut off industry-specific criminal opportunities; improvement of enforcement effectiveness with the emerging knowledge of the offender's characteristics and with the increase of technical training etc.). It is certainly true that strengthening the guardianship of the environment also depends on the relationship of regulator and regulatee. As will be shown, situational crime-prevention does not contain the ambition of eliminating the so called root causes of (environmental) crime, but is addressing to solving of these issues very pragmatically and with concrete actions. The contribution presents the use of five above-mentioned theories of situational crime-prevention as possible measures of preventing and reducing environmental threats. Situational crime-prevention has, in combination with rising of people's awareness on crime and other forms of threats, proves to be quite an effective form of prevention.

3.2 Situational Crime-Prevention in the Field of Environmental Criminality

Crimes against nature are far-reaching, dangerous and complex crimes that destroy the environment and consequently affect our society and world. These crimes may directly affect our health today as well as hurt future generations. Therefore, efforts to detect and deter environmental crimes have to be effective and culminate in prosecution to the fullest extent permitted. To achieve this, measurable data, records, information, case studies and surveys, and useful suggestions formed on the basis of obtained results are important.

When talking about harms against the environment, the term environmental crime is broadly used. Environmental crime refers to every temporary or permanent legally defined deviant act or resigned activity, which causes an artificial change, worsening, burdening, degeneration or destruction of (human) environment or breaking its natural changes. The perpetrator of environmental crime could be anyone, or every one of us (corporations, companies, groups, individuals, etc.). Their common characteristics are the victims, because by harming of the environment (plants, animals, and natural elements) they harm people as well. Dealing with issues of environmental criminality, within the social sciences, green criminology has best responded to the phenomena of environmental crime. Green criminology has developed as a branch of criminology, a science about criminal acts and their committers, which researches the forms of a deviant behaviour, investigates the causes of such behaviour and describes such phenomena and observes them in their development. After all, criminology is not a legal, but an empirical science, which uses comprehensions of empirical researches and results of the experience. Green criminology has developed in the mid 1970s of the twentieth century. Although with a huge lack of clarity, lack of necessary knowledge from the naturalistic field and with a small number of scientific research work and publications (White 2009), it succeeded to set a basis for further work. In just three decades the development of green criminology had numerous attempts at defining environmental crime and environmental criminality. As mentioned above, situational crime-prevention has proven to be quite an effective form of prevention, which addresses the solving of crime issues very pragmatically and with concrete actions. In combination with the rise of people's awareness on crime and other forms of threats, situational crime-prevention will be presented in the next sections as possible measures of preventing and reducing of the environmental threats phenomena.

Falling under the auspices of the "primary" crime-prevention (Hughes 1998: 20), situational crime prevention is focused on crime and situations in which crime occurs, rather than on the perpetrator. This can be seen as a consequence of the fact, that the theory of situational crime prevention is built on the basic idea that the situations are more predictable than the individuals (Weisburd 1997: 8). According to Rosenbaum et al. (1998: 153–154) situational crime-prevention is significant in two respects – the already mentioned shift of emphasis away from the criminal to the environment, and its crime specific focus. The concrete measures are targeting very specific crime, based on the assumption, that the situational determinants of each type of crime are different. Therefore individual measures can be very different one from another, ranging from being fairly simple (supplying toughened drinking glasses to reduce injury from fights in bars, or training in conflict management) to complex (installing CCTV, or appropriate architectural design of whole neighbourhoods). The main agent of situational crime prevention is not the criminal justice system, but a variety of public and private organizations such as schools, trading companies, transportation businesses and others (Clarke 1997: 2).

By designing and managing the immediate environment, where crime is most likely to occur, the situations in which potential perpetrator and potential victim get together are altered to the point, that this specific changes influence the offender's *decision* or *ability* to commit crime (Ekblom 2006: 384). As Ekblom (2006: 384) continues, the interventions can also act on prior 'scenes', in which offenders prepare, or become primed for crime, and they can be implemented directly or indirectly.

People have used situational crime-prevention measures in a form of door locks and other physical barriers long time before they were ever labelled. However, government officials, police and academics started to pay greater attention to this approach to crime-prevention in the late 1970s and early 1980s, beginning with Crime Prevention trough Environmental Design movement in USA (Ekblom 2006: 384) and Home Office Research Unit's 'administrative criminologists' in UK (Hughes 1998: 59). They put their effort into a search for a more theoretical foundation built around a concept of crime as opportunity (Ekblom 2006: 384). The results of their work are best seen in the writings of Ronald V. Clarke, who initially presented the theory of three ways of approaching the prevention of crime: (1) increasing the effort, (2) increasing the risks, and (3) reducing the rewards (Clarke 1992), which were later supplemented by two more – reducing provocations and removing excuses (Clarke and Eck 2008: 91).

To overcome a common critique that situational crime-prevention is only concerned with offences and situations and that it neglects offenders along with their unpredictable response when confronted to prevention measures, authors usually combine the theory of situational crime prevention with the theories of rational choice theory, routine activity theory, control theory and lifestyle theory (Rosenbaum et al. 1998: 155; Hughes 1998: 63–65; Ekblom 2006: 384). There are many other critiques of situational crime prevention, of which displacement is most often emphasised (Ekblom 2006: 384–385), but inconsistent results of different researches do not give a comprehensive answer to whether these critiques are justified or at least to what point (Rosenbaum et al. 1998: 164–167).

The use of situational crime-prevention aspects of the response to environmental threats represents a novelty in practice of crime-prevention methods in the environment protection field. As will be shown in the continuation, the five types of crime-prevention measures can be used in the planning of responses to environmental threats. Furthermore, these measures represent a very useful starting point for all further research and development of programs, resolutions and other documents about responding to environmental threats.

Criminological research in the area of environmental pollution and environmental crime is almost as new to criminologists as penal sanctions for environmental crimes are to criminal lawyers. When they enter the sphere of environmental criminality, researchers indeed come to an unfamiliar territory. For this reason, proposed changes and solutions must be based on previous reliable research results and analysis. Furthermore, solutions should be implemented progressively and the studied area closely observed. By all means, the problems that occur must be handled by experts and addressed the next section presents an analysis of the crimes against the environment in the Republic of Slovenia. The results of performed analysis are presented as the basis for the use of a method of reflection by analysis of theories of situational crime-prevention and the development of proposals of useful measures.

3.3 Crimes Against the Environment in the Republic of Slovenia

When considering the idea of environmental crime, a majority would agree that the environment needs to be treated with respect. People need clear air, good food, and clean water to drink. They may even agree that some aspects of the natural environment need to be protected. Yet a good way of protecting natural resources has evaded our understanding since the question of preservation versus conservation first emerged. Protecting the natural environment is a complicated task, and the issues associated with determining violations and imposing sanctions are controversial

(Clifford 1998: 252), as there are different forms of environmental crime, different types of environmental criminality exist. In addition, the *modus operandi* of these offenders can be different as well. Carter (1998: 172) notes that the reasons why people commit environmental crime are very heterogeneous. Some people may accidentally, unknowingly, mistakenly, negligently or unconsciously violate environmental law. Others may be motivated by the reduction of their operating costs, due the financial costs involved in illegally disposing of waste or sewage discharges into the nature. The most dangerous and the most wide-spread are environmental criminals who are business associated or members of organized crime.

The criminal justice system plays a very important, or better to say crucial, role in responding to the environmental criminality. First of all, environmental justice represents the basis for the responding to this problem. Furthermore, enforcement, prosecution, and sentencing of the environmental crime offences have to be appointed. Last, but not least, social sciences and their cooperation with criminal justice, especially in the field of research and testing, represent an important part of the efficient response against environmental crimes and the resulting consequences. For environmental crime, responsible state agencies still employ reactive approach to addressing environmental crimes instead of proactive enforcement that should be, by Carter's (1998: 170) opinion, initiated into the reactive approach.

"The environmental law system is an organized way of using all of the laws in our legal system to minimize, prevent, punish, or remedy the consequences of actions which damage or threaten the environment, public health, and safety". (Steinway and Botts 2005: 3–4).[1] In the legal protection of nature and environment it is the subsystem in the legal system of each country, which is properly responsible for setting up this legal system and ensure its operation. This function is implemented by the adoption of various laws and executive acts, as well as with the ratification of various international legal sources (conventions, declarations, etc.). In this way, environmental protection law is created and formed. Selinšek (2003: 659) defines environmental law as a community of actions, necessary to ensure the situation in which there is no health risk and where there is a decent standard of living. Furthermore, the protection of soil, air, water, plant and animal life from harmful human interventions and repairing of the damage and the harmful consequences, because of the protection of the human, is ensured. Environmental legislation must be balanced, while it should also allow the normal functioning of society, as well as restrict (respect for fundamental human rights) by virtue of its functioning. Likewise, environmental law (protection of the environment) must combine prevention, enforcement and reparation functions of environmental protection.

---

[1] Steinway and Botts (2005: 4) explain that their definition illustrates how the environmental law system is often interpreted to encompass the protection of public health and workers' safety in addition to the environment. Furthermore, what makes a law or regulation a part of the environmental law system is not its label or original function but the purpose for which is used (for example, the criminal code play important role in the environmental law system, although, its original function has almost nothing with the environmental protection).

From the constitution and laws arise certain principles for effective and harmonized legislation on environmental protection. Among the most essential are four principles: the principle of prevention,[2] the principle of causality,[3] the principle of cooperation[4] and the principle of shared burdens.[5] These are principles associated with the legal norms, which can ensure the functionality of the entire environment. The objectives of the rule or regulation for protection of the environment against environmental crime, and other harmful practices, are as follows (Faure and Visser 2003; Selinšek 2003; Šinkovec 1994; Strojin 1987): prevention of hazards and risks to the environment (reducing aggravation); elimination of environmental damage; regenerating ability to restore the environment; reduction of consumption of natural resources and energy; maintaining qualitative and renewable natural resources; and preservation of local characteristics and environmental values.

An individual, as a member of society, has the right to a clean and healthy living environment. In the Republic of Slovenia a fundamental human right to live in a clean and healthy living environment is guaranteed by the Constitution of the Republic of Slovenia (Ustava Republike Slovenije [URS] 1991) (in Article no. 72, entitled *Healthy Living Environment*[6]). As already mentioned, environmental protection in the Slovenian legal process has its basis in the Constitution of the Republic of Slovenia. On the systemic level it is governed by the laws. The legal basis for action in the Republic of Slovenia is the Environmental Protection Act (Zakon o varstvu okolja [ZVO-1] 2006),[7] some 20

---

[2] The legislation prohibits any adverse action as a timely prevention works long term, in the benefit of future generations (Šinkovec 1994).

[3] The legal rule provides that any person, which causes damage or delayed hazard and cause injury to the environment, must suffering the removal costs or the costs of removal of negative effects (Strojin 1994).

[4] In the protection of the environment a state as well as society or individual operators must be involved. Šinkovec (1994: 164) adds that this is a shared responsibility and cooperation of all concerned in all stages of environmental protection (from planning to confrontation with the consequences of decisions).

[5] Individual costs incurred in protecting the environment, are financed from the state budget (Šinkovec 1994: 170). This group also includes specific taxation (potential) agents of environmental disturbance.

[6] "Everyone has in the accordance with the law the right to a healthy living environment. The state is in charge for a healthy living environment. For this purpose, the law defines the conditions and ways for the performance of economic and other activities. The law stipulates under what conditions and to what extent the provoker of the damaged in the living environment is obliged to pay the damages. The protection of animals from cruelty shall be regulated by law".

[7] The Act represents a basic legislation in the field of environmental protection in the Republic of Slovenia. It sets out general principles and guidelines for a comprehensive approach, in addition to introducing a series of implementing regulations for the regulation of more specific relationships and concrete examples (Viler Kovačič 2007).

3 Situational Crime-Prevention Measures to Environmental Threats

other acts,[8] concerning the nature and animals, and a number of other executive acts (regulations, strategies and rules).

The growing technological and economic development requires more and more complex, sophisticated and, unfortunately, environmentally dangerous technology. Striving for profit, or as Pečar (1981: 36) wrote, economic expansion, a struggle for survival, competition, rivalry, giving priority to fetishism of economic goods prior to the spiritual, medical and other goods, are more often followed by the abuses and violations of laws and regulations, consequences of which are often disastrous (mass fish mortality, loss of forest cover, leakage of noxious liquid substances, radiation, radioactive waste, rinse sprays into rivers, streams and lakes, and toxic waste incineration, etc.). Problem solving, and sanctions for violations of environmental protection, is based on a combination of administrative, civil and criminal law. Selinšek (2003: 652) notes that the latter is even more limited, and the reason for this is its *ultimo ratio*. Although, the criminal law in environmental protection is being regarded as accessory, today at both the national and international levels, the need to protect the environment through criminal law measures is highlighted. The reasons for this are more frequent and the extensive "attacks" on the environment occur with increasingly severe consequences. Environmental crime can occur within the Slovenian legal system as all known forms of criminal activities, such as disciplinary offenses, misdemeanours and criminal offenses. Selinšek (2003) points out that the environmental awareness of most people is still at a very low level, therefore it is necessary to resort to various means, including the criminal law, to prevent these harms. In doing so, Dežman (2004) stresses, that fundamental function of modern criminal law is in the protection against the most serious forms of deviant behaviour in society. Environmental crime has become just that, the most dangerous form of deviant behaviour, in a global perspective, as an increasingly serious threat to lives and livelihoods of people and their types, as well as all other forms on Earth.

For Slovenia, most typical forms of deviations against the environment are (IRSOP 2008; Policija 2002, 2003, 2004, 2005, 2006, 2007, 2008): water pollution (intensive agriculture; industrial emissions, unregulated discharge of waste water); air pollution (traffic; industrial emissions); soil pollution (hazardous waste, illegal dumps, intensive agriculture, waste oils and motor fuels); waste management[9] (communal and

---

[8] Nature Conservation Act (Zakon o varstvu narave [ZON] 1999), Waters Act (Zakon o vodah 2002), partly Animal Protection Act (Zakon o varstvu živali 1999), Forests Act (Zakon o gozdovih 1993), partly the Spatial Planning Act (Zakon o prostorskem načrtovanju 2002), Chemicals Act (Zakon o kemikalijah 2003), Act on Protection Against Natural and Other Disasters (Zakon o varstvu pred naravnimi in drugimi nesrečami [ZVNDN] 2006), Act on Inland Waterway Transport (Zakon o plovbi po celinskih vodah 2002), Fire Protection Act (Zakon o varstvu pred požari 2007), Act on the Triglav National Park (Zakon o Triglavskem narodnem parku 2001), Act on the Mountain Trails (Zakon o planinskih poteh 2007), Nature Conservation Act (Zakon o varstvu narave [ZON-1] 2004), Act on the Safety on Ski Slopes (Zakon o varnosti na smučiščih 2002), Road Traffic Safety Act (Zakon o varnosti cestnega prometa [ZVCP] 2004), Act on the Protection of Public Order and Peace (Zakon o varstvu javnega reda in miru 2006), etc.

[9] Waste management is one of the worst resolved problems within the framework of environmental protection in the Republic of Slovenia (IRSOP 2008).

industrial waste, hazardous waste, waste oils and motor fuels, asbestos waste); illegal trafficking with animal and plant species and minerals and fossils; import, export and unlawful acquisition or use of radioactive or other hazardous substances; environmental degradation and destruction of forests; animal torture and illegal hunting; pollution with the electromagnetic radiation; light pollution; and driving in the natural environment with bicycles and motor vehicles, and motor slides.

In the Penal Code of the Republic of Slovenia (Kazenski zakonik Republike Slovenije [KZ-1] 2008) the field of environmental crime is defined in the section 32 as *Crimes against the environment, space and natural resources*. Section includes 16 following environmental crime offenses (articles 332–347):

- Nuisance and environmental degradation (Article 332)
- Marine and water pollution by ships (Article 333)
- Import and export of radioactive substances (Article 334)
- Unlawful acquisition or use of radioactive or other hazardous substances (Article 335)
- Contamination of drinking water (Article 336)
- Contamination of food or feed (Article 337)
- Unlawful occupation of property (Article 338)
- Destruction of crops with harmful substance (Article 339)
- Destruction of forests (Article 340)
- Animal torture (Article 341)
- Illegal hunting (Article 342)
- Illegal fishing (Article 343)
- Illegal handling with protected animals and plants (Article 344)
- The transmission of infectious diseases at animals and plants (Article 345)
- Production of harmful medications for animals treatment (Article 346)
- Unconscionable veterinary assistance (Article 347).

Most environmental crimes are blanket in nature and have their origins outside the criminal law, in other regulations in the field of environmental protection, especially in the Environmental Protection Law. The specialty of environmental crime law is in the fact, that it is a relatively new field of action, which is still constantly changing. Environmental law protects, in a special way, particular values (e.g., human life and health). Environmental crimes are acts that leave behind unpredictable consequences, therefore it is difficult to prove the existence of specific imminent threats, and even harder to prove causation and guilt to the perpetrator. Prosecution of environmental crime is still in its infancy. With such a continuous and rapid growth of environmental protection, the question arises whether such a prosecution will not always be too late, when the violations already have serious and irreversible consequences. For now, stays the focus on the preventive affect of penal law in the field of environmental crime predominantly only wish. Undoubtedly is important to have good cooperation between law enforcement and other departments for the response to this kind of criminality. Furthermore, the role of each person as an applicant or as non-polluter is increasing and by authors opinion the most easiest, effective and successful way to solve the above exposed issues is the pass over from reactive to proactive responding against the environmental threats.

The above analysis of environmental criminality in the Republic of Slovenia confirms the need to move from reactive to proactive responses to crimes against the environment. One of these proactive responses is the use of crime-prevention theory. Situational crime-prevention has, in combination with an increased public awareness on crime and other forms of threats, proved to be quite an effective form of prevention, as it will be shown in the continuation of the chapter.

3.4 Situational Crime-Prevention Methods and Environmental Criminality

Since Ronald V. Clarke had a privileged institutional location of his theory in the Home Office (Hughes 1998: 61), with an extensive, well-funded and well publicized programme of research, he could develop one of the most comprehensive, detailed and evidence-based models of situational crime prevention. Initial three clusters containing 12 techniques (Clarke 1992), have been gradually modified and expanded to 5 clusters of 25 techniques by 2003 (Clarke 1997; Cornish and Clarke 2003), which is due to the developments in theory, practice and technology surrounding the prevention of crime.

The first cluster of techniques is titled *increasing the effort*, under which Cornish and Clarke (2003) list five techniques: (1) Target hardening – criminal opportunities are reduced by changes in design, by which physical barriers prevent potential offenders from easily committing a crime (Clarke 1997: 17). (2) Access controlling – as a central component of the defensible space theory, is all about measures intended to exclude potential offenders from places where crime is likely to occur (Clarke 1997: 17). (3) Screening exits – is not about excluding potential offenders, but about increasing the likelihood of detecting those who are trying to remove objects that should not be removed from protected area (Clarke 1997: 19). (4) Deflecting offenders – as a technique suggested by the routine activity theory, it is about redirecting offenders away from crime targets (Clarke 1997: 18). (5) Controlling tools/weapons – is about controlling a range of crime facilitators ranging from measures such as implementing and inspecting photos on credit cards, to surrendering any potentially dangerous objects before entering museums or courts (Clarke 1997: 19).

The second cluster of techniques is titled *increasing the risks* and consists of the following techniques: (6) Extending guardianship – is about enabling potential victims to have guardians at hand (Cornish and Clarke 2003: 90). (7) Assisting natural surveillance – is about enhancing natural surveillance provided by people, who go about their everyday business (Clarke 1997: 21). (8) Reducing anonymity – is about making potential offenders more visible, individually recognizable (Cornish and Clarke 2003: 90). (9) Utilizing place managers – is about providing employees with means and motivation to better control the place where they work (Clarke 1997: 21). (10) Strengthening formal surveillance – is about making police and security guards more effective by means of advanced modern technology and by making them more accessible to potential victims (Clarke 1997: 20).

The third cluster of techniques is titled *reducing the rewards* and consists of the following techniques: (11) Concealing targets – is about making potential targets for crime less visible or at least less obviously attractive to offenders (Cornish and Clarke 2003: 90). (12) Removing targets – is about relocating targets to a safer location, or their modification to the point where they become useless to the potential offender (Clarke 1997: 22). (13) Identifying property – is about making very transparent who the rightful owner of the property is, or marking it in a way that it can be easily traced back to its owner (Clarke 1997: 22). (14) Disrupting markets – is about making it difficult for the offender to convert crime products into money (Cornish and Clarke 2003: 90). (15) Denying benefits – is about preventing the offenders to enjoy in the results of the crime they have committed (Clarke 1997: 23).

The fourth cluster of techniques is titled *reducing provocations* and consists of the following techniques: (16) Reducing frustrations and stress – is about managing environment, procedures and attitude of employees in a way that people do not get unnecessarily agitated and nervous (Cornish and Clarke 2003: 90). (17) Avoiding disputes – is about preventing people from getting into disputes in situations where quarrels are expected (Cornish and Clarke 2003: 90). (18) Reducing emotional arousal – is about preventing people from being confronted with emotionally intense situations (Cornish and Clarke 2003: 90). (19) Neutralizing peer pressure – is about preventing, for instance, troublemakers at school to exercise pressure on each other and others, and about driving an efficient anti-propaganda on common peer pressure (Cornish and Clarke 2003: 90). (20) Discouraging imitation – is about making results of committed crimes or information about how to commit these crimes unnoticed and incomprehensive (Cornish and Clarke 2003: 90).

The fifth cluster of techniques is titled *removing excuses* and consists of the following techniques: (21) Setting rules – is about introduction of new rules or procedures, which are intended to remove any ambiguity concerning the acceptability of conduct, while not all such rules require the backing of the law (Clarke 1997: 24). (22) Posting instructions – is about making rules and expected behaviour clearly known to everyone entering a designated area (Cornish and Clarke 2003: 90). (23) Alerting conscience – is about focusing on specific forms of crime, by stimulating feelings of conscience at the point of considering the commission of a specific kind of offence (Clarke 1997: 24). (24) Assisting compliance – is about managing the procedures and situations in a way that "doing the right thing" is just as easy, or even easier, than committing an offence (Clarke 1997: 25). (25) Controlling drugs and alcohol – is about controlling psychological influence of legal and illegal drugs, which weaken the usual social or moral inhibitions, or impair perception and cognition (Clarke 1997: 25).

On the basis of previous research in Slovenia the analysis of the detected forms and consequences of environmental threats were made. Furthermore, the analysis of the threats was made on the basis of five theories of situational crime-prevention using the method of reflection. The developed proposals of situational crime-prevention measures of environmental threats are presented in the continuation.

3.4.1 Modification of Situational Crime-Prevention Measures for Preventing Crimes Against the Environment

As already mentioned, a thorough analysis has to be conducted on every concrete case or a problem, to be resolved, before it can designed and applied any sort of preventive measure, and environmental crime is no exception in this manner. After a detailed analysis a selection and adjustment of individual techniques follows, which are according to Clarkes' classification sorted into five clusters.

3.4.1.1 Increasing the Effort

1. Harden targets – perhaps the simplest application of this measure can be found in the field of preventing of destruction of natural heritage in national parks. White (2008: 236) suggests the use of creative architecture and strategic planning of roadways and footways through national parks. Appropriate marking of these paths, systematic placing of toilets and viewpoints across the park can help to divert tourists away from the most pristine parts of wilderness, which are most sensitive about human presence and therefore require the highest level of protection.[10] Very concrete architectural measures can be similarly used in the field of preventing the pollution of drinking water. This planning begins with placing the well as far away from potential sources of pollution as possible. With the construction having a deeper well casing, enhanced grouting, redirection of surface runoff, installation of fences and alarms and similar measures. This technique is useful in preventing most of the previously mentioned criminal offences stated in the Slovenian penal code, particularly nuisance and environmental degradation, and contamination of drinking water.

2. Controlling access – most widely used for the general public probably the best known measure dealing with controlling access is customs. Customs have several tasks when it comes to preventing environmental crime. For instance, a radiation portal monitor can detect any intake of radioactive substance into a country.[11] Closely linked to the concept of customs is quarantine, which protects many countries from diseases entering into their natural space.[12] Another example of controlling access is maritime inspection, which surveillances ships entering Slovenian see in

---

[10]Arrangement or physical protection of natural values, which can be threatened by sightseeing, is included into the Resolution on National Program of Environmental Protection 2005–2012 (Resolucija o nacionalnem programu varstva okolja [ReNPVO] 2006), by which at least ten natural value sites a year should be protected.

[11]For instance, in May 2009 customs in Obrežje border crossing (between Slovenia and Croatia) a truck was discovered to transport a source of radioactive radiation among scrap metal. It was denied entry into Slovenia/European Union (Carina 2010a).

[12]For instance, Slovenian prime minister has received a camel couple as a gift from Libyan leader Gaddafi, but they can not be imported into Slovenia without a lengthy quarantine in Canada.

accordance with the sailing safety inspection decree.[13] Other possible measures include fences around significant locations, entry ramps, and a limited number of access routes in national parks. This technique is particularly useful in preventing nuisance and environmental degradation, marine and water pollution by ships, import and export of radioactive substances, contamination of drinking water, contamination of food or feed, unlawful occupation of property, destruction of crops with harmful substance, illegal handling with protected animals and plants, and transmission of infectious diseases at animals and plants.

3. Screening exits – is another area where customs play a major role. Customs prevents import and export of plant and animal products and other substances which could have an effect on the reduction of natural sources and degradation of natural environment. And in Slovenia as well customs often prevent smuggling of poached animals.[14] Maritime inspection also has a mandate to inspect ships before they exit Slovenian ports with the authority to prevent these ships from sailing out until they have not fixed any deficiencies which pose a serious threat to maritime environment. A similar effect as customs, is inspections at the exits from national parks, where entry/exit points are carefully monitored.[15] This technique is especially useful in preventing nuisance and environmental degradation, import and export of radioactive substances, destruction of forests, illegal hunting, illegal fishing, and illegal handling with protected animals and plants.

4. Deflecting offenders – just as diverting "street racers" off the streets to appropriate racing grounds restricts the danger they pose in everyday traffic, a similar effect can be achieved by establishing proper polygons for dirt-bike riders, off road four-wheelers, snowmobiles and similar offenders who break the prohibitions regarding motor-vehicle presence in the natural environment.[16] A measure which could assist in deterring offenders from illegal dumping is creation of walking paths in areas where illegal dumps are often emerging. As stressed by White (2008: 248), in some parts of Canada a program of guaranteed income for indigenous hunting people deflected hunters from excessive hunting in overexploited areas. This technique is especially useful for preventing nuisance and environmental degradation, contamination of drinking water, destruction of forests, illegal hunting, and illegal fishing.

5. Controlling tools/weapons – radioactive materials are, among substances, which due to their nature pose a great threat to the environment, are probably the

---

[13] If a ship does not meet certain conditions, of which those concerning see pollution are fundamental, its entry into Slovenian ports can be denied by the inspection service.

[14] For instance, poachers from Italy are often discovered by customs inspection to be smuggling endangered birds which they have poached in Balkan countries (Carina 2010b).

[15] In Slovenia Triglav national park has no surveillance on entry and exit points, which renders impossible to exert such control, however some national parks in other parts of the world do have a limited number of controlled entry/exit points.

[16] This measure should be accompanied by other measures (such as strengthen formal surveillance) so that offenders find it easier to comply and visit this tracks instead of natural environment.

3 Situational Crime-Prevention Measures to Environmental Threats

most strictly controlled.[17] Similar strict control is being exerted over chemicals, especially those which can have significantly negative effect on human and environment. In Slovenia, this sort of control is carried through the inspection of chemicals, a duty of the ministry of health.[18] In the field of controlling firearms used by poachers in Africa when hunting for endangered species, Lemieux and Clarke (2009: 465) suggest the use of special sensors which can detect gunshot – that way park rangers and the police can be alarmed in time to intervene. These techniques are particularly useful for preventing nuisance and environmental degradation, import and export of radioactive substances, unlawful acquisition or use of radioactive or other hazardous substances, contamination of drinking water, contamination of food or feed, destruction of crops with harmful substance, illegal hunting, illegal fishing, the transmission of infectious diseases at animals and plants and production of harmful medications for animal treatment.

3.4.1.2 Increasing the Risks

6. Extending guardianship – the essence of this technique is in broadening the number or range of 'guardians'. One possible measure is a distribution of pollution measurement systems and qualification of local residents for handling with this technology. In that way, it is necessary to ensure scientific measurement supported surveillance over potential polluters, not just by competent state officials, but the local people also become supervisors of large companies. This measurement is perhaps particularly useful in undeveloped countries where large corporations are situating their heavy industry, often not minding their devastating effect on the environment. Local residents could then alert the authorities and general public in the developed world where these corporations are selling their very affordable products. This technique is particularly useful for preventing nuisance and environmental degradation, marine and water pollution, contamination of drinking water, contamination of food or feed, and destruction of crops with harmful substance.

7. Assisting natural surveillance – one measurement falling under this technique has already been mentioned earlier, that is, setting walking paths, jogging tracks and similar in areas where illegal dumps often appear. This way walkers, runners, sportsmen and other also become supervisors in a certain area, where there was no natural surveillance earlier. Voluntary actions such as *Cleaning up Slovenia in one day*, when thousands of volunteers join in an effort to clean illegal dumps, is also an example of expanding natural surveillance. A result of this action is also a register of illegal dumps (Geopedia.si 2010) which works similarly as Geographic Information System

---

[17] Use, transport and trade are under a very strict surveillance on national and also on international level, where International Atomic Energy Agency – IAEA plays a major role (IAEA 2010).

[18] In the last few years CFC gases received a fair share of attention, the goal is to completely ban the use of these gases in near future, since they have a devastating effect on ozone layer. Commercially named gas "Freon" is one of the last of these gases to be removed from production.

(GIS) – another tool for using natural surveillance.[19] Anonymous call numbers, on which residents can report pollution, or maybe tormenting or molesting of animals, also assist natural surveillance. This technique is especially useful for preventing nuisance and environmental degradation, marine and water pollution by ships, contamination of drinking water, contamination of food or feed, destruction of forests, animal torture, illegal hunting, and illegal fishing.

8. Reducing anonymity – along with the list of companies, which severely burden the environment, the public should also be presented with the names and faces of these companies' directors and owners, and, with an appropriate campaign, their anonymity could be diminished. White (2003: 501) also advocates this sort of approach, saying that it is important to publicly expose information on environmental vandals as a part of a public accountability process. Lewis (2009: 237) concurs with this idea, pointing out that naming, shaming and public outrage can have positive consequences due to the 'embarrassment effect' on the side of governments. Adaptation of "How's my driving?" stickers for dirt-bikes, off road four-wheelers, snowmobiles and similar is one more measure, and White (2008) is also suggesting obligatory ID badges for fishermen during fishing, which could be used also for hunters. This technique is especially useful for preventing nuisance and environmental degradation, marine and water pollution by ships, destruction of forests, illegal hunting and illegal fishing.

9. Utilizing place managers – foresters and forestry inspectors are most certainly representing professions which require workers to operate in the natural environment. Appropriate stimulations could encourage them to be more attentive to signs of environmental crime, such as illegal dumps and their "authors", or violators of the regulations about motor vehicle presence in the natural environment. The same goes for hunters, which also have to observe each other to possible prevent poaching. Farmers are also in a workplace position which enables them to spot instances of crime against natural environment. Appropriate government stimulations could bring them more benefit than letting their neighbours suppressing pests by illegal means, and thus enable them as place managers. Veterinaries are just another example of certain "place" managers which could be utilized to prevent ecological crime. This technique is especially useful for preventing nuisance and environmental degradation, marine and water pollution by ships, destruction of forests, illegal hunting, illegal fishing, production of harmful medications for animal treatment and unconscionable veterinary assistance.

10. Strengthening formal surveillance – police and inspection services are the primal representatives of formal surveillance in the context of crime against natural environment. In New Zealand there are some areas where fishing with drift nets are

---

[19] Anyone who comes across an illegal dump (when hiking, picking mushrooms or alike) can add the description and location of the dump into register over the internet, and the dump is later cleared.

3 Situational Crime-Prevention Measures to Environmental Threats   57

only allowed when two ministry appointed observers are aboard, and net position is monitored by an electronic system (White 2008: 249). White (2008: 240) also suggests the use of satellite surveillance, as a CCTV is used in the streets, to monitor clearing of forests or illegal fishing.[20] Other, simpler ways of performing formal control over illegal fishing are patrolling by ships and aeroplanes, and ship checking schemes on the docks (White 2008: 250). United bases of electronic data about large companies, regarding their production in undeveloped countries, could provide another tool for more effective formal surveillance. The nature of this technique is such that it is useful in preventing every criminal offence stated in the Slovenian penal code.

3.4.1.3 Reducing the Rewards

11. Concealing targets – placement of already mentioned measure of well weighted design of walking paths and tourist infrastructure in national parks also under this technique since it directs visitors away from the most sensitive areas of the park in which way the natural heritage is concealed and protected from the danger of destruction. An establishment of a national park can be a measure by itself, if the park regime forbids any exploitation of natural sources (hydro-potential, animals, wood, ores, oil), or even researching the terrain with an intention to find ore or oil. Another example can be found in caving tourism, where tourist are often allowed to visit only a part of a cave, while the most beautiful and at the same time most vulnerable parts are closed to public, or the visits are very restricted.[21] This technique is particularly useful for preventing nuisance and environmental degradation, contamination of drinking water, and illegal hunting.

12. Removing targets – local authorities have used this technique in Zimbabwe and Namibia to protect endangered black rhinos, which were on the verge of extinction by sedating them and then sawing their horns off. Poachers find no use in the rhinos themselves, they only kill them for their horns, therefore such rhinos do not represent a target any more – the target is removed. Another measure proposed also by White (2008: 251) is relocation of an endangered species to an area where it is out of danger. In Slovenia a butterfly species is being relocated out of an area where a motorway will be constructed to a newly adjusted environment appropriate for the butterfly species to survive.[22] Another example of target removal can be seen in

---

[20]Obligatory equipping of fishing vessels with GPS systems can also assist in monitoring the position of these vessels when fishing, thus preventing them to fish in the prohibited areas without being noticed, for instance in the "green areas" of the Australian great coral reef (Australian Government 2010).

[21]Examples in Slovenia are Križna cave (Križna jama 2010) and Županova cave (Županova jama 2010).

[22]Another example is from Malavi, where an elephant herd will be relocated to a newly established national park, to protect it form killing by the local people (African Conservation Foundation 2010).

the authorisation of veterinary inspection, which can dispossess an animal, forbid breeding or fine the owner of the animal in the case of animal protection law violation.[23] This technique is particularly useful for preventing animal torture, illegal hunting and illegal handling with protected animals and plants.

13. Identifying property – countries and international organizations are already taking advantage of such sophisticated technology as DNA testing in the endeavour to stop the illegal ivory trade. Programs of identifying animals by their DNA are already being performed in certain parts of Asia and Africa, and when an ivory of suspicious origins is being confiscated anywhere in the world (mostly in Japan), it can be DNA analysed and its true origin determined.[24] Identifying property is a useful measure also when it comes to large objects containing hazardous substances, which pose a threat to be illegally discarded in a natural environment. Therefore every car has a VIN number, which has, along with some other measures, contributed to fewer cars being simply left in forests or rivers of the Slovenian countryside. This technique is especially useful for preventing nuisance and environmental degradation, contamination of drinking water and illegal hunting.

14. Disrupting markets – an especially clear case representing this technique is the disruption of the international ivory market as a consequence of the 1989 Convention on International Trade in Endangered Species of Wild Fauna and Flora. With extensive media coverage the burning of 2,000 elephant tusks in Kenya has become a symbol of disruption of ivory markets (Lemieux and Clarke 2009: 254).[25] White (2008: 251) sees chances to disrupt markets in the field of illegal fishing by interfering with distribution channels, and implementation of strict control over permit and licenses issuing. Yet another example is the absolute ban on any waste import from abroad to Indonesia, which was imposed after the authorities have realised, that almost half of the garbage brought to country to be recycled were in fact non-recyclable, a good part even hazardous (Lewis and Chepesiuk 1994: 2). This technique is especially useful for preventing nuisance and environmental degradation, destruction of crops with harmful substances, destruction of forests, illegal hunting, illegal fishing, illegal handling with protected animals and plants and production of harmful medications for animal treatment.

15. Denying benefits – one of these measures is green public procurement, where a company with bad environmental impact is being denied benefits, since it is not selected to do business with the state. A similar measure, which takes effect on a

---

[23] See more about veterinary inspection in (Veterinarska inšpekcija 2010).

[24] This measure is supported by Lemieux and Clarke (2009: 465) who say that it is vital to equip authorities in the countries of origin and especially the customs services in countries of destination with appropriate equipment and knowledge.

[25] However, Lemieux and Clarke (2009: 264) stress that it is not always enough to disrupt international markets; often the internal markets are also very strong and need to be handled with. They offer a case of parrots market in Mexico, which was not really affected by the disruption of international ban on export, since the poaching of parrots remained almost the same due to local demand.

3 Situational Crime-Prevention Measures to Environmental Threats

little different level, is labelling various marking products as "eco" or "green". In EU and also in Slovenia a new way of labelling products and services is being implemented by 'eco label' which is uniform for the whole EU.[26] Labelling of products can also have the opposite effect, when a label provides the products' "carbon footprint" and thus informs the consumer which of the products have a more negative effect on the environment.[27] And labelling domestic appliances with energy class, or stressing the CO_2 output of a car, have a similar goal and effect. This technique is particularly useful for preventing nuisance and environmental degradation, contamination of drinking water, destruction of crops, and illegal fishing.

3.4.1.4 Reducing Provocations

16. Reducing frustrations and stress – when the procedures regarding toxic, construction or other waste removal are too complicated, too lengthy, or too expensive for a citizen, then this unpleasant experience at trying to do things right can cause him to simply dump those waste products on an illegal dumping ground. Simplification of procedures (regular, free and well advertised cartages) can thus be an effective measure. Disappointed and angry Slovenian peasants often explain to the media, that the wolfs have more protection by the state than human, since killing them is prohibited even when they tear apart whole flock of sheep. Frustration can lead to illegal hunting of these wild animals. And it is similar with bears. State help at providing adequate fencing is one measure, alternative food sources provision (by hunters) is another one, GPS tracking of bears combined with alarms when they get too close to human yet another one.[28] This technique is particularly useful for preventing nuisance and environmental degradation, marine and water pollution by ships, destruction of crops with harmful substance, animal torture, and illegal hunting.

17. Avoiding disputes – destruction of foreign plants, fruit trees, or other crops with harmful substances is a criminal offence, which can be a result of a dispute between land owners. Measure for prevention of such offence is an effective and accessible legal assistance (mediation). Ineffectual cohabitation of an animal and human can also lead to a sort of dispute. Lemieux and Clarke (2009: 466) are mentioning a very simple situational measure for this instance, coming from African savannah. Local residents have been experiencing serious disputes with authorities who prohibited any elephant killing, while these animals have been causing a lot of damage on local peoples' farming grounds. The solution came in

---

[26] Products and services with this label have an advantage on the market, but companies have to adjust their production to acquire it.

[27] The third largest international retailer Tesco has calculated carbon footprint of 20 of its products (the calculation includes all of the life cycle of the product: production, distribution and consumption) and in that way offered its customers one more criteria by which they decide what to buy (Umanotera 2010).

[28] Some bears in Slovenia have already been equipped with GPS collars with an intention of their study; however the use of those collars could be upgraded (Medvedi 2010).

the form of hedgerows around the fields, which contained certain bushes with special characteristics unpleasant to elephants, for which they have avoided them and consequently also the surrounded fields. This technique is especially useful for preventing unlawful occupation of property and destruction of crops with harmful substance.

18. Reducing emotional arousal – strong emotional arousal is rarely a cause of crime against natural environment, though there are exceptions. One of those is, for instance, pyromania or the irresistible urge to start a fire, which can have devastating fires as a consequence.[29] A principal measure is timely detection, for which reason medical examinations of children should also include an interview with a psychologist (OrganizedWisdom 2010). Sexual molestation of animals is yet another form of ecological crime, stipulated by strong emotional arousal. A complete ban and internet blockade of violent pornography is one appropriate measure intended to prevent such acts (POP Center 2010). This technique is especially useful for preventing nuisance and environmental degradation, destruction of forests and animal torture.

19. Neutralising peer pressure – juveniles riding motorbikes, all terrain four wheelers, or snowmobiles in the natural environment are usually not alone. They encourage each others as in the case of street traffic offences. Therefore the preventive measures are also similar, for instance, media campaigns specifically aimed at this behaviour.[30] But there is also peer pressure among directors of large companies, who compare one to another and compete for success, which leads to one-way race for profit at any cost. With appropriate public campaign, rewarding 'green companies', and by awarding managers who achieve the most substantial reduction of negative impact of its company to environment, a change can be achieved.[31] When economic profit is not the only measurement of success, the pressure will also decline. This technique is particularly useful in preventing nuisance and environmental degradation, animal torture, and illegal hunting.

20. Discouraging imitation – Singapore is a pocket county with extraordinarily strict environmental legislation. Its streets are not polluted, nor there are any problems with illegal dumps. Insufficiently maintained ships are avoiding its harbours, since they are aware of the fact, that they can expect harsher sanctions than elsewhere. The "broken windows theory" would suggest that illegal dumps can be avoided by removal of all little signs of disorder – in the form of roadside waste. Therefore, effective communal service by taking care of little garbage also contributes to the absence of large (illegal) dumps. And this also holds true for harbours.

---

[29] We can see the consequences of large fires every year over the media [Croatia, Spain, Karst (TimesOnline 2010; Delo 2010; BBC 2010)].

[30] For instance with a slogan: "Don't let your maid be the one!".

[31] Planet Zemlja association annually awards the car manufacturer, which has achieved the greatest reduction of CO_2 of its products (Planet Zemlja 2010).

3 Situational Crime-Prevention Measures to Environmental Threats

Thus strict environmental standards need to be implemented and enforced.[32] This technique is especially useful for preventing nuisance and environmental degradation, marine and water pollution by ships, animal torture, illegal hunting, and illegal fishing.

3.4.1.5 Removing Excuses

21. Setting rules – it seems that many people are not familiar with the rules of behaviour in the natural environment, since there are a lot of people who visit nature with motor vehicles, destroy natural heritage, and ignore animal rights. If vagueness of rules is the cause of this, then the first job for authoritative services is to simplify these rules and make them more understandable. If the reason lies in ignorance, than informing the people by media campaigns, in schools and similar could be the answer. Perhaps any buyer of a motorbike should also conduct a short course and/or exam to prove his awareness of the fact that these machines do not belong in the natural environment. From the companies' perspective, there are too many rules already, so they should be simplified and priorities determined. This technique is particularly useful for preventing nuisance and environmental degradation and illegal handling with protected animals and plants.

22. Posting instructions – this is a widespread technique in the field of preventing environmental harm. Warning signs forbidding waste disposal are quite common and so are warning signs at the entries of natural parks, which warn visitors in letters and by symbols what is forbidden in the park and what is allowed to do. Since there are piles of waste often surrounding warning signs, which ban waste dumping, an improvement of this measure would be needed. Perhaps signs similar to modern radar-signs in traffic, which offer the information about the threatened penalty for the offence committed, would have a better effect.[33] White (2008: 251) also suggests the use of warning signs in harbours, drawling attention to the problem of illegal fishing. This technique is particularly useful for preventing nuisance and environmental degradation, marine and water pollution by ships, contamination of drinking water and illegal fishing.

23. Alerting consciousness – some aircraft companies include short warning messages, regarding dangers of illegal trade in endangered animal species and their products, into instructions movie concerning rules of behaviour on the plane in case of emergency before takeoff or landing.[34] White (2008: 251) suggests strengthening

---

[32] An example of good practice is an introduction of environmental manager [Port Koper has one in accordance with regulations since 2004 (Luka Koper 2004)].

[33] Warning signs could include information about the penalty and even about modern ways of detecting offenders – that would serve as a better deterrent than just a notice of forbiddingness.

[34] Such warnings serve as first information for the ignorant passengers, while at the same time have a deterrent effect, and also alert consciousness about the problem.

of moral condemnation of illegal fishing, which can be achieved by well thought out media campaigns. Further, he proposes distribution of pamphlets with information regarding fish stock in the oceans, which can be accompanied by information boards in harbours. Directors of large companies should be confronted with victims of the polluted environment they cause on a regular basis, which would perhaps influence their decision making, when it comes to protecting the environment. This technique is especially useful for preventing nuisance and environmental degradation, illegal fishing, and illegal handling with protected animals and plants.

24. Assisting compliance – subventions are most certainly a measure, which assists compliance. It is often the municipality or the state, who is the one who gives subventions; however it is the European Union who is becoming a major source for environmental subventions and other forms of financial stimulations.[35] Another example of assisting compliance is regular collection and removal of large and hazardous waste.[36] Lemieux and Clarke (2009: 465) state, that local people in Africa will only stop killing elephants, if they find elephant survival is also beneficial to themselves. Therefore eco-tourism could be a good way, when it is assured that the local people are really benefiting from it. This technique is especially useful for preventing nuisance and environmental degradation, illegal hunting and illegal fishing.

25. Controlling drugs and alcohol – consideration of a total ban on alcohol and other drugs consumption in natural parks would mean that alcoholic beverages, cigarettes and similar products could not be bought in mountain cabins and tourist centres. Personal intake of these substances would also have to be prohibited and controlled. Other areas, where human error resulting from inappropriate psychophysical condition can lead to ecological disaster, are various work places, where error can result in severe pollution. Such work places exist in nuclear power plants, in maritime traffic, on oil rigs and refineries, chemical factories and alike. Every work place should have a risk assessment, and for all of them, where environmental hazard would be established, a strict prohibition of any drug usage should be enforced. This technique is particularly useful for preventing nuisance and environmental degradation, animal torture and unconscionable veterinary assistance.

3.5 Conclusion

Crimes against nature are far-reaching, dangerous and complex crimes that destroy the environment and consequently affect our society and world. Therefore, efforts to detect and deter environmental crimes should be effective. To achieve this, measurable

---

[35] More about European Unions' subventions available on http://ec.europa.eu/environment/funding/intro_en.htm.

[36] An example of this is an annual activity of pharmacology students association, which organizes collection and removal of old medicines posing threat to environment (Farma društvo 2010).

data, records, information, case studies and surveys, and useful suggestions formed on the basis of obtained results are important.

The role of prevention environmentally harmful activities is attached to criminology, because it was criminology that realized humans cannot conquer nor control nature. People are much better off constructing our lives within ecological constraints; therefore, they have to pass over from reactive to proactive (preventive) behaviour towards the environment.

Situational crime-preventive aspects of the response to the environmental threats represent useful ideas and methods to improve the current situation in Slovenia, in the South Eastern European region, and elsewhere in the world. Situational crime-prevention does not have the ambition of eliminating the so called root causes of (environmental) crime, but rather its goal is to solve these issues very pragmatically and with concrete actions. Transferring situational crime prevention techniques to crimes against the environment involves the basic situational crime prevention principle: reduce the commission of crime by designing models that eliminate the crime opportunities (e.g. redesigning enforcement strategies to cut off industry-specific criminal opportunities; improvement of enforcement effectiveness with the emerging knowledge of the offender's characteristics and with the increase of technical training etc.).

On the basis of previous research and analysis of the use of situational crime-prevention methods in Slovenia, an analysis of the detected forms and consequences of environmental threats was made. Furthermore, the analysis of the threats was made on the basis of three theories of situational crime-prevention using the method of reflection.

The analysis of the environmental criminality in Slovenia confirms the need to move from reactive to proactive responses to the crimes against the environment. One of these proactive responses is the use of crime-prevention theory. A thorough analysis has to be conducted on every concrete case or a problem needing resolution before it is designed and applied any sort of preventive measure, and environmental crime is no exception in this manner. After a detailed analysis of the detected forms and consequences of environmental threats, a selection and adjustment of individual techniques from Clarkes' situational crime-prevention model, containing 25 techniques in five clusters, follows.

To prepare the effective situational crime-prevention measures, any kind of generalization must be avoided. A thorough analysis has to be conducted on every concrete crime problem. Therefore, it is before the use of the situational crime prevention techniques in practice, according to standard methodology for each project of situational crime prevention necessary to carry out all five steps of action research, as proposed by Clarke (1997: 15): (1) collecting data on the nature and dimensions of the problem; (2) analysis of the situational conditions that allow or even enable the enforcement of the offense; (3) systematic study of the resources available to prevent opportunities for these crimes, including cost analysis; (4) implementation of the most promising, feasible and cost-effective action; and (5) monitoring of results and dissemination of experience.

Presented situational crime-preventive aspects of the response to environmental threats represent a novelty in the Slovenian practice of crime prevention methods in the environment protection field. The findings of the debate can be used in the planning of crime-preventive measures with regard to environmental threats. The idea of opportunity, social control and elements of 'intimidation' – the probability (reliability), the speed and severity of the response stand out. Furthermore, these measures represent a very useful starting point for all further research and development of programs, resolutions and other documents about responding to environmental threats.

Situational crime-prevention has, in combination with rising of people's awareness on crime and other forms of threats, proven to be quite an effective form of prevention. Nevertheless, it is necessary to be aware that situational crime prevention is just one piece of the jigsaw of modern societies responding to environmental threats. However, situational crime-prevention is no insignificant part in this process, especially if it is used as an element in problem solving-oriented way of responding to problems and not just as a political attempt to publicly likeable rapid elimination of current problems (White 2008: 237). Given the previous failure of other approaches to addressing issues of environmental crime, development and application of situational prevention measures is a promising attempt to achieve tangible results and may protect the environment in which people live up to the moment when humanity realizes the need for radical social changes.

References

Adler F (1996) Offender-specific vs. offence-specific approaches to the study of environmental crime. In: Edwards SM, Edwards TD, Fields CB (eds) Environmental crime and criminality: theoretical and practical issues. Garland Publishing, New York, pp 35–54

African Conservation Foundation (2010) Resource document. http://www.africanconservation. org/content/view/1516/409/. Accessed 12 May 2010

Australian Government (2010) Great barrier reef. http://www.gbrmpa.gov.au. Accessed 5 Mar 2010

BBC (2010) Dubrovnik fire threat subsiding. http://news.bbc.co.uk/2/hi/europe/6932578.stm. Accessed 5 Mar 2010

Biderman AD, Reiss JR (1976) On exploring the "Dark Figure" of crime. Ann Am Acad Pol Soc Sci 374(1):1–15

Carina (2010a) Radioaktivni kos kovine med odpadnim železjem (Eng. Radioactive Piece of Metal among Iron Waste). http://www.carina.gov.si/si/novosti/sgd/radioaktivni_kos_kovine_ med_odpadnim_zelezjem/. Accessed 12 May 2010

Carina (2010b) Carina v službi okolja – varujemo naravno dediščino (Eng. Customs in the Service of Environment – Protection of Natural Heritage). http://www.carina.gov.si/si/novosti/sgd/ carina_v_sluzbi_okolja_varujemo_naravno_dediscino/. Accessed 12 May 2010

Carter T (1998) Policing the environment. In: Clifford M (ed) Environmental crime: enforcement, policy, and social responsibility. Aspen Publishers, Gaithersburg, pp 169–203

Clarke R (1992) Introduction. In: Clarke R (ed) Situational crime prevention: successful case studies. Harrow & Heston, New York, pp 3–38

Clarke R (1997) Introduction. In: Clarke R (ed) Situational crime prevention: successful case studies, 2nd edn. Harrow & Heston, New York, pp 2–44

Clarke VR, Eck JE (2008) Priročnik za policijske (kriminalistične) analitike – v 60 korakih do rešitve problema (Eng. Crime analysis for problem solvers – in 60 small steps). Fakulteta za varnostne vede, Univerza v Mariboru, Ljubljana

3 Situational Crime-Prevention Measures to Environmental Threats 65

Clifford M (1998) Environmental crime: enforcement, policy, and social responsibility. Aspen Publishers, Gaithersburg

Cohen LE, Felson M (1979) Social change and crime rate trends: a routine activity approach. Am Sociol Rev 44(4):588–608

Cornish D, Clarke RV (1986) Introduction. In: Cornish D, Clarke RV (eds) The reasoning criminal. Springer, New York, pp 1–16

Cornish DB, Clarke RV (2003) Opportunities, precipitators and criminal decisions: a reply to Wotley's critique of situational crime prevention. In: Smith MJ, Cornish DB (eds) Theory for practice in situational crime prevention. Criminal Justice Press, New York, pp 41–96

Delo (2010) Komenski požar pod nadzorom (Eng. Fire on Komen under control). http://www.delo.si/clanek/o150770. Accessed 5 Mar 2010

Dežman Z (2004) Nekaj kritičnih pogledov na novi model slovenskega kazenskega postopka (Eng. Some critical views on the new model of Slovenian criminal procedure). Podjetje in delo 30(6/7):1575–1589

Ekblom P (2006) Situational crime-prevention. In: McLaughlin E, Muncie J (eds) The sage dictionary of criminology. Sage, London, pp 383–385

Farma društvo (2010) Resource document. http://www.farma-drustvo.si/project.php?id=163. Accessed 5 Mar 2010

Faure GM, Visser M (2003) Law and economics of environmental crime: a survey. http://www.hertig.ethz.ch/LE_2004_files/Papers/Faure_Environmental_Crime.pdf. Accessed 29 Jan 2008

Geopedia.si (2010) Register divjih odlagališč (Eng. Register of illegal dumps). http://www.geopedia.si/?params=T1199_vT_b4_x483479_y102619_s9#T1199_x483479_y102619_s9_b4_vT. Accessed 12 May 2010

Heckenberg D (2009) Studying environmental crime: key words, acronyms and sources of information. In: White RD (ed) Environmental crime: a reader. Willan Publishing, Cullompton, pp 9–24

Hirschi T (1969) Causes of delinquency. University of California Press, California

Homer-Dixon TF (1994) Environmental security and violent conflict. Int Secur 19(1):5–40

Hughes G (1998) Understanding crime prevention: social control, risk and late modernity. Crime and Justice, Open University Press, Buckingham

IAEA (2010) International Atomic Energy Agency. Resource document. http://www.iaea.org/OurWork/index.html. Accessed 5 Mar 2010

IRSOP (2008) Ministrstvo za okolje in proctor: Poročilo o delu Inšpektorata Republike Slovenije za okolje in prostor za leto 2008 (Eng. Ministry of the Environment and Spatial Planning: Report on the Work of the Inspectorate of the Republic of Slovenia for the Environment and Spatial Planning for 2008). Resource document. Inspectorate of the Republic of Slovenia for the Environment and Spatial Planning. http://www.iop.gov.si/fileadmin/iop.gov.si/pageuploads/IRSOP_dokumenti/POROCILO_IRSOP_ZA_2008.pdf. Accessed 23 July 2009

Kazenski zakonik Republike Slovenije [KZ-1] (2008) (Eng. Penal Code of the Republic of Slovenia). Uradni list Republike Slovenije, št. 55/2008, 66/2008

Keel RO (1997) Rational choice and deterrence theory. Resource document. http://www.umsl.edu/~keelr/200/ratchoc.html. Accessed 10 May 2010

Križna jama (2010) Resource document. http://www.krizna-jama.si/uvod_ENG.htm. Accessed 12 May 2010

Lemieux AM, Clarke RV (2009) The international ban on ivory sales and its effects on elephant poaching in Africa. Br J Criminol 49(4):451–471

Lewis H (2009) Race, class, and katrina: human rights and (Un)natural disaster. In: Steady FC (ed) Environmental justice in the new millennium. Palgrave Macmillan, New York, pp 233–251

Lewis DL, Chepesiuk R (1994) The international trade in toxic waste: a selected bibliography of sources. Electron Green J 1(2):1–15

Luka Koper (2004) Odnos do okolja (Eng. Environmental management). http://www.luka- kp.si/slo/o-podjetju/odnos-do-okolja. Accessed 10 May 2010

Medvedi (2010) Resource document. http://www.medvedi.si/index.php?option=com_content&view=article&id=79:telem-projekt-podrobno&catid=33:telemetrijasplosno&Itemid=90. Accessed 12 May 2010

OrganizedWisdom (2010) Resource document. http://organizedwisdom.com/helpbar/index. html?return=http://organizedwisdom.com/Pyromania&url=www. usfa.dhs.gov/downloads/ pdf/publications/fa-239.pdf. Accessed 17 Apr 2010

Pečar J (1981) Ekološka kriminaliteta in kriminologija (Eng. Environmental crime and criminology). Revija za kriminalistiko in kriminologijo 1(34):33–45

Planet Zemlja (2010) Planetu Zemlja prijazna vozila (Eng. Planet earth friendly vehicles). http://www.planet-zemlja.org/planetu-zemlja-prijazna-vozila/. Accessed 17 Apr 2010

Policija (2002) Poročilo o delu policije za leto 2002 (Eng. Report on the Work of the Police in 2002). Resource document. Police. http://www.policija.si/images/stories/Statistika/ LetnaPorocila/PDF/lp2002.pdf. Accessed 15 May 2010

Policija (2003) Poročilo o delu policije za leto 2003 (Eng. Report on the Work of the Police in 2003). Resource document. Police. http://www.policija.si/images/stories/Statistika/ LetnaPorocila/PDF/lp2003.pdf. Accessed 15 May 2010

Policija (2004) Poročilo o delu policije za leto 2004 (Eng. Report on the Work of the Police in 2004). Resource document. Police. http://www.policija.si/images/stories/Statistika/ LetnaPorocila/PDF/lp2004.pdf. Accessed 15 May 2010

Policija (2005) Poročilo o delu policije za leto 2005 (Eng. Report on the Work of the Police in 2005). Resource document. Police. http://www.policija.si/images/stories/Statistika/ LetnaPorocila/PDF/lp2005.pdf. Accessed 15 May 2010

Policija (2006) Poročilo o delu policije za leto 2006 (Eng. Report on the Work of the Police in 2006). Resource document. Police. http://www.policija.si/images/stories/Statistika/ LetnaPorocila/PDF/pol-lp2006.pdf. Accessed 15 May 2010

Policija (2007) Poročilo o delu policije za leto 2007 (Eng. Report on the Work of the Police in 2007). Rersource document. Police. http://www.policija.si/images/stories/Statistika/ LetnaPorocila/PDF/pol-lp2007.pdf. Accessed 15 May 2010

Policija (2008) Poročilo o delu policije za leto 2008 (Eng. Report on the Work of the Police in 2008). Rersource document. Police. http://www.policija.si/images/stories/Statistika/ LetnaPorocila/PDF/LetnoPorocilo2008.pdf. Accessed 15 May 2010

POP Center (2010) Resource document. http://www.popcenter.org/25techniques/. Accessed 5 Mar 2010

Resolucija o nacionalnem programu varstva okolja 2005–2012 [ReNPVO] (2006) (Eng. Resolution on National Program of Environmental Protection 2005–2012). Uradni list Republike Slovenije, št. 2/2006

Rosenbaum DP, Lurigio AJ, Davis RC (1998) The prevention of crime. Wadsworth Publishing, Belmont

Seis M (1996) A native american criminology of environmental crime. In: Edwards SM, Edwards TD, Fields CB (eds) Environmental crime and criminality: theoretical and practical issues. Garland Publishing, New York, pp 121–146

Selinšek L (2003) Kazenskopravno varstvo okolja in naravnih dobrin – izziv ali nuja? – (Eng. Criminal law protection of the environment and natural assets – a challenge or a necessity?). Pravnik 58(9–12):651–672

Šinkovec J (1994) Pravo okolja: načela in mednarodnopravni prikaz (Eng. environmental law: principles and international law relation). Uradni list Republike Slovenije, Ljubljana

Steinway DM, Botts B (2005) Fundamentals of environmental law. In: Sullivan TPF (ed) Environmental law handbook, 18th edn. The Scarecrow Press, Lanham, pp 1–10

Strojin T (1987) Uvod v pravo okolja (Eng. Introduction to environmental law). Delavska enotnost, Ljubljana

Strojin T (1994) Osnove prava okolja (Eng. Basics of environmental law). ČZ Uradni list Republike Slovenije, Ljubljana

TimesOnline (2010) Fire hits Spain's La Palma in Summer of Blazes. http://www.timesonline. co.uk/tol/news/world/europe/article6735877.ece. Accessed 5 Mar 2010

Umanotera (2010) Resource document. http://www.umanotera.org/upload/files/Ogljicni_odtis___ primeri_DRAFT.pdf. Accessed 5 Mar 2010

Ustava Republike Slovenije [URS] (1991) (Eng. Constitution of the Republic of Slovenia). Uradni list Republike Slovenije, št. 33/1991

3 Situational Crime-Prevention Measures to Environmental Threats

Veterinarska inšpekcija (2010) Zakonodaja (Eng. Legislation). Resource document. http://zakonodaja.gov.si/rpsi/r08/predpis_ZAKO4018.html. Accessed 5 Mar 2010
Viler Kovačič A (2007) Spremembe ZVO. *Pravna praksa*, 26 (44): 17–18
Weisburd D (1997) Reorienting crime prevention research and policy: from the causes of criminality to the context of crime, Research Report. National Institute of Justice, U.S. Department of Justice, Washington
White R (2003) Environmental issues and the criminological imagination. Theor Criminol 7(4):483–506
White R (2008) Crimes against nature: environmental criminology and ecological justice. Willan Publishing, Devon
White RD (2009) Environmental crime: a reader. Willan Publishing, Cullompton
Zakon o gozdovih (1993) (Eng. Forests Act). Uradni list Republike Slovenije, št. 30/1993
Zakon o kemikalijah (2003) (Eng. Chemicals Act). Uradni list Republike Slovenije, št. 110/2003
Zakon o planinskih poteh (2007) (Eng. Act on the Mountain Trails). Uradni list Republike Slovenije, št. 61/2007
Zakon o plovbi po celinskih vodah (2002) (Eng. Inland Waterway Transport Act). Uradni list Republike Slovenije, št. 30/2002
Zakon o prostorskem načrtovanju (2002) (Eng. Spatial Planning Act). Uradni list Republike Slovenije, št. 110/2002
Zakon o Triglavskem narodnem parku (2001) (Eng. Law on the Triglav National Park). Uradni list Republike Slovenije, št. 35/2001, 110/2002
Zakon o varnosti cestnega prometa [ZVCP] (2004) (Eng. Law on Road Traffic Safety). Uradni list Republike Slovenije, št. 83/2004
Zakon o varnosti na smučiščih (2002) (Eng. Act on the Safety on Ski Slopes). Uradni list Republike Slovenije, št. 110/2002, 98/2005, 3/2006, 17/2008, 52/2008
Zakon o varstvu javnega reda in miru (2006) (Eng. Act on the Protection of Public Order and Peace). Uradni list Republike Slovenije, št. 70/2006
Zakon o varstvu narave [ZON] (1999) (Eng. Nature Conservation Act). Uradni list Republike Slovenije, št. 56/1999, 31/2000, 119/2002
Zakon o varstvu narave [ZON-1] (2004) (Eng. Nature Conservation Act). Uradni list Republike Slovenije, št. 96/2004, 8/2010
Zakon o varstvu okolja [ZVO-1-UPB-1] (2006) (Eng. Environment Protection Act) Uradni list Republike Slovenije, št. 39/2006
Zakon o varstvu pred naravnimi in drugimi nesrečami [ZVNDN] (2006) (Eng. Act on Protection Against Natural and Other Disasters). Uradni list Republike Slovenije, št. 51/2006
Zakon o varstvu pred požari (2007) (Eng. Fire Protection Act). Uradni list Republike Slovenije, št. 3/2007
Zakon o varstvu živali [ZZŽiv] (1999) (Eng. Animal Protection Act). Uradni list Republike Slovenije, št. 98/1999, 126/2003, 14/2007
Zakon o vodah [ZV] (2002) (Eng. Waters Act). Uradni list Republike Slovenije, št. 67/2002, 110/2002
Županova jama (2010) Resource document. http://zupanova-jama.dolenjska.com. Accessed 12 May 2010

Chapter 4
The Role of Economic Instruments in Eco-Crime Prevention

Radmilo V. Pešić

Abstract The chapter deals with economic policy instruments for environmental policy with an emphasis on the instruments' potential to prevent eco-criminal activities. Certain types of economic instruments, frequently used in environmental policy, can serve as prevention tools in a struggle against organized eco-crime. The purpose of the chapter is to put forward a proposal for an adequate instrument design, and is aimed to help policy-makers to set up not only efficient public policy tools, but to create an economic support for the functioning of the legal system. The chapter is based upon the rational polluter model (Spence 2001; Emery and Watson 2004). The Emery–Watson model is elaborated and extended in a dynamic context, introducing discount rates and present value calculations. It is used to analyze the economic impacts of environmental policy instruments on polluters' behaviour. Finally, a set of policy recommendations obtained from the extended model is confronted with the lessons learned from the Republic of Serbia environmental policy.

4.1 Introduction

The concept of rationality is deeply rooted in the essence of economic science. Either a definition of economics as a "science of choice" is accepted, or as a "science of contract" (Buchanan 1975), it has actually been presupposed that economic agents are rational, and that all decisions are made rationally, based upon available information. If economics is considered as "a science of contract", it has been assumed that both sides involved in contracting are, to the extent of available information, rational. Otherwise, no contract would have ever been made. In various schools of economic thinking, economic choice is considered rational, but not

R.V. Pešić (✉)
Department of Agricultural Economics, Faculty of Agriculture,
University of Belgrade, Belgrade, Serbia
e-mail: radmilo@sbb.rs

G. Meško et al. (eds.), *Understanding and Managing Threats to the Environment in South Eastern Europe*, NATO Science for Peace and Security Series C: Environmental Security, DOI 10.1007/978-94-007-0611-8_4, © Springer Science+Business Media B.V. 2011

always to the same degree. Neoclassical economics is based upon an absolute rationality premise. Neo-institutional or transaction cost economics relies on "bounded rationality". This means that economic actors are assumed to be "*intendedly rational, but only limitedly so*" (Simon 1961) because of their limited cognitive competence. However, limited rationality is in fact rationality under incomplete information, and the limits of rationality are not to be mistakenly interpreted as non-rationality or irrationality (Williamson 1985). Therefore, it is not a surprise that economists usually regard themselves as "*guards of rationality*" (Arrow 1974).

In this chapter an attempt has been made to use a Rational Polluter Model (Spence 2001) as a starting point in analyzing of economic instruments and their role in environmental policy. A broad concept of environmental economic instruments, as institutional and regulatory arrangements that produce economic effects on involved agents and on society, has been accepted. The concept has included not only direct regulation and economic incentive instruments, but discount rates, eco-fines and penalties as well.

4.2 The Rational Polluter Model

A common approach in the theory of environmental regulation is to use the Rational Polluter Model which describes the response of organizations and individuals to environmental legislation. The Rational Polluter Model has had a profound impact on American environmental regulation (Spence 2001). Although, it is found valid for an individual's behaviour it offers only a partial explanation of organizational behaviour. The virtue of the model in terms of its explanatory role has been linked with individual behaviour. It is well known that organizations including business firms do not react in the same way as individual actors do (Williamson 1985). However, behind any decision of an organization stands an individual either as an owner or as a manager. Coleman (1990), while recognizing the difference between organizations and individuals, does not see this as sufficient reason to refute the rational choice perspective (Emery and Watson 2004). Therefore it is theoretically justified to accept the rationality concept as a general pattern of economic behaviour. Starting from such premises David Spence in his seminal article (2001) has developed a simple mathematical model of the rational polluter's behaviour based on expected value of an eco-crime, or of a non-compliance with environmental legislation. Using original notation the model can be described as follows:

$$E(NC) = [S - pF] \tag{4.1}$$

Where:

$E(NC)$ = the expected value of non-compliance,

S = the economic benefit (or savings) associated with non-compliance, such as money saved by taking fewer steps to minimize pollution, failing to monitor, or failing to report as required by law,

4 The Role of Economic Instruments in Eco-Crime Prevention

pF = the expected costs of non-compliance,
p = the probability that a violation will be detected,
F = the expected penalty (or fine) if detected.

Conclusions from the model say that if expected value of non-compliance $E(NC)$ is negative, then the rational polluter will comply with the legal regulation, and if $E(NC)$ is positive, the law will be violated. The conclusions are consistent with micro-economic theory, predicting that the polluter will continue to pollute up to the point where marginal cost equals marginal benefit of doing so.

Emery and Watson (2004) state that the Rational Polluter Model holds for small firms where decision-making power remains in the hands of a single person, but the complexity of larger organizations makes the model inadequate. In order to find an explanation of why organizations may wish to stay within the legal framework Emery and Watson have turned attention to a theory suggesting that organizations, if they want to survive and to develop, have to be legitimized by society (Weber 1996). In spite of being focused on the legitimacy theory, the authors have concluded that "the rational polluter model is not without value but at best it may offer a partial explanation of organizational behaviour and at worst an inappropriate premise for legislation" (Emery and Watson 2004).

4.3 The Extended Rational Polluter Model

Previous argument about the Rational Polluter Model and its potential validity is based on a simple and rudimentary version of the model given by Spence (2001). If the model is to be properly assessed, it has to be developed in a more complex and dynamic sense. Assuming that the Rational Polluter Model has a high theoretical value rooted in the essence of economics, an upgrade has been made by introducing a distinction between direct and indirect benefits of non-compliance. It has also been assumed that the benefits are not static, but come as a series of events, or as a flow in a time dimension, up to the moment of detection. In order to make the model more realistic, a time dimension of envisaged penalty is included. The extended version of the model consists of a sum of net present values of benefits over time, minus a net present value of the expected non-compliance costs, obtained by multiplying the probability of detection p, by the net present value of the penalty in the moment of detection. The extended Rational Polluter Model is:

$$E(NC) = \sum_{t=0}^{K} \frac{S_1 t + S_2 t}{(1+r)^t} - p \frac{F}{(1+r)^K} \qquad (4.2)$$

Where:

$E(NC)$ = the expected value of non-compliance,
$S_1 t$ = the direct economic benefit of eco-crime or of a non-compliance with environmental legislation (e.g. economic benefit of illegal toxic waste dumping, or of illegal wastewater discharge into a river body, or polluting the atmosphere),

S_2t = the indirect benefit of non-compliance from unpaid eco-taxes or pollution charges in case of the environmental law violation, (e.g. whenever the waste is illegally dumped or wastewater discharged, no eco-charges are being paid),

r = discount rate,

t = period of violation (time from the beginning of an illegal practice, t=0, to the moment of detection, K),

K = moment in time when eco-crime or non-compliance with eco legislation is being detected, prosecuted and sentenced (end of a violation period),

p = the probability that a violation will be detected, prosecuted and sentenced,

F = the envisaged penalty (or fine) if detected.

The same conclusions as from the original model can be drawn from the extended model. If the expected value of non-compliance $E(NC)$ is negative, then the rational polluter will comply with the environmental regulation, and if $E(NC)$ is positive there is a high probability that the law is to be violated.

What are the other conclusions that can be made from the extended model? First of all, the higher the eco-standards imposed by a regulator, the higher the direct economic benefits of non-compliance. This means that whenever society wants to implement high standards of environmental protection and quality, there is more danger of eco-crime or of non-compliance with the legislation. Secondly, the longer the period of violation t, the higher the sum of benefits obtained from non-compliance. Also, the longer the violation period t, the higher the value of K, so the net present value of the expected penalty, F, is lower. Thirdly, the higher the eco-taxes and eco-charges, the higher the indirect benefits of non-compliance, created by the non-payment. All the mentioned parameters increase the expected value of non-compliance $E(NC)$. However, the envisaged penalties are pushing in the opposite direction. The higher the expected penalties defined by regulators, the less attractive the eco-crimes, so the expected value $E(NC)$ is lower.

4.4 Implications

All these conclusions have direct policy implications. In order to set up a good environmental policy it is essential to diminish the expected value of non-compliance $E(NC)$. The list of policy recommendations consists of:

1. Shortening the period of time t, from "now" to the moment of detection K. There is no doubt that long periods of an undetected illegal behaviour will produce long streams of benefits coming from the non-compliance. Also, the net present value of any penalty will be low, creating an additional motive for eco-crimes. Therefore, it is necessary for an effective environmental policy to make the time period t as short as possible, under the existing legal and institutional framework.
2. Impose moderate eco-taxes and eco-charges. High eco-taxes and charges make non-compliant behaviour even more attractive because no charges or taxes are

paid on the basis of illegal operation. Generally, a non-compliant behaviour is always non-taxable. It will not be a surprise that high eco-charges may create strong additional motives for eco-crimes.

3. Unrealistically high eco-standards increase the expected value of non-compliance, too. Generally, all the eco-standards that create public costs above a socially-sustainable level or above the level that can be achieved without a significant sacrificing of other social goals are to be considered unrealistic. High eco-standards demand high costs of monitoring and enforcement. Over ambitious eco-targets in developing and transitional economies are unlikely to be achieved. Therefore, realistically-defined environmental standards, particularly the standards that can be accomplished with moderate financial resources are to be implemented. Instead of highly ambitious environmental quality requirements, imposed on poor developing communities and households, it would be better to promote a gradual policy approach, with realistically-defined standards that are subject to frequent revisions, simultaneously with re-evaluations of eco-policy instruments. It is always good to start with a "low-hanging fruit" and to reach higher standards later. It appears to be better to impose moderate eco-standards and frequently assess their implementation, than to start with far reaching high goals that would only increase the expected value of non-compliance.

4. Imposition of high penalties for non-compliant behaviour is essential for a sound environmental policy. From the extended model, it can be seen that high F would diminish the expected value of non-compliance $E(NC)$, and consequently decrease the motivation for the eco-crime.

5. Rational eco-offenders regularly do not make their judgments upon the envisaged levels of penalty, but on the present value of penalty. This means that discount rate levels play a significant role in decision-making about the eco-crime rationality. It becomes obvious that high discount rates would have the same effect on the expected value of non-compliance as protracted periods of time t, from the beginning of the non-compliant behaviour to the moment of detection. This gives an additional rationale for minimising discount rates. In environmental economics literature (Perman et al. 1996), it has been recognized that high discount rates are against the activities and aspirations of conservationists. If the Hotelling (1931) rule reasoning is to be accepted, low discount rates will be found favourable for natural resource preservation and will be recommended as a protection policy instrument. The conclusion based upon the Extended Rational Polluter Model fortifies the protectionists' claim for low discount rates.

6. One of the most important barriers against eco-crime is the high probability of detecting, prosecuting and sentencing eco-offenders, denoted as p in the model. There is no need to explain that the higher the probability p, the lower the expected values of non-compliance. In order to make the probability of detection and sentencing high, it is essential to provide cost-effective and efficient legal procedures against the eco-offenders. Costly and slow procedures diminish p. Emery and Watson (2004) described a well-documented tendency for regulators to concentrate on relatively trivial offenders or on small-size operators rather

than on the more serious offences committed by large corporations (Diver 1980; De Prez 2001). Similar cases can be found in most of developing and transitional economies. Therefore, high investments in capacity building of the judicial system are needed, rather than investing in expensive high-tech monitoring devices. It is more effective to create efficient courts than to invest in costly equipment for detection. Detection is just the first step in a legal combat against the offenders; the court is the place where they are to be defeated.

7. The high probability of detecting, prosecuting and sentencing is related not only with the legal system but with an appropriate design of policy instruments. In environmental economics theory there are two main categories of instruments: economic incentive instruments (EII) and regulatory instruments often referred as command and control instruments (CACI). In recent years, EII are becoming more and more popular on both sides of the Atlantic Ocean. However, in a survey of professional opinions gathered in Europe and in the United States (Frey et al. 1985), it has been shown that support for EII was stronger in academic circles than amongst businesses and policy-makers. Compared to regulatory policies based upon CACI, economic incentive policies are proved to be more efficient, both in static and in dynamic senses (Harrington et al. 2004). EII require less information than CACI, and therefore are less costly to be implemented. One of the advantages of EII can also be found in lower administrative costs. Regulatory policies have higher costs because establishing a regulation system requires setting specific tasks for each of the regulated entities, whereas charges and fees would be set uniformly to all the polluting sources. The multiplicity of regulation standards involved in CACI could open space for lobbying and informal organizing of polluters which might compromise the entire policy efforts. Monitoring and enforcement costs are expected to be lower with EII, particularly with tradable permit systems, because all of the participants have a strong motivation for mutual monitoring. In order to make the policy system efficient it is good to encourage reliable self-monitoring and self-reporting activities. However, there is a broad consensus that regulatory instruments achieve policy objectives faster and with greater certainty than incentives (Harrington et al. 2004). Better effectiveness of CACI is particularly linked with highly differentiated spatial and temporal effects of pollution. It is easier to impose more stringent pollution reductions at those plants where the emissions cause greater damage (Rose-Ackerman 1973). Therefore a selective approach to the instrument design and a gradual approach to their calibration are to be recommended.

If the Extended Rational Polluter Model is applied to a specific country, the policy implications are becoming more valid than theoretical argumentation. After analyzing legal documentation it becomes clear why eco-policy outcomes in the Republic of Serbia can be considered as unsatisfactory. In Article 116 of the Environmental Protection Act (Zakon o zaštiti prirode 2009) it has been envisaged that a legal entity is obliged to pay fines ranging from RSD 150,000 to RSD 3,000,000 (about €1,500–€30,000) if it commits one or other of the predefined commercial offences.[1] For these offences a responsible person within the legal entity shall also pay a fine ranging from RSD 30,000 to RSD 200,000 (about €300–€2,000). It is more than obvious that such low level penalties have no or negligible effect on the polluters' behaviour. Compared to the potentially high benefits from a non-compliant behaviour, such low level fines are not fulfilling their task of eco-crime prevention. Even if it is assumed that eco-crimes will be detected instantly ($t=0$), and with absolute certainty (probability $p=100\%$), which is practically impossible, the expected level of non-compliance costs remains too low. In 2009, the Environmental Protection Act was amended and new higher fines have been introduced, ranging from RSD 1,500,000 to RSD 3,000,000 (€15,000–€30,000) for legal entities, and from RSD 100,000 to RSD 200,000 (€1,000–€2,000) for the responsible persons. However, if the Rational Polluters' reasoning is applied, it appears that even recently increased penalties are still too low to serve as an obstacle for eco-crime.

The problem of low level penalties on eco-crime in Serbia is exacerbated by high discount rates. As in the other transitional economies, in Serbia discount rates above 10% are frequently used making future effects of any investment or disinvestment unrealistically low.

A set of problems with the functioning of the judicial system in Serbia seems to have even more deteriorating effects on the prevention of eco-crimes than the previously mentioned low level of penalties. Long periods of time from detection to sentencing are one of the most serious deficiencies of the national judicial system.[2] Problems of the efficacy of the legal system may lead to very low values of p (the probability that a violation will be detected, prosecuted and sentenced) and create additional motives for eco-offenders. Therefore, the recently initiated reforms of the judicial system in Serbia may be seen as inevitable preconditions for a better functioning of society as a whole, including the environmental regulation sector.

---

[1] Amongst the mentioned commercial offences are: (1) natural resources used without the Ministry consent, (2) failure to implement preventive measures, (3) not undertaking recultivation or rehabilitation of degraded environments, (4) collecting and selling of certain types of wild flora and fauna, (5) importing and exporting of endangered and protected species of wild flora and fauna, (6) if dealing with dangerous substances, not undertaking all the necessary protective and safety measures, (7) not making risk assessments from a potential accident, (8) releasing polluting and dangerous substances into the air, water or soil in a way and in quantities that are exceeding those prescribed, (9) producing and selling vehicles that do not fulfil conditions regarding emissions for mobile sources of pollution, (10) using devices for pollution elimination for which no domestic standards have been prescribed, (11) producing, importing and exporting substances that are depleting the ozone layer, (12) importing hazardous waste, (13) importing, exporting or transferring waste without a Ministry permit, (14) not insuring against damage made to third parties in an accident.

[2] According to a high executive of the Environmental Protection Ministry, in 2007 from over 600 accusations made by the National Environmental Inspection Office, only 18 were brought to court and prosecuted in the same year. Most of the others exceeded the deadline for court proceedings, and so were dropped. In 2008 and 2009 some improvements of the judicial system efficacy were made.

If focused on the economic tools for environmental policy in Serbia, it can be said that a gradual introduction of certain eco-charges[3] is absolutely in accord with "sound-policy" recommendations that can be drafted from the Extended Rational Polluter Model. In Article 27 of The directive on types of pollution, criteria for eco-charge calculations and their application, known as The Eco-Charge Directive, (Uredba o vrstama zagađivanja, kriterijumima za obračun naknade i obveznicima, visini i načinu obračunavanja i plaćanja naknade za zagađivanje životne sredine 2005), a full amount of payment is expected to start from January 1, 2016. Before that a gradual introduction of charges is envisaged so that from the end of 2005 to December 31, 2008 polluters were obliged to pay only 20% of the charges. From January 1, 2009 to December 31, 2011 only 40% of charges are expected to be paid, and from January 1, 2012 to December 31, 2015 charges amounting to 70% are expected to be paid. Finally from the beginning of 2016 the entire amount of charges is expected to be collected from the polluters. Having in mind that one of the recommendations from the extended model is to introduce moderate eco-charges, it can be concluded that economic instruments envisaged by the Serbian Eco-Charge Directive are well designed and properly calibrated in a time dimension. Therefore the imposed eco-charges are not expected to have adverse effects on the policy outcomes.

4.5 Conclusion

The Extended Model of a Rational Polluter gives a deeper insight into the motivation for the behaviour of eco-offenders. The expected value of non-compliance $E(NC)$ is considered to be a crucial argument in polluters' reasoning and can be used as an explanatory tool for the behaviour of both individuals and organizations. The Extended Model consists of a sum of net present values of direct and indirect benefits, minus net present value of the expected non-compliance costs, obtained by multiplying detection probability by the net present value of the penalty at the moment of detection. In order to create a proposition for an effective environmental policy that prevents eco-crime, potential determinants of the polluters' behaviour have been analyzed. Among the most important are: (a) the time period from "now" to the moment of detecting an eco-crime being shortened as much as possible (b) moderate eco-taxes and eco-charges, alongside realistically-defined eco-standards (c) prohibitively high penalties for a non-compliant behaviour (d) low discount rate levels (e) high probability of detecting, prosecuting and sentencing eco-offenders and (f) an adequate set of economic instruments for environmental policy, carefully-designed and calibrated to nation-specific conditions and circumstances.

---

[3]In the Eco-Charge Directive industrial waste charges plus SO_2, NO_2 and particle matter charges are envisaged (Pravilnik o utvrđivanju usklađenih iznosa naknade za zagađivanje životne sredine 2009).

4 The Role of Economic Instruments in Eco-Crime Prevention   77

Not only can all the above mentioned be analyzed in a theoretical context but it can
be applied to policy-making practice and can serve as a guidepost for a specific
country's efforts to make sound environmental policy.

Acknowledgments The author is grateful to Professor Steve A. Quarrie, from the Newcastle
University for a peer review and English language interventions.

References

Arrow KJ (1974) The limits of organization. W.W. Norton, New York
Buchanan J (1975) A contractarian paradigm for applying economic theory. Am Econ Rev
   65(May):225–230
Coleman JS (1990) Foundations of social theory. Belknap, Cambridge
Diver C (1980) A theory of regulatory enforcement. Public Policy 28(3):257–299
De Prez P (2001) Biased enforcement or optimal regulation? Reflections on recent parliamentary
   scrutiny on the environmental agency. Environ Law Manage 13(3):145–150
Emery A, Watson M (2004) Organizations and environmental crime – legal and economic
   perspectives. Manage Audit J 19(6):741–759
Frey B, Schneider F, Pommerehne WW (1985) Economists' opinions on environmental policy
   instruments: analysis of a survey. J Environ Econ Manage 12(1):62–71
Harrington W, Morgenstren RD, Sterner T (eds) (2004) Choosing environmental policy:
   comparing instruments and outcomes in the United States and Europe. Resources for Future,
   Washington, DC
Hotelling H (1931) The economics of exhaustible resources. J Polit Econ 39(2):137–175
Perman R, Ma Y, McGilvray J (1996) Natural resource and environmental economics. Longman,
   London and New York
Pravilnik o utvrđivanju usklađenih iznosa naknade za zagađivanje životne sredine (2009) (Eng.
   Eco-Charge Directive). Službeni glasnik Republike Srbije, br. 7/2009
Rose-Ackerman S (1973) Effluent charges: a critique. Can J Econ 6(4):512–528
Simon H (1961) Administrative behaviour, 2nd edn. Macmillan, New York
Spence D (2001) The shadow of the rational polluter: rethinking the role of rational actor models
   in environmental law. Calif Law Rev 89(4):917–998
Uredba o vrstama zagađivanja, kriterijumima za obračun naknade i obveznicima, visini i načinu
   obračunavanja i plaćanja naknade za zagađivanje životne sredine (2005) (Eng. The Directive
   on Types of Pollution, Criteria for Eco-Charge Calculations and their Application). Službeni
   glasnik Republike Srbije, br. 113/2005, 6/2007
Weber M (1996) The theory of social organizations. Free Press, New York
Williamson O (1985) The Economic institutions of capitalism. Free Press, New York
Zakon o zaštiti prirode (2009) (Eng. Environmental Protection Act). Službeni glasnik Republike
   Srbije, br. 36/2009

Chapter 5
International Waste Trafficking: Preliminary Explorations

Ana Klenovšek and Gorazd Meško

Abstract This chapter concentrates on the problem of international waste trafficking, one of the dark sides of technological development in the late twentieth century. The first section frames illegal waste trafficking within the field of green criminology and cross-national environmental crime. The article goes onto define waste and the classification of waste in different categories. The first section concludes with the definition of the consequences of illicit waste dumping for the environment, people, and seeing waste trade through a human rights perspective. The second section looks into the reasons for international waste trafficking and the methods used; it discusses destination countries, phases of trafficking and the traffickers themselves. International legislation is the essential element of the third section, taking a closer look at the Basel, Bamako and Lomé IV Conventions. Domestic waste law in a sample of countries is further examined. To get a better insight into the seriousness of the problem, some of the most notorious cases are summarized in the fourth section. The final section presents some of the international illicit waste trafficking combating efforts, and the results of different national and international programs are discussed.

5.1 Understanding the Problem of Waste

Illicit waste trafficking, as a part of green criminology, has been gaining in prominence among criminologists during the past few years (Eman et al. 2009). Environmental crime is not new, but growing awareness has extended its scope (Carrabine et al. 2009).

A. Klenovšek (✉)
Faculty of Criminal Justice and Security, University of Maribor,
Kotnikova 8, 1000 Ljubljana, Slovenia
e-mail: ana.klenovsek@gmail.com

G. Meško
Faculty of Criminal Justice and Security, Institute of Security Strategies, University of Maribor,
Kotnikova 8, 1000 Ljubljana, Slovenia
e-mail: gorazd.mesko@fvv.uni-mb.si

G. Meško et al. (eds.), *Understanding and Managing Threats to the Environment in South Eastern Europe*, NATO Science for Peace and Security Series C: Environmental Security, DOI 10.1007/978-94-007-0611-8_5, © Springer Science+Business Media B.V. 2011

Many activities detrimental to the environment were not covered by the law in the past, but new environmental trends have promoted many changes in legislation.

Carrabine et al. (2009) classify green crimes into two categories: primary and secondary. Actions that negatively affect the environment are green crimes of the primary type. Secondary or symbiotic green crimes are criminal actions promoted by legal regulation; most of the hazardous waste trafficking belongs in this category.

Waste trafficking is further classified as a form of transnational environmental crime. Elliot (2009) defines transnational environmental crime as trafficking and smuggling of plants, animals, natural resources and contaminants in violation of multilateral international agreements and domestic law. Transboundary dumping of hazardous waste, illegal wildlife trade, timber trafficking and the ozone depleting substances black market are four categories of transboundary criminal activities where risks are low and profits high (Elliot 2009).

When it comes to waste, there is no single internationally recognized definition. Waste can be defined in numerous ways, depending on its environmental impact, its form, its properties, or its legal definitions (Buckingham and Turner 2008). Defining waste as hazardous or non-hazardous is usually controversial, and definitions of hazardous waste vary widely from country to country, creating loopholes and making it difficult to measure the volume of hazardous waste trade (Burns and Fuchs 2004). For example, the Basel Convention on the Control of Transboundary Movements of Hazardous Wastes and Their Disposal, instituted in 1989, defines waste by disposal destination or recovery processes (Puckett and Smith 2002). Two lists were established: List A consists of hazardous, and List B of non-hazardous waste. Electrical components that are intended for direct re-use are a part of List B (Puckett and Smith 2002). It will be later shown how this classification allows waste traffic.

A large proportion of today's exported waste is electronic waste or "E-waste", including computer components, cell phones and cathode ray tubes (CRTs). It is estimated that about 50 million tons of E-waste is produced annually (Nordbrand 2009). Because of continuous technological development the amount of obsolete electronics is expected to rise, with the global market growing by almost 9% annually. In 2007 there were approximately 140 million cell phones and 205 million computer products ready for disposal. There are estimates foretelling a huge increase in the numbers of obsolete TVs by 2011, because of the switch to digital broadcasting in North America and the EU (Isarin and Whitehouse 2009).

E-waste contains many toxic substances. Lead was removed from gasoline in the 1970s, but it was still present in all computer monitors and circuit boards until recently, causing potential damage to the nervous system, blood, kidneys and reproductive organs. Cables and computer housings, which are made of plastic that includes poly-vinyl-chloride (PVC), form dioxins when burned. Mercury is used in sensors, thermostats and switches and can cause damage to the brain, kidneys and foetus. Barium is used in CRTs to protect users from radiation but it can affect the brain, heart, liver and causes muscle weakness. Beryllium, widely used in semiconductors and lighting, is a known carcinogen. Studies have shown that black toner pigment (carbon black) is possibly carcinogenic to humans (Puckett and Smith 2002). These toxic substances are dangerous when not handled in an environmentally sound manner.

5 International Waste Trafficking: Preliminary Explorations

There seems to be more than enough reasons for defining E-waste as hazardous, but there is opposition to this from Canada and the United Stated of America (USA) which do not consider E-waste as hazardous. The Resource and Recovery Act (RCRA) is an important piece of federal legislation in the USA that regulates waste. It classifies E-waste sometimes as non-hazardous and sometimes not a waste at all; electronic devices that are planned to be re-used or recycled are classified as non-waste. The rationale behind this is that materials destined for recycling should not be classified as hazardous waste (Puckett and Smith 2002). They are categorized as products or commodities instead of waste (Interpol 2009). The problem is the propensity of this to be misused. Export of old or broken computers to developing countries has no positive effects.

E-waste is not the only type of hazardous waste that is often exported, whether legally or illegally. Radioactive waste, substances that are the result of processes that involve radioactive materials, is often trafficked. It is classified, depending on activity, as low-, intermediate- or high-level. Because exposure to highly radioactive waste can cause serious health problems, it is hard to find a place for its disposal (Buckingham and Turner 2008). Many cases of radioactive waste dumping at sea are acknowledged, especially dumping of broken nuclear submarine reactors or radioactive waste containers by the Russian navy in the Barents and Kara Seas. A successful action of Greenpeace in 2000 found 28,500 corroded containers of nuclear waste near the Channel Islands that were dumped by the United Kingdom (UK) during the 1950s and 1960s (Walters 2007). Besides producing significant danger to the environment and people, radioactive waste can be used for producing weapons and this should be taken into account when disposing of it.

How do all these toxic substances affect us? The south of Italy has long been known as a landfill for the entire country; more than 1,200 illegal dumping sites were identified in Campania alone. Uncontrolled disposal of waste has had devastating consequences for agriculture. In 2002 and 2003, toxins in sheep and cattle milk reached dangerous levels and more than 10,000 animals had to be euthanized and 9,000 l of contaminated milk had to be destroyed (Arie 2004). Massari (2004) notes a dramatic rise in the incidence of cancer with a 400% increase in some types of cancer in the same area.

A study carried out in Jinghai County, Tianjin, China provides yet another example. The DNA of 171 villagers who were constantly exposed to E-waste was compared to a control group of people from a neighbouring town who were not exposed. The results showed considerable differences in the DNA damage in the exposed group, which had drastically higher chromosome aberration rates. In women the amount of genetic damage was even higher than in men (Qiang et al. 2009).

The town of Guiyu, China, located near Hong Kong, has changed from a rural community to an E-waste processing centre by 1995. The way E-waste is handled is primitive and harmful to people working there and to the local ecosystem. In the year 2000 a water sample, a sediment sample, and three soil samples were taken along the nearby Lianjiang River to be analysed. The banks of the river had been used as a dumping ground for the disposed E-waste. The test revealed shocking levels of heavy metals. Water lead levels were 2,400 higher than the World Health

Organization (WHO) deems acceptable, and concentration of other toxic metals as well, exceeded recommended levels (Puckett and Smith 2002).

Illegal waste trafficking and dumping also has human rights implications. Gwam (2000) observed that economic, social and cultural rights are less protected in comparison to political and civil rights. Rights to life, health and a sound environment are often ignored by Western nations when trying to get rid of waste in a developing country. Developing countries lack proper facilities, trained, educated and well equipped staff that could manage waste in an environmentally sound manner. People, including children, are exploited while working in poor conditions, which sooner or later affect their health (Schmidt 2006). Public opinion in developed nations is against building recycling facilities, incinerators or having landfills in their proximity (O'Neill 1998). This is known as the NIMBY syndrome: "*Not In My Backyard*" (Gwam 2000). Such public opposition rarely happens when garbage is sent elsewhere and becomes someone else's problem.

5.2 Waste Trafficking

The 1970s marked the beginning of the international trade of toxic waste. Technological development progressed and the amount of generated waste was increasing (Krueger 1998). This continued to grow during the 1980s and the 1990s, and was also encouraged by globalization and the price reduction in transportation and communication networks (Clapp 1994). Those were also the times when environmental legislation in developed Western countries was becoming more and more stringent, costs of handling waste increased, disposal capacity decreased and export to countries with looser environmental regulation became very attractive (Brack and Hayman 2002). It is four times more expensive to incinerate waste in the EU than to illegally send it elsewhere (Rosenthal 2009). While in Africa a company needed and still needs to pay between US$2.50 and US$50 for disposal of 1 ton of their waste, this amount could reach US$2,000 in a developed country (Gwam 2000). The export of waste to less developed or developing countries was an easy solution to the problem. Typically, such countries do not only lack proper legislation that would protect their inhabitants and environment from large amounts of western garbage, but are desperately in need of money (Burns and Fuchs 2004). The example of Guinea Bissau illustrates the problem: they were offered US$600 million for storing and disposing hazardous waste, which is twice the amount of the country's foreign debt and about four times of its Gross National Product (Ibitayo 2008). Besides poorly implemented environmental policies, health-based standards are minimal, and there are little or no obstacles to building a landfill next to a very populated area. Residents are not aware of the dangers that are threatening them and the widespread corruption of government officials in many African countries compounds the problem. Western companies do not even need large sums of money to bribe an official; a small bribe opens the doors for toxic waste movement into the country. When government or customs officials are not bribed, waste could enter the country because of improper training and lack of expertise for the unmasking of disguised waste shipments. Unskilled and inexperienced officials are not qualified to recognize them and present an easy target for clever exporters (Ibitayo 2008).

This begs the question: who is involved in the illegal waste trade? Conceptions of violent and aggressive criminals are in most cases inaccurate. Actors are not typical criminals of the classic criminal underworld; they have different backgrounds, with some of them being members of organized crime groups while others belong to more mainstream organizations. Even when companies are in compliance with the law most of the time, many are prepared to step out from the legal side and into illegal business for the purpose of saving or even earning money. Legal businesses are in that way co-responsible for the proliferation of organizations specialised in the illegal trade in waste that otherwise would not exist without the demand for such services (Massari and Monzini 2004).

Many would be surprised to know that most of the legal recyclers actually do more waste trading than actual recycling (Puckett and Smith 2002). Independent companies that specialize in recycling are not always what they appear to be. There were cases when a large international parent company owned both exporting and importing facilities and had, as a matter of fact, exported to itself (Interpol 2009). The correlation between the business structure and amount of exported waste is weak. Large companies are not necessarily the major exporters of waste; it is possible that smaller recycling facilities have in fact greater chances to, intentionally or unintentionally, act outside the law, while they often need to rely on third parties in the process of recycling. In the same manner, smaller enterprises are more likely to rely on doubtful recyclers when they want to lower their waste disposal costs (Interpol 2009). For example, the costs of simply sending CRT monitors to China instead of properly recycling them in the US are ten times lower. Besides that, waste brokers are paid twice: for taking someone's obsolete device and secondly when selling it in a developing country (Puckett and Smith 2002).

People involved in waste trafficking are in most cases white-collar criminals, most of them businessmen or brokers, but in many instances chemical engineers and analysts. But they could not run the business without the help of blue-collar workers, such as drivers and guards. One of the most important links in the chain of trafficking are the so-called "middle men", who contact companies and offer them services of waste handling at low prices (Massari and Monzini 2004). The term "E-waste tourists" is used when talking about people, mainly from developing countries, who come to the UK to purchase electronic waste with the intention to sell it or to extract valuable materials and afterwards dump it illegally (Interpol 2009; Gray 2009).

There is some doubt as to whether the name "organized crime" is the one that should be used to best describe actors in waste trafficking (UNODC 2009). The organizational structure of such criminal groups tends to be quite simple, with a maximum 3–4 people forming the heart of the business. The network of links with companies and firms is broad, with only a few of them being previously involved in criminal activities (Massari 2004). When trading with African countries, sources say, these groups are also small with only 5–6 members. In most cases, at least one of them has ethnic connections with the country of destination, but these groups seem to be less professional than those trading with Asian countries (UNODC 2009).

The position of the Italian mafia in the process of waste trading needs to be described more precisely. Profits gained from environmental crimes in Italy are estimated at US$8.8 billion a year (Liddick 2009), while €2.6 billion was gained only from illegal waste management in 2002 (Massari and Monzini 2004). It is not hard for them to gain new customers when their prices are 400 times lower than the prices of legal companies (Colombo 2003). Estimates about the quantity of waste that "disappears" every year in Italy vary from 15 to 50% (Massari and Monzini 2004; Liddick 2009). This phenomenon of environmental crimes across Italy led to the introduction of a new term – *Ecomafia*, which could be today found in any Italian dictionary (Massari 2004). The waste traffic is mainly oriented from the North to the South of Italy, where the trafficking routes were first noticed in the early 1990s. The investigation of such crimes is even harder while the many mafia-type organizations also control the legal waste sector and the legal landfills. The Camorra controls most of the waste trading in the region of Campania, while "ndrangheta" occupies Calabria. The Italian environmental organization Legambiente has also named all 22 Italian ecomafias that are involved in waste trafficking and are control-ling the entire waste trade, while their links are spread across the entire country, across many companies (Massari and Monzini 2004) and even among public offi-cials (Colombo 2003).

Like any other process, waste trafficking also consists of different phases. Massari and Monzini (2004) outline three stages of waste trafficking. The first includes transferring waste from its origin to a waste management firm. The most important decision in this stage is the choice of service. In some cases producers are not aware that the service they choose is illegal, but they could assume so if an offer is much lower than the rival's. But producers are in many cases completely aware that their actions are in contradiction with the law and are therefore as responsible as the firm that manages waste.

During the phase of storage and transit, the second in the process, waste is often treated in different ways to deceive further inspection. Traffickers use different methods while trying to get rid of the waste, where the main aim is still to gain as much money possible and to take little risk. There are many situations where the shipment is destined for recycling, when actually such shipment will be dumped – whether legally or illegally (Brack and Hayman 2002). Waste laundering is also one example of trafficking waste. It could be laundered by selling it as fertilizers or soil conditioners, by making bricks out of it or using it for constructing roads and high-ways (Massari 2004) or even as a donation of humanitarian aid (Clapp 1994). Such incidents of "charity" happened when radioactive milk was sent to Jamaica from the EU in 1987, and when illegal pesticides were sent to Albania from Germany in 1992 (Clapp 1994). Co-mingling hazardous waste with non-hazardous waste is another model of trafficking. In that way hazardous waste "disappears" and new, "non-hazardous" waste occurs, that can be easily dumped in landfills or even laun-dered as fertilizer or fuel oil (Clapp 1994; Liddick 2009). In the very similar manner shipments of E-waste are mixed with functioning second-hand electronic goods

5 International Waste Trafficking: Preliminary Explorations

and are labelled as material for re-use (UNODC 2009). Another method also turns hazardous waste into non-hazardous without any treatment. This time waste is not even mixed with less contaminated waste, but is simply taken from the producer to a storage centre, which becomes the new producer of waste. Documents are easily modified and waste is reclassified from toxic to non-toxic waste. This is carried out by using false or old certificates with new a date but the same identification code (Massari and Monzini 2004).

The final step in the cycle is dumping of waste at the destination sites. Waste could be taken into recycling centres, where it could be recycled, but is used instead in construction or agriculture as mentioned above. Other possibilities are incineration or illegal dumping that appears almost anywhere; waste can be found in lakes, caves, rivers, canals, and in the seas (Massari and Monzini 2004). There were cases when farmers were even paid for accepting waste on their own land where they raised their dairy livestock (Colombo 2003). Destinations of waste are mainly developing nations that lack environmental legislation, have lower ecological standards, are in desperate need of money and do not have adequate knowledge and skills to prevent themselves from such harm. Most common targets are Asian and African countries, but also East Europe and South America are not immune to such criminal activities (Clapp 1994).

Africa is becoming a major importer of the world's E-waste, especially the West African countries of Nigeria and Ghana. Estimates are that about 95,000 tons of E-waste enters West Africa each year (UNODC 2009). Every month about 500 containers of used electronics pass through the port in Lagos, Nigeria (Schmidt 2006). There is indeed a demand for electronic equipment in Africa, but shipments usually consist of only 25% of useful second-hand equipment and about 75% of E-waste (UNODC 2009). One container is usually filled up with 800 computer monitors or 350 TV sets. The price of shipping such container from US to Africa is about US$5,000, while the price of one used Pentium III computer in Africa is approximately US$130. Only 40 such computers can pay the transport of the entire container (Schmidt 2006). But what happens to the other 75% of the shipment? Because African countries lack facilities and knowledge, such toxic materials are treated in an environmentally unsound manner.

Equally hazardous is illegal dumping of toxic and radioactive waste that has already been seen in the case of Somalia and Ivory Coast. Containers of such waste were dumped near the shoreline of Somalia and a tsunami in 2004 has revealed this dirty secret hidden under the sea, and many people died from breathing toxic dust and smoke (White 2008b).

Asia is another continent that is turning into a graveyard for obsolete Western technologies. China is one of the major importers of E-waste mainly from the US, but also from South Korea, Japan and Europe, with many cities transforming from rural communities to recycling centres. But those centres cannot compete with modern Western facilities. Computers and its components are dismantled using bare hands and a screw driver, valuable processors and chips are resold, plastic and wires are burned. Circuit boards contain various components and valuable materials that are heated on home-made wok-grills with melted lead-tin solder, which helps remove the chips. These chips contain precious materials, such as gold, which are removed by a primitive process using acid baths and the remaining electronic equipment that is not burned is simply dumped in open fields, along riverbanks or ponds that are highly contaminated with high levels of heavy metals. All of this is done by unskilled people, with many of whom are women and children (Puckett and Smith 2002).

E-waste is not the only type of waste destined for Asia. The costs of dismantling outmoded ships in Europe and the US have risen and legislation became more stringent. Developing countries such as India, Bangladesh, China, and Pakistan seemed like a perfect destination for ships no longer functional. While ships and aircraft carriers on the one hand are a source of raw materials, they also carry many hazardous materials and substances. Those are even more dangerous if workers are uneducated about the possible risks, are ill equipped and work in poorly regulated conditions (Sonak et al. 2008). Bangladesh currently dominates the market, breaking down more than half of the world's old ships; the Indian yard of Alang is also one of the top destinations (White 2008b). Until October 2006, they dismantled 4,327 ships in a period of 20 years; most of the 15,000 workers that worked there in 2003 were immigrants from undeveloped States of Northern India, with many of whom were illiterate. They were paid approximately US$6 per 6 h shift, with no bonus for working overtime. Accidents occurred on a daily basis and workers had no health or life insurance (Sonak et al. 2008). But the future of this industry does not look more promising. The number of out-of-date ships and aircraft will increase dramatically; there are estimations that predict about 25,000 outdated civil aircraft in the next 10–15 years. The situation is very similar with ships – many of them will be soon destined for disposal (White 2008b).

While much of the waste from Western countries is destined to poorer areas of the world, that is not always the case. Members of the EU, such as the Czech Republic, are not completely safe from illegal shipments of waste from their neighbours. Since the Czech Republic joined the EU, border checks with other EU members became a thing of the past. In the years 2005 and 2006, significant amounts of waste were found in Bohemia, with some 30,000 tonnes of waste distributed around the country, in 26 documented illegal dump sites, most of them along the German border. Because of the difficulty in demonstrating its origin, only 7,000 tonnes were proven to originate from Germany, even though it is suspected that all the waste belonged to this neighbouring country. Germany has a long reputation for advanced environmental legislation and waste management techniques since the early 1990s (Vail 2007), but on the other hand it is also known to be a great exporter of its waste. Before the fall of communism they exported waste from the Federal Republic of Germany to the German Democratic Republic, later to other Eastern European countries (Vail 2007). Free waste trade among EU countries puts new members at risk while they still lack appropriate recycling facilities and they often serve as a handy backyard for Western Europe's waste, which puts them in an even less legally protected situation than developing and non-OECD countries. European Environmental Agency (EEA) data, which is based on the reports of member states to European Commission (EC) and to the Basel Convention

Secretariat, indicates that the most significant importers in the EU are not new member states, but Belgium, Germany and Norway, while the largest exporters are the Netherlands, Ireland, Luxembourg and Belgium (EEA Report 2009). Exports from and imports into the new member states are not a large part of the entire EU waste trade, but the research based on data gathered between 1997 and 2005 could raise doubts about its validity today.

One of the main problems in investigating international waste trade is the insufficient and inaccurate data, most of which is self-reported. Waste that is legally dumped in landfills or is recycled represents only a small part of the enormous pile of produced garbage; it is often unclear what happens to the rest. There are assumptions that the seized amounts are only the tip of the iceberg and that the problem of exported waste is even larger. Analysed data, gathered from the Basel Convention website in 1998, showed a relatively large difference between reports made by importers and exporters of waste and raise critical reliability issues (Burns and Fuchs 2004). Recent data were gained by inspections made by the IMPEL-TFS, the informal network of environmental authorities of EU members, but they lack the cooperation of three significant countries: Spain, Greece and Italy, which reduces the value of the research. Annual reports about illegal shipments vary between 6,000 and 45,000 tonnes, but reported numbers may only be a fraction of the total (EEA 2009).

5.3 Incidents

Since the 1970s, there have been many incidents that involved waste, and the most devastating involved illegally transported waster later dumped in Third world countries. Probably the most serious waste incident in recent years occurred in August 2006 on the coast of Abidjan, the capital city of Ivory Coast. Although the consequences of the disaster were enormous, there were almost no media reports about it. About 600 tonnes of caustic soda and petroleum residues were dumped across Abidjan, causing nosebleeds, nausea and vomiting by five million inhabitants. Sixteen people lost their lives, over 100,000 sought medical attention and around 75 were hospitalized. The toxic cargo originated in the Netherlands and was produced by the global oil and metals trading company Trafigura, which has its offices in the UK and Switzerland. The ship used for transport, was Probo Koala, a Korean-built tanker that was registered in Panama, owned by Greeks with a Russian crew. The final disposal of waste in Ivory Coast was held by a company named Tommy, which was established only for the purposes of this special shipment and charged 16 times less than Trafigura would pay for disposal in the Netherlands. Trafigura denied responsibility for the disaster in Abidjan, but still agreed to pay US$198 million to the Ivorian Government (White 2008a). Trafigura even hired a team of 20 experts that were looking for links between exposure to chemicals and health problems, but they were not able to find such (CNN 2010). They have been constantly changing the Dutch version of Wikipedia on the entry of Probo Koala to clear their name (White 2008a).

Several serious incidents occurred prior to the Abidjan case. More than 20 years ago in 1986 the ship Khian Sea, loaded with toxic ash, left the port of Philadelphia in the United States. The ash contained aluminium, copper, lead, mercury, zinc and toxic dioxins and was intended to be disposed somewhere around the globe. But after being turned away by at least 11 countries, they decided to dump the ash illegally. They sailed around the globe for 2 years, looking for a suitable and hidden place to dump its cargo. The ship was re-named twice during that period, first to Felicia and then to Pelicano, to made tracking them more difficult. In 1988, after 2 years of sailing, an empty ship anchored in Singapore (AP 1988). Four thousand tonnes of about 14,000 tonnes of its cargo were dumped on a beach in Haiti, while another 10,000 tonnes somewhere between the Suez Canal and Singapore (Brack and Hayman 2002).

Another incident in the late 1980s is known as Karin B, after the name of the ship that illegally transported 2,100 tonnes of undefined industrial waste from Italy to Koko, Nigeria. Environmentalists and citizens of Nigeria protested and demanded that Italy take their waste back, but Italy had no interest in doing so. Karin B eventually loaded Italian waste once again, left Nigeria and wandered around the shores of France, Spain, West Germany and the Netherlands with the intention of finding a location for unloading its cargo, but without success. In the end Italy finally agreed to accept its waste back (Greenhouse 1988).

A more recent event did not involve a ship filled with waste, but the ship itself being classified as waste. Le Clemenceau was a French aircraft carrier that was sent to Alang, India to be dismantled. The ship was built in 1957 and because of its age, 500 tonnes of the entire weight of 27,307 tonnes was asbestos in addition to other toxic materials. Because of the international ban on the import of asbestos many developing countries refused to accept this French waste. The French authorities have argued that the ship was decontaminated and thus safe to recycle. The Indian Supreme Court subsequently prohibited the entry of Le Clemenceau into Indian waters (Sonak et al. 2008).

5.4 International Legislation

The Basel Convention on the Control of Transboundary Movements of Hazardous Waste and their Disposal is the international response to scandals regarding the trade in hazardous waste that were occurring during the late 1980s. It was adopted in Basel, Switzerland in 1989, and came into force in 1992 (Interpol 2009). Its precursors were the UNEP Cairo Guidelines that were non-binding and adopted in 1984 (Krueger 1998). The Basel Convention calls for self-sufficiency in waste management practices of its signatories and strives for the general reduction of hazardous waste generation (Puckett and Smith 2002). A member of the agreement should not export its waste to a non-member without a formal bilateral agreement (Burns and Fuchs 2004). Member countries are also not allowed to export their waste to Antarctica (Lipman 2002), and to countries that have banned the import, or lack adequate treatment facilities (Gwam 2000). In the case of illegal transportation of waste, the exporter is liable to accept the waste back and to pay for the damage made or for the cleaning of the waste (White 2008b). By the end of September 2009, 172 states had signed the Basel Convention, but the USA, as one of the biggest exporters of waste, is still avoiding ratifying the Convention (Waste Trade Ban Agreements 2009).

The stated purpose of the convention is the protection of developing countries from the hazardous waste of developed countries. The system of "prior informed consent" (PIC) gives the members of the Basel Convention the right to refuse imports of hazardous waste (Clapp 1994). PIC requires the exporting member state to inform the importing member state about the type of waste, reasons for the export and the method of disposal. The importing member needs to reply to the notification assenting the shipment, rejecting the shipment or requesting more information. All of these reports need to be forwarded to the national competent authority that is responsible for the administration as well, and the exporter needs to wait for the response before transporting its waste. Member transit states need to be informed about the shipment in the same way as the possible importer and have the same abilities for permitting, denying or requesting more information. Status of the transit states that are not parties to the Convention is less clear (Krueger 1998). These guidelines seem like a good solution for the control and supervision of the international movement of waste; sadly in practice things are often a bit different. Falsification of documents and mislabelling of shipments are common, and furthermore the national authorities are not informed about every movement made, which disables monitoring and presents limitations for the successful functioning of the Convention (Krueger 1998).

In 1994, developing countries showed their intention to update the Convention by banning the movement of hazardous waste from the members of the OECD, the EU and Liechtenstein to other countries. The ban was adopted as an amendment to the Convention in 1995 (Lipman 2002), but six of the largest producers and exporters of waste did not support the idea of total prohibition. The USA, the UK, Germany, Japan, Australia and Canada decided not to sign the amendment (Clapp 1994). However, Denmark was a strong supporter of the ban, and forced the entire EU to adopt it, which was confirmed by adopting EC Regulation 259/93 that prohibited export of hazardous waste from EU to non-EU countries in 1993 (Sonak et al. 2008). The UK and Germany were also forced to ratify the ban. In the same manner, some developing countries did not agree that ban would bring any advantages to them and so Ivory Coast, Pakistan, Bangladesh, Philippines, India and others still have not ratified the ban (White 2008b). There are 67 countries that have ratified the Basel ban to date, which is still not enough for it to finally get in force. There are many supporters of the ban, but many doubt of the effectiveness of such prohibition.

Besides becoming the basic document in regulating the movement of waste across international borders, the Basel Convention also became a severely criticized international agreement. The definition of hazardous waste was criticised as to lack clarity that was improved with the development of three lists of waste (Clapp 1994; Vander Beken and Balcean 2006). Radioactive waste was not included in the lists

(Clapp 1994). Other critics have been exposing persistent growth of waste exports to developing countries, which could be an indicator that the implementation of the Convention has not been that successful (Liddick 2009). Increasing waste trade could be the outcome of allowing bilateral waste agreements between Parties and non-Parties of the Basel Convention and not adopting the Basel Ban amendment (Clapp 1994). But there are also doubts about the effectiveness of the total ban on waste movement. Krueger (1998) argues that adaptation of the ban would simply turn the movements into illegal and hidden ones, and would not have an impact on reducing the number of exportations. Other critiques are related to the lack of knowledge and lack of training in developing countries, absence of proper and standardized reporting procedures, need for an international cooperation and information exchange, and lack of financial aid to developing countries that have implemented the Convention (O'Neill 1998; Krueger 1998; Ibitayo 2008).

Because of such weaknesses, many countries began to form their own regional initiatives that are usually more stringent than the Basel Convention such as Bamako and Lomé IV Conventions (Ibitayo 2008). Lomé IV offered the first regional waste trade protection to 69 African, Caribbean and Pacific (ACP) countries from the EU and came into force in 1991. The intention of the EU states was to export toxic waste to ACP countries under certain conditions, but ACP countries resisted. The final agreement was that EU countries would not export their toxic waste to ACP countries and that ACP countries will not import toxic waste from non-EU countries (Clapp 1994). The Lomé IV was replaced by new ACP-EU partnership known as the Cotonou Agreement in 2000 which is intended to run for 20 years (Babarinde and Faber 2003).

In the meantime, members of the Organization of African States (OAS) decided to adopt their own convention in 1991, the Bamako Convention, that would prohibit the importation of radioactive and hazardous waste into African countries (Clapp 1994). The convention follows the idea of the Basel Convention, but also shows improvements and is much stricter when it comes to banning the waste trade. Counselling and help from the Greenpeace organization is obvious, while African countries decided for the total ban on the import of waste, have prohibited ocean dumping and incineration and gave support to cleaner production practices (Clapp 1994). The convention entered into force in 1996 (Sonak et al. 2008) after it was ratified by many African countries, and to date 24 nations have done so (African Union 2010). Ghana and Nigeria appear to be the biggest West African importers of waste, and have not ratified the Convention yet (UNODC 2009).

The European Union, on the other hand, has a long reputation for adopting strict and rigorous environmental legislation that is sometimes hard for its members to abide. Although such legislation offers a healthier place to live, it also creates loopholes for illegal actions to happen. The number of those has risen in the last couple of years, but there are doubts if the reason for this is true increase or just due to increased number of inspections made and better monitoring (EEA Report 2009). But contrary to the Bamako or Lomé IV Conventions, the EU legislation not only protects the EU itself, but also does the same for developing countries.

5 International Waste Trafficking: Preliminary Explorations 91

The most important waste management act for the purposes of this chapter is the Waste Shipment Regulation (WSR). EU Regulation 259/93 supervised the movements of waste among EU members and between EU and the rest of the world and was replaced by Regulation 1013/2006 that came into effect in July 2007. The aim of both WSRs is the protection of the environment by requiring effective processing of waste and prevention of hazardous waste shipments to developing nations. The new WSR 1013/2006 is based on the Basel Convention (1989), the OECD Decision (2001) and the Council Directive 2006/12/EC on waste. While the Basel Ban is still not in effect because it has not been ratified by a sufficient number of countries, the EU implemented the Ban Amendment in its WSR back in 1993 (Collins 2009; EEA Report 2009) and banned the export of hazardous waste to non-OECD countries. However, non-hazardous waste can still be exported to non-OECD countries with prior notification. It is possible to ship all kinds of waste within EU borders, except in cases when a member country implements a specific ban on import. The prior notification with detailed information on the shipment is needed, while only general data is later on forwarded to the EC (EEA Report 2009). The new WSR also brings obligation to EU members to perform inspections of ships on their territory and permits further examinations of the interior of the containers where the waste is stored (White 2008b).

With the EU legislation being very strict, its implementation still received many negative critiques. Lack of proper infrastructure, high numbers of illegal shipments, and reliance on landfills all make the regulations hard to implement. The EU started to encourage its members to exchange information, to carry out joint actions and cooperation, and even tried to achieve this with lawsuits against member states before the European Court of Justice (IHS 2009).

As all the other waste, WSR also prohibits export of E-waste to developing countries, but on the other hand allows export of second-hand goods for re-use or as a charitable donation. The problem is that about 50–70% of such shipments present devices that do not function (Nordbrand 2009). By analysing the shipments of second-hand goods, investigators were surprised that electronic devices had such low values, which suggested that shipments were as a matter of fact waste (EEA 2009). Because checking and testing these huge amounts of shipments is a time-consuming and complicated effort, violations of ban may be happening daily.

Several principles have been introduced for better guidance through the environmental law in the EU. The proximity principle is related to dealing with environmental problems as close to its origin as possible and achievement of self-sufficiency of the member states in the case of waste management practices. Another principle sets the producer as the responsible partly for cleaning its products after use and is known as polluter pays principle (PPP). Similar to PPP is extended producer responsibility (EPR) which also makes producers responsible for financing collection and treatment of waste after consumers stop using it (Vail 2007; Nordbrand 2009).

One might be a bit surprised about the differences in severity between the laws of the EU and the US. The USA remains the only developed nation in the world that still has not ratified the Basel Convention and is at the same time the world's greatest waste exporter. Federal regulation of waste movements in the US is much less rigorous than in the EU and it is not illegal to export E-waste as long as the goal of an export is recycling (UNODC 2009). The Resource Conservation and Recovery Act (RCRA) is the main federal document that manages hazardous waste and surprisingly it does not address E-waste as hazardous. By not ratifying the Basel Convention it disabled the US from being able to freely export their waste to countries that have ratified it. It is also not legal to sent waste to non-OECD countries without bilateral agreements. The absence of effective federal regulation, has forced many US states to adopt their own directives on the cleaner waste management practices (Interpol 2009).

Unlike the USA and the EU, China is one of the greatest importers of waste from developed countries. Rural parts of the country have become recycling centres with very low health and safety standards. One would think that China's policies regarding waste are even looser than those in the US but in reality, that might not be the case. China was actually one of the first proponents of international ban on toxic waste exports and also one of the first signatories of the Basel Convention. China's national legislation on the import of waste is in theory quite severe and since 2000 it prohibits import of most categories of E-waste. But the problem of totally fine legislation and policies lies in its improper implementation and enforcement (Puckett and Smith 2002).

5.5 National and International Operations and Actions

During the last couple of years the amount of legally and illegally exported waste increased. The fact is that we are producing more and more waste and electrical and electronic devices are becoming obsolete very rapidly. It is not certain if the increased amounts of exported waste truly reflect conditions on the market, or are the increased amounts only the outcome of more intense monitoring procedures and inspections (EEA Report 2009). International cooperation has become an important part of fighting against international waste trafficking which reflects in numerous national and international joint actions and operations.

An example of a national programme against waste trafficking is that of Italy. They have a long reputation for being a country with high levels of environmental crimes. That was also the reason for organizing a special unit that would combat such crimes back in 1986. This unit is still active today, and constantly cooperates with Europol and Interpol and has significant success, for instance, in 2007, more that 170 people were arrested. The unit is divided into two sections: Radioactive Special Section and Information System for Environmental Protection (S.I.T.A.). The first deals with sources of radioactivity, including illegal trafficking of contaminated waste, while the S.I.T.A. system is EU sponsored and helps to locate places where environmental crimes occurred by interpreting photos taken by satellite or aircraft (Contri 2009).

Another example is the Sky-Hole-Patching project, a joint project among Asian and Pacific countries that was in force from September 2006 until November 2007.

It was performed with the World Customs Organization (WCO), with the help of United Nations Environmental Programme – Regional Office for Asia and Pacific (UNEP ROAP), Regional Intelligence Liaison Office for Asia and Pacific (RILO A/P), customs and environmental agencies of member countries. One of the aims of the project was to change the whole idea of illegal trade that still mostly includes images of drugs and arms, and is not aware of the size of the international waste trafficking problem. Twenty Asian-Pacific countries joined the programme that was separated into two phases (UNEP Press Release 2006). The subject matter of the first 6 months period was the monitoring of ozone-depleting substances (ODS) that were made illegal with the Montreal Protocol on Substances that Deplete the Ozone Layer in 1989. For the purposes of this chapter the second period that concentrated on the trafficking of hazardous waste is more significant. Intense monitoring reflected 98 seizures of waste in port of the Hong Kong that originated from 25 different countries. Most of the seizures presented E-waste, while most of the waste was exported from Mexico.

It is important to interpret such data with scepticism: many developed countries are using developing countries as exporters of their waste only to avoid customs attention while crossing borders. In the same matter Hong Kong is usually not the real destination for most of waste shipments, but an important centre of transhipment, primarily because of its geographical location and free port status. If not seized, the waste is re-exported to other developing countries such as Cambodia, India, Vietnam and China. While the project was promoted all around the A/P Region, more countries joined the programme. The follow-up investigations of waste from Europe were enabled with the help of European Union Network for the Implementation and Enforcement of Environmental Law (IMPEL) which later became an example of cooperation between two continents. The final evaluation of the project showed that it was successful, because it improved cooperation between customs, environmental agencies and other parties. What the project lacked were more detailed reports from the participants that were important for the overall analysis and for future comparisons (WCO RILO A/P 2007).

One of most recent projects between the Asia/Pacific region and Europe is Demeter. An international joint operation on fighting illegal waste trade between 65 countries of Europe, the Asia/Pacific region and Africa was held under the World Customs Organization (WCO) between March and May 2009 and it resulted in more than 30,000 tonnes of waste taken during 57 seizures. The operation included more than 300 seaports, with most of the seizures taking place in Europe, before shipments were sent to their destination in developing nations (Isarin and Whitehouse 2009). The primary goal of the operation was not only the increased number of successful inspections, but above all the improvement of exchanged information among member countries that is often missing. Coordination, cooperation and communication are, according to WCO (World Customs Organization) Secretary General, Kunio Mikuriya, the biggest enemies of waste traffickers and essential elements in fighting waste trafficking (WCO Press Release 2009). The operation gave information on the most common types of waste and the most desired destinations, while it emphasized the importance of collaboration and early intervention.

Among the first European joint projects were the Seaport projects I and II. They were carried out under the IMPEL-TFS; a network of representatives of authorities from different EU member states and neighbouring countries that are involved in the inspection of transnational shipments of waste. It is a part of the EU network for the Implementation and Enforcement of Environmental Law (IMPEL). The first project was carried out in 2003 and 2004 in six major seaports around Europe. The aim of the project was to improve the enforcement of Regulation 259/93 which was in force at that time and was differentially enforced across the EU. To improve such practice, inspectors were included into the project and a manual to assist them was developed, which standardized phases of inspections and reporting of the results. These reports showed that 1,230 shipments were checked and about 40% of them were waste shipments, 20% of those turned out to be illegal. The main characteristic of the project was cooperation and knowledge exchange between cooperating nations and its customs, police and other authorities. Because of its success, the project was extended for a second phase with 12 participating countries, some of them not members of the EU (Isarin 2005).

Because these projects proved to be effective, there was a desire for further cooperation and continuation of successful practices. The new project Enforcement Actions I began in 2006, whose aim was cooperation, and better implementation of the new EU waste regulation. The first part of the project lasted from September 2006 to June 2008 and linked together 25 European countries. During this period, three conferences were organized, four inspection periods were carried out and 34 inspectors took part in an international inspector exchange programme. Four inspection periods resulted in almost 14,000 shipment inspections, of which more than 2,000 contained waste. More than 300 of these 2,000 shipments (15%) were in violation of the EU Waste Shipment Regulation. Most of detected waste was actually E-waste that was destined for non-OECD countries. Cooperation seemed to be one of the most important parts of the project, with the Inspector Exchange Programme promoting knowledge and experience sharing. Inspectors who joined this programme reported with enthusiasm about the improvements in quality and effectiveness. Joint inspections of two or more countries were usually held on the same day and at the same border crossing, which enabled further cooperation. In more than 90% of the cases, other authorities, such as the police, also took part in the inspections (IMPEL-TFS 2008).

Another important aspect was improved implementation of the existing legislation by the member states. During the Enforcement Actions I, European legislation changed and EC Regulation 259/93 was replaced with Regulation 1013/2006, which also reflected changes in practice. After implementation, the amount of violations was reduced significantly, probably because of improper and insufficient enforcement. Therefore, standardized inspection procedures were introduced, and national coordinators were appointed to gather reports for further analysis (IMPEL-TFS 2008).

Most inspections were performed in seaports and on the highways, while trains and inland-water ports were less common places for inspections to occur. Administrative violations (61%) reflected in incomplete, incorrect or missing documentation, while illegal shipments (39%) were defined as the ones which did not obtain the needed authorisation, did not accurately describe the contents or were just prohibited for transport (IMPEL-TFS 2008).

The project was a success, but there were still chances for improvement. International cooperation showed not to be a part of everyday practice, which indicated that such a project should be continued. Countries that did not take part, were to be involved in the next project, especially Italy and Spain which have been attractive countries for waste origin with easy access to international waters (IMPEL-TFS 2008).

The project that followed Enforcement Actions I and is still being performed is Enforcement Actions II. It started in October 2008 and is planned to last until March 2011. So far three inspection periods were carried out in 22 EU countries that joined the project and four non-members. The goals have stayed more or less the same and desired to improve inspection practices and international cooperation between countries in the light of better waste regulation enforcement (IMPEL-TFS 2009).

During a period of 8 months, 10,481 transports were inspected, which includes containers, trains, trucks and documentation checks. In 75% of these transports physical inspection has taken place and about 1,935 shipments or 25% contained waste. Out of 1,935 shipments about 19% turned out to be in violation of the WSR, most of whom (46%) were administrative violations, followed by illegal transports (37%) (IMPEL-TFS 2009).

Because the primary intent of such project is international cooperation between countries and cooperation at the local level between local authorities, the increased number of participating countries is positive. Twenty-two out of 26 countries produced reports about their activities, which is satisfactory, but still indicates that additional guidance is needed to achieve more homogeneous reporting. In general, port inspections proved to be more successful at discovering violations. Many countries just started to learn how to perform inspections and the outcome could be a lower number of uncovered violations, while more experienced countries could achieve better results because of this experience, but on the contrary traffickers could be already aware of their practice and could change their routes. In some countries, inspections were performed during basic inspections of transportation by the police or customs. Only six countries have reported company inspections, but violations were observed in more than 50% of those cases. Generalisations cannot be made due to the small number of reports. On the other hand it can be generalised that destinations of waste shipments are Asia and Africa, while the countries of origin are European states (IMPEL-TFS 2009).

Cooperation and exchange of inspectors was carried out in the same way as in project EA I. There were 12 collaborating countries in the exchange programme with 30 participating inspectors with the most popular hosting country being the Netherlands.

When comparing Enforcement Actions I with its successor, it is obvious that some progress was achieved. The number of countries performing inspections increased from 17 to 22; the number of physical inspections also increased, as have detection rate of violations from 15 to 19%. The participation rates of customs and police in joint inspections also increased, as did the participation of other enforcement agencies and local authorities. However, the success and improvements need to be examined and analyzed carefully. Three major European members (Italy, Spain, and Greece) still have not joined the project, which definitely presents a major deficit in reported violations. In the same way, problems occurred in larger EU countries where some regions did not take an active part in the project. On the contrary, some countries did actively collaborate in the project, but have mostly concentrated on import rather than on the export side of waste trafficking. Enforcement Actions II are planned to end in March 2011, but this would not present the ending of united European intention of fighting illegal waste trade, while there are already plans for its continuation (IMPEL-TFS 2009).

5.6 Concluding Remarks

The scale of international waste trafficking is enormous, yet mostly hidden and under-represented in the media. This low profile promotes more waste being trafficked, usually from rich countries into poorer countries, with all the consequences it has on people and environment.

While developed nations are adopting new laws on waste that are even stricter from previous ones, developing countries are becoming desired destinations due to poor legislation that do not offer them protection. Laws adopted by developed countries are, as a matter of fact, not bad for the third world, as they aim to protect weaker members of the international community. The problem lies in the lack of enforcement and political will. European laws do prohibit and limit the movements of waste between nations, but lack of adequate implementation of such regulations puts developing countries in a sometimes helpless position. For waste exporter nations, it is much easier to simply ship their waste elsewhere without harming their own environment and avoiding public opposition, than to take care of that waste at its origin. On the other hand, poorer countries are sometimes willing recipients of waste and hazardous shipments as there is money to be made. Leaders of both sides need to be aware of the harm caused by waste traffickers and should prioritize actions to stop this.

Because waste trade is an international form of crime, no country can successfully deal with it on its own. Cooperation is the essential element of all the actions aiming to combat this illicit trade. Such collaboration enables experience and intelligence sharing and has proven to be an effective method in detecting and seizing illegal shipments of waste.

Reports about waste shipments are the basis for further research in the field of waste trafficking, but data is often missing or incomplete. There are only estimates about the real quantities of waste produced worldwide annually and the assumptions about the quantities of trafficked waste are even more approximate. Based on the available data, there is an increase in waste shipments. This increase could be

5 International Waste Trafficking: Preliminary Explorations

due to intensified monitoring performed during the last few years or due to a true increase of international waste movements. Better data is needed for further analysis that could be used for the identification of traffic routes, seeing changes in trafficking trends, making presumptions about future situation and so on.

Ideally the entire problem should be examined from a different perspective from the very beginning. Waste production rates are high and are getting even higher because of endless technological progress. The consumer society has created an artificial need for brand new, always modern, fashionable, throw away goods. To achieve an actual decline in the production of waste, we need to first change our behaviour. New products require raw materials, energy and labour – not only for production or for their usage, but also for their disposal. The life cycle of products should not end with its shipment to "somewhere" but with complete recycling.

References

African Union (2010) Resource document. http://www.africa-union.org/root/au/Documents/Treaties/List/Bamako%20Convention.pdf. Accessed 2 Feb 2010

AP (1988) After 2 years, ship dumps toxic ash. Resource document. The New York Times. http://www.nytimes.com/1988/11/28/us/after-2-years-ship-dumps-toxic-ash.html. Accessed 2 Feb 2010

Arie S (2004) Toxic scandal in mozzarella country. http://www.guardian.co.uk/world/2004/oct/14/italy.sophiearie. Accessed 2 Feb 2010

Babarinde O, Faber G (2003) From Lomé to Cotonou: business as usual? In: European Union Studies Association, Biennial conference, European Union Studies Association, Nashville, TN, pp 27–29

Brack D, Hayman G (2002) International environmental crime: the nature and control of environmental black markets. Background paper for RIAA workshop, Royal Institute of International Affairs, London. http://www.chathamhouse.org.uk/files/3049_environmental_crime_background_paper.pdf. Accessed 2 Feb 2010

Buckingham S, Turner M (2008) Understanding environmental issues. Sage, London

Burns TJ, Fuchs J (2004) The international transport of hazardous waste: some preliminary findings from the Basel Convention data. Resource document. http://www.allacademic.com//meta/p_mla_apa_research_citation/1/1/0/8/3/pages110830/p110830-1.php. Accessed 7 April 2010

Carrabine E, Lee M, Plummer K, South N, Iganski P (2009) Criminology: a social introduction. Routledge, London

Clapp J (1994) The toxic waste trade with less-industrialized countries: economic linkages and political alliances. Third World Q 15(3):505–518

CNN (2010) Firm offers to settle toxic waste case in Ivory Coast. Resource document. http://edition.cnn.com/2009/WORLD/africa/09/21/ivory.coast.toxic.waste/index.html. Accessed 12 April 2010

Collins KJ (2009) Europe exporting more waste to less developed nations. Resource document. http://greenlegals.com/2009/09/europe-exporting-more-waste-to-less-developed-nations. Accessed 18 April 2010

Colombo F (2003) Environment – Italy: 'Eco-mafia' reaps billions in waste disposal. Resource document. http://proquest.umi.com.nukweb.nuk.uni-lj.si/pqdweb?did=351211121&sid=1&Fmt=3&clientId=16601&RQT=309&VName=PQD. Accessed 7 Feb 2010

Contri M (2009) Illegal trafficking of waste in the light of national and international legislation. In: De Amicis R, Stojanovic R, Conti G (eds) GeoSpatial visual analytics: geographical information processing and visual analytics for environmental security. Springer, Dordrecht, pp 499–508

EEA (2009) Not in my back yard – international shipments of waste and the environment. Resource document. http://www.eea.europa.eu/articles/international-shipments-of-waste-and-the-environment. Accessed 7 April 2010

EEA Report (2009) Waste without borders in the EU? Transboundary shipments of waste. EEA, Copenhagen

Elliot L (2009) Combating transnational environmental crime: "joined up" thinking about transnational networks. In: Kangaspunta K, Marshall IH (eds) Eco-crime and justice: essays on environmental crime. United Nations Crime and Justice Research Institute, Turin, pp 56–77

Eman K, Meško G, Fields CB (2009) Crimes against the environment: green criminology and research challenges in Slovenia. J Crim Justice Secur 11(4):574–592

Gray L (2009) Environment agency to crackdown on "waste tourists". Resource document. http://www.telegraph.co.uk/earth/earthnews/6502305/Environment-Agency-to-crackdown-on-waste-tourists.html. Accessed 18 April 2010

Greenhouse S (1988) Toxic waste boomerang: Ciao Italy! Resource document. http://www.nytimes.com/1988/09/03/world/toxic-waste-boomerang-ciao-italy.html?sec=&spon=?pagewanted=1. Accessed 7 Feb 2010

Gwam CU (2000) Toxic wastes and human rights. Brown J World Aff 7(2):185–196

Ibitayo O (2008) Transboundary dumping of hazardous waste. Resource document. http://www.eoearth.org/article/Transboundary_dumping_of_hazardous_waste. Accessed 7 April 2010

IHS (2009) EC calls for better implementation of EU waste law. Resource document. http://engineers.ihs.com/news/environment/2009/eu-waste-management-112009.htm. Accessed 7 April 2010

IMPEL-TFS (2008) Enforcement actions I. Enforcement of EU waste shipment regulation. Resource document. http://www.vrom.nl/docs/IMPEL-TFS.pdf. Accessed 27 Mar 2010

IMPEL-TFS (2009) Enforcement actions II. Enforcement of EU waste shipment regulation. Resource document. http://impeltfs.eu/wp-content/uploads/2009/05/Interim-Report-IMPEL-TFS-EA-II-final-word.pdf. Accessed 27 Mar 2010

Interpol (2009) Electronic waste and organized crime – assessing the links. Trends Organize Crime 12:352–378

Isarin N (2005) IMPEL-TFS project: European Enforcement Initiative to detect waste shipments. In: Seventh International Conference on Environmental Compliance and Enforcement, pp 249–252

Isarin N, Whitehouse T (2009) The international hazardous waste trade through seaports. Resource document. http://www.inece.org/seaport/SeaportWorkingPaper_24November.pdf. Accessed 7 April 2010

Krueger J (1998) Prior informed consent and the Basel Convention: the hazards of what isn't known. J Environ Dev 7(2):115–137

Liddick D (2009) The traffic in garbage and hazardous waste: an overview. doi: 10.1007/s12117-009-9098-6

Lipman Z (2002) A dirty dilemma: the hazardous waste trade. Harvard Int Rev 23(4):67–71

Massari M (2004) Ecomafias and waste entrepreneurs in the Italian market. Resource document. http://www.cross-border-crime.net/pdf/CCC-2004-Massari.pdf. Accessed 17 Jan 2010

Massari M, Monzini P (2004) Dirty businesses in Italy: a case-study of illegal trafficking in hazardous waste. Glob Crime 6(3–4):285–304

Nordbrand S (2009) Out of control: e-waste trade flows from the EU to developing countries. SwedWatch, Stockholm

OECD Council (2001) Council Decision C(2001)107/FINAL. Paris: OECD. Resource document. OECD. http://www.oecd.org/dataoecd/37/49/30654501.pdf. Accessed 12 March 2010

O'Neill K (1998) Out of the backyard: the problems of hazardous waste at a global level. J Environ Dev 7(2):138–163

Puckett J, Smith T (eds) (2002) Exporting harm: the high-tech trashing of Asia. http://www.ban.org/E-waste/technotrashfinalcomp.pdf. Accessed 12 Dec 2009

Qiang L, Cao J, Ke Qiu L, Xu Hong M, Guang L, Fei Yue F et al (2009) Chromosomal aberrations and DNA damage in human populations exposed to the processing of electronics waste. Environ Sci Pollut Res 16(3):329–338

5 International Waste Trafficking: Preliminary Explorations

Rosenthal E (2009) Smuggling Europe's waste to poorer countries. Resource document. http://www.nytimes.com/2009/09/27/science/earth/27waste.html. Accessed 7 Feb 2010

Schmidt CW (2006) Unfair trade: e-waste in Africa. Environ Health Perspect 114(4):A232–A235

Secretariat of the Basel Convention (1989) Basel Convention on the control of transboundary movements of hazardous wastes and their disposal. Resource document. UNEP, Basel. http://www.basel.int/text/17Jun2010-conv-e.pdf. Accessed 5 May 2010

Sonak S, Sonak M, Giriyan A (2008) Shipping hazardous waste: implications for economically developing countries. Int Environ Agreements Polit Law Econ 8(2):143–159

UNEP Press Release (2006) Project sky hole patching goes into operation: tackling illegal trade in ozone-depleting substances and dangerous waste. Resource document. http://www.unep.fr/ozonaction/information/mmcfiles/4819-e-PRunep_roap1906.pdf. Accessed 2 March 2010

UNODC (2009) Transnational trafficking and the rule of law in West Africa: a threat assessment. United Nations Office on Drugs and Crime, Vienna

Vail BJ (2007) Illegal waste transport and the Czech Republic: an environmental sociological perspective. Sociologicky časopis/Czech Sociol Rev 43(6):1195–1211

Vander Beken T, Balcean A (2006) Crime opportunities provided by legislation in market sectors: mobile phones, waste disposal, banking, pharmaceuticals. Eur J Crim Pol Res 12(3–4):299–323

Walters R (2007) Crime, regulation and radioactive waste in the United Kingdom. In: Bierne P, South N (eds) Issues in green criminology – confronting harms against environments, humanity and other animals. Willan, Devon, UK, pp 186–205

Waste Trade Ban Agreements 2009 (2010) Resource document. http://www.ban.org/country_status/country_status_chart.html. Accessed 2 April 2010

WCO Press Release (2009) Operation Demeter yields tons of illegal shipments of hazardous waste. Resource document. http://www.wcoomd.org/press/default.aspx?lid=1&id=187. Accessed 12 Feb 2010

WCO RILO A/P (2007) Evaluation report on project sky-hole-patching. Resource document. http://www.greencustoms.org/reports/workshop/Sky_hole_patching.pdf. Accessed 7 Feb 2010

White R (2008a) Toxic cities: globalizing the problem of waste. Soc Justice 35(3):107–119

White R (2008b) Transnational environmental crime. In: White R (ed) Crimes against nature. Willan, Devon, UK, pp 115–143

Chapter 6
Primary Categories and Symbiotic Green Crimes in Bosnia and Herzegovina

Elmedin Muratbegović and Haris Guso

Abstract The aim of this chapter is to highlight the problem of the environmental degradation in Bosnia and Herzegovina from a criminological perspective and to define two types of "Green Crime" in contemporary criminology. There are four "primary categories" of green criminology in which the environment becomes degraded through human action – all of which have become the subject of legislative efforts in recent years. These new categories are as follows: (a) crimes of air pollution; (b) crimes of deforestation; (c) crimes of species decline and against animal rights; and (d) crimes of water pollution. The secondary categories of green criminology can be found under the name of symbiotic green crimes. These are defined as crimes that grow out of the flouting of rules that seek to regulate environmental disasters. They include, for instance, state violence against oppositional groups, hazardous waste and organized crime. "Green Criminology" in South Eastern European Countries is still a very strong taboo. On paper, there are many legal and policy-related issues raised by green crimes, but in reality, these crimes are not considered important by the institutions of formal social control. This chapter is mainly based on the studies available in the field of "Green Criminology" in Bosnia and Herzegovina. Several case studies from Bosnia and Herzegovina will also be presented, which will illustrate the real extent of the problem of environmental crime.

E. Muratbegović (✉)
Faculty of Criminal Justice and Security, University of Sarajevo, Sarajevo,
Bosnia and Herzegovina
e-mail: emuratbegovic@fkn.unsa.ba

H. Guso
Department of Public Relations, Criminal Policy Research Center, Sarajevo,
Bosnia and Herzegovina
e-mail: hguso@cprc.ba

G. Meško et al. (eds.), *Understanding and Managing Threats to the Environment in South Eastern Europe*, NATO Science for Peace and Security Series C: Environmental Security, DOI 10.1007/978-94-007-0611-8_6, © Springer Science+Business Media B.V. 2011

6.1 Introduction

6.1.1 Research Problem

Ecological and environmental protection certainly represents the most important strategic interests of modern civilization. Knowledge about the transience of the benefits of nature caused by human mistakes naturally occurs in response to the daily industrial race, which determines the predatory nature of capitalism in general (Mujanović 2009). People are already infected by the basic goals of modern living that start with an instinct for self-preservation and, gradually, through the process of accumulation of capital, are becoming more "eager for additional profits" for themselves and to ensure a secure future for the coming generations (Modly 1998).

It is in pursuit of this goal that a secure future is now already at stake. Water and air grow more and more polluted, waste piles up all around, industrial toxic gases triumph over nature and everything is slowly melting down. Nature, of course, reciprocates with the same measures and if we continue "to ride this road" then it is for sure: "As always, nature wins". All these are only superficial reasons why man should finally return to his original motto: "Living with nature".

In this chapter, we will try to show the conservation status of individual natural resources in Bosnia and Herzegovina, a country far well less known for its natural resources which mark it out as an oasis of the Balkans, than for being a site of war at the end of the twenthieth century, involved in one of the greatest horrors of recent European history.

Criminological thought, even for the analysis of conventional crimes, is an extremely new subject in Bosnia and Herzegovina, and therefore it should be made clear in advance that the field of environmental protection or environmental crime has not been sufficiently researched as yet. Although modest, this study will make an initial contribution to the development of a new criminological orientation in Bosnia and Herzegovina, namely as "Green Criminology".

The term "Green Criminology", first coined by Lynch in 1990, as a useful paradigm for analyzing the two faces of eco-crime, sociological and legal definitions. This phenomenon was established as terminological umbrella under which to theorize and critique the emerging terminology related to environmental harm (McLaughlin and Municie 2006: 147). Like other parts of contemporary criminology, "green criminology" is a social construction, influenced by: social locations, power relations in society and definitions of environmental crimes (Lynch and Stretsky 2003: 217).

As a new field of critical analysis, green criminology involves both new sets of concerns – particularly harms to other species and violations of "animal rights" and harms to ecological systems – and what have more generally been understood as environmentally destructive conduct usually by business corporations and states. These are important issues but what, if anything is gained by relating these to criminological discourse and what is lost by doing so (Pearce 2007).

As few years ago, Carabine et al. (2004: 28) identified four main tasks of contemporary green criminology. Those tasks include: (1) documentation of the existence of green crimes in all their forms, (2) charting the ways in which the laws have been developed around area of green crimes, (3) the connections of green crimes to social inequalities and (4) the assessment of the role of green social movements in bringing about social change. Issues in Green Criminology: confronting harms against environments, humanity and other animals aims to provide, if not a manifesto, then at least a significant resource for thinking about green criminology, a rapidly developing field (White 2003: 490). This article demonstrates the complexities in determining the character, extent and impact of environmental harm. It furthermore identifies diverse, and at times, competing approaches to environmental regulation and to the prevention of environmental harm.

The field of criminology has historically shown relatively little interest in the topic. The emergence of environmental, or green criminology, over the past decade marks a shift in this trend, but attempts to define a unique area of study have been extensively criticized (Gibbs et al. 2010: 126; South 1998: 213).

In the next part of this chapter we will outline the significance of green issues for Bosnian criminology, thought legal, sociological and other criminological relevant views.

6.1.2 Environmental Crime in Bosnia and Herzegovina

Under environmental crime in Bosnia and Herzegovina, criminal justice practitioners consider a group of offenses that threaten the environment, that have been defined by the Criminal Code and special laws governing the particular area of environmental protection, and that have been incriminated within the Criminal Code provisions as the worst forms of threat to the environment as criminal acts (Sućeska 2008; Ramljak and Petrović 2005).

Legislation protecting and improving the environment in Bosnia and Herzegovina is treated by four main Laws: (a) Environmental Protection Act (Zakon o zaštiti okoliša 2009); (b) Nature Protection Act (Zakon o zaštiti prirode 2003); (c) Air Protection Act (Zakon o zaštiti zraka 2003); and (d) Water Protection Act (Zakon o zašitit vode 2006). These offenses are particular forms of crime which result in environmental pollution on a large scale or affecting a broad area, thus endangering human health or life or causing the large-scale destruction of plant or animal life (Ramljak and Petrović 2005). These crimes have been especially acute in recent years in Bosnia and Herzegovina, due to the rapid development in science and technology and the introduction of new technologies, which use new, powerful energy sources, and the construction of a large number of industrial plants followed by the development of large urban areas (Modly 1998). We will start our analysis with the conservation of the status of water and air in Bosnia and Herzegovina.

6.2 Main Primary Categories of Environmental Crimes in Bosnia and Herzegovina

6.2.1 Water Pollution in Bosnia and Herzegovina

People are now living through an extensive period of globalization, in which they reap all its benefits but suffer its "side effects" too. One such side effect is that the water flowing in the world is increasingly becoming more polluted. In particular, the European Union, for example, has had to adopt certain conventions that contest and mitigate the consequences of harmful actions against nature. So, the Convention on the Protection of Water has been passed in the ECE, e.g. the Helsinki Convention on the Protection and Use of Transboundary Watercourses and International Lakes. Certainly in Bosnia and Herzegovina the greatest burden of pollution of natural watercourses comes from industrial wastewater, which is being discharged without any purification at all or an insufficient degree of purification. If one sentence could describe the state of water in Bosnia and Herzegovina, it would be roughly as follows: "Water in Bosnia and Herzegovina is used to mitigate the effects of the discharges of household and industry wastewater, and given the state of the system for flood control it represents a potential threat to the population" (Ramljak 2002).

The increase in water consumption in Bosnia and Herzegovina is attributed to a large number of reasons: a higher social–material status, urbanization and partial industrialization, all of which lead to increased amounts of wastewater being discharged into the natural water systems (Mujanović 2009). In fact, these water systems are the most common sets of wastewater, which cannot tolerate pollution without serious disruption of the natural balance. Such practice has resulted in the intensive pollution of water resources and loss of water quality. The pollution of water resources is largely caused by materials which are a product of human activity (Korajlić 2008). It is obvious that the environmental crime of water pollution exists in Bosnia and Herzegovina, because the quality of drinking water is already poor in a number of cities because the city and the city's water resources are polluted (Sućeska 2008).

There are several types of water pollution caused by humans or which involve human factors. They are mainly microbiological contaminants and chemical pollutants. Water may be contaminated by a large number of microorganisms which come from the air, soil and plants, mostly from feces of human or animal origin. These forms of microbiological pollution are in most cases directly or indirectly encouraged by humans. Germs that can be transmitted through water include bacteria and viruses. Water used for drinking should not contain pathogenic microorganisms or any of the indicators of fecal bacteria contamination. The latest test of water on the largest water source of the capital of Bosnia and Herzegovina, symbolically named "Source of Bosnia", showed the presence of bacteria *Escherichia coli* (Ramljak 2002). These findings are devastating because the source has long been reputed for being one of the best water sources in the Balkans: Sarajevo is famed for being the city whose inhabitants drink the highest quality water.

Unfortunately, the construction of "tourist facilities" at the Olympic mountain Igman and Bjelašnica, site of the above-mentioned sources, without previously solving the wastewater treatment in an environmentally correct way, is effectively leading to the "self-poisoning of Sarajevans".

Microbiologically-contaminated drinking water leads to diseases, either by the direct use of polluted water through drinking or through watering vegetables, washing fruit or preparing food using contaminated water. On the other hand, chemical pollution is caused by the wastewater of most industrial complexes (in the production of oil, mining, etc.), as well as erosion from agricultural lands (in which case groundwater is most often affected) (Sućeska 2008). Chemical pollutants have been proven to lead to water pollution. Chemical pollutants that are usually found in the water in Bosnia and Herzegovina include ammonia, nitrates, detergents, mineral oils, fats and oils, pesticides and heavy metals (including arsenic, mercury, lead and copper).

6.2.2 Air Pollution in Bosnia and Herzegovina

Industrial plants in Bosnia and Herzegovina affect the air with the following gases: SO_2, NO_2, CO_2, CO, etc. These listed gases affect the air with increased emissions of stone and other dusts, as well as an increased amount of ash and soot. The air quality is not only affected by different industrial sectors but also by private furnaces, used for heating, which burn around a thousand tons of coal annually (Mujanović 2009). In addition to problems with water pollution, Bosnia and Herzegovina also has problems with air pollution and the presence of hazardous substances in the larger cities of our country. The chimney is both a means of expelling unwanted, harmful gases from the household and a way of getting rid of unwanted gases in industry. In the latter case, its dimensions are far greater as are the harmful, clearly visible effects of the gases on the environment. Toxic gases rise higher due to the physical laws of hot air rising in the upper classes and the law of flow in the chimney (Đonlagić 2005).

These two physical/chemical acts, facilitated by individuals to resolve the problem of unwanted gases as the current problems, in the second "act" two new acts are directly turned involved. The first is gravity that brings the particles of harmful substances back to the earth (heavier and bigger in the immediate vicinity of the chimney, other a bit further). The second is the extremely limited space of the atmosphere and its physical–chemical composition, as well as the function that this composition has had in the formation of the life on the planet Earth and its role in the further maintenance of eco-system. Direct air pollution in Bosnia and Herzegovina is also caused by the increase in the use of motorized transport. Motor vehicles allow greater mobility, but in turn require high amounts of oxygen (combustion in the engine) and return carbon dioxide and carbon monoxide to the atmosphere (Mujanović 2009).

Bosnia and Herzegovina has the legal framework that, in general, enforces preventive measures related to the problems of water and air protection. The legal framework is not in the hands of state level authorities but in those of authorities at the entities and Brčko District, which, in principle, resolve the matter in a similar manner (Golić 2005). Therefore, for the purposes of this article, we will give an example of the Federation where the following laws are in effect: Air Protection Act (Zakon o zaštiti zraka 2003); Waters Act (Zakon o vodama 2006); Environmental Protection Act (Zakon o zaštiti okoliša 2003); Nature Protection Act (Zakon o zaštiti prirode 2003); Act on Fund for Environment Protection of Federation of Bosnia and Herzegovina (Zakon o financiranju zaštite prirode Federacije Bosne i Hercegovine 2003); Act on Amendments to the Law on the Protection of Water (Zakon o izmjenama i dopunama Zakona o zaštiti voda 2003); Amendments to the Regulations on the type, mode, volume measurements and tests on used water and extracted material from the waterway (Zakon o izmjenama i dopunama Pravilnika o vrsti, načinu rada, volumen mjerenja i ispitivanja upotrijebljene i iskorištene vode i izvađenog materijala iz plovnog puta 2004); the Rules of plants and facilities which stipulates that environmental impact assessment is mandatory and that plants may only be constructed and operated if they have an environmental permit (Pravila o biljkama i postrojenjima, koji propisuje procjenu utjecaja na okoliš 2004); the Republic of Srpska (RS) and Brčko District have identical laws and regulations in their positive-legal framework as an integral part of Bosnia and Herzegovina administration.

A positive example in Bosnia and Herzegovina is certainly the Hydro-meteorological Institute of RS, which is the institution at the entity level for monitoring air quality. Although it does not yet have an extensive network of stations to measure air quality in all the cities in the RS, it has quickly achieved the tasks envisaged in the strategy to protect air, which is currently under the construction and development by the Ministry of Ecology of RS. In the RS, there are currently several stations for the measurement of air quality, following the level of primary pollutants in the air, SO_2, NO, NO_2, $(NO)_x$, CO, ozone and suspended particles to 10 µm. The regulation on the monitoring of air quality requires that the analysis should be conducted year-round so that a detailed assessment of air quality can be given annually.

6.2.2.1 Some Ecological Specificity in Bosnia and Herzegovina

The problem with ignoring of air and water pollution in Bosnia and Herzegovina is best illustrated with the following example. During the year 2008, when a state of emergency was declared in Budapest after the value of airborne particles in the air exceeded the value of 100 µg/m³, the authorities in Sarajevo, where the pollution was five times higher (500 µg), did nothing except issue useless appeals.

Most experts in Bosnia and Herzegovina have the opinion that the main air pollutants in Bosnia and Herzegovina are the power plants burning domestic coal with a high rate of sulphur dioxide. Bosnia and Herzegovina power plants have filters for dust, but at the same time do not have filters for sulphur dioxide. Incorporation of these systems to remove sulphur dioxide is very expensive for Bosnia and Herzegovina, which is practically the only official reason for their absence. Pollution is particularly prominent in the winter period when stable weather conditions occur after snow or rainfall, while in summer, the situation is much better because there is more wind, then the smog, which consists of substances such as sulphur dioxide, carbon monoxide, nitrogen oxides, disappears more easily.

In addition to thermal power plants in Bosnia and Herzegovina the motor vehicles are the second largest polluter. This situation is especially significant in Sarajevo where the greatest amount of pollution comes from old motor vehicles. The media in Bosnia and Herzegovina call this phenomenon the "European waste management". The cars are on average 15 years old and a large number of them have no catalysts and cause emission of higher level of toxic gases than newer cars. However, even if the cars did have catalysts, due to heavy congestion in Sarajevo experts predict air pollution at alarmingly high levels (Sućeska 2008).

According to research by Mujanović (2009), a large number of cars in Sarajevo (and specifically only this city) are not in technically acceptable condition. The problem is that the technical centres for vehicles certification have a lot of shortcomings. Permits are obtained in different ways and exhaust gases are poorly tested. According to the regulations, vehicles throughout Bosnia and Herzegovina should not emit more than one percent (1%) of carbon monoxide. However, there is almost no verification of the exhaust level in Bosnia and Herzegovina by law enforcement agencies. It is obvious that poverty resulted in people driving really old cars, and, in the context of our topic, can be concluded that the situation will not change until the old cars ruminated (Mujanović 2009).

6.2.2.2 Goals and Prevention: What Bosnia and Herzegovina Plans to Do?

Every country does its best to protect its water resources and air quality, aiming to prevent, or limit, pollution through the implementation of regular, systematic, efficient measures. Further, the text will present some goals which are mentioned in BH legislation and outline the preventive measures that would ensure the transfer of these laws from paper to reality. That is, the paper offers measures such as raising awareness among people about the importance of protection, reduction of air pollution and improvement of air quality by establishing regional networks for monitoring of air and creating strategies for improvement in areas where monitoring will detect deficiencies in air quality, making air quality management plans to reduce air pollution and reduced emissions below permitted limits prescribed by law enacted by Bosnia and Herzegovina, as well as the regulation in the EU; reducing the percentage of air pollution from sources of low combustion (private houses, buildings, furnaces for heating), reducing air pollution and enabling households to use natural gas for heating, having clean, high-quality, hygienically proper drinking water; reducing the pollution of city water and solving the problem of sewer network; adequately supplying the population with water; working to improve water quality and informing the public on that issue; cleaning rivers; reducing the pollution of rivers with nitrogen fertilizers and finally awakening a sense of responsibility among water pollutants (Šator 2001).

If contemporary civilization considers water scarcity as a key problem faced by humanity today, then the people of Bosnia and Herzegovina are very wealthy people. However, they still need to appreciate this fact and pay close attention to the quality of their water.

6.2.3 Crimes of Deforestation

It is generally known that trees maintain the balance of the eco-system. The forests play an extremely important role in the circulation of water in nature, and the consequences of their excessive felling or deforestation may be long-term and fatal. Deforestation is a process of the destruction of forests by cutting and burning.

Deforestation occurs for many reasons: the trees are sold as a fuel or as a commodity, while the devastated land is used as pasture for livestock, plantations and settlements. The destruction of forests without reforestation leads to landslides, the destruction of natural habitats and the destruction of biological diversity. Deforestation has a great influence on the amount of carbon dioxide in the atmosphere (Đonlagić 2005).

Carelessness, ignorance, lack of skilled personnel for the management of forests and ineffective environmental laws are some of the factors that lead to deforestation. In many countries deforestation is causing the extinction of plant and animal species, changes in climatic conditions, desertification and the displacement of indigenous peoples.

There are many causes of deforestation including corruption in government institutions, the unjust distribution of wealth and power, population growth, over-population and urbanization. Forests, given their self-reproducibility, natural structure, mixed composition and natural regeneration, represent one of the basic resources in the Development Strategy of Bosnia and Herzegovina for the future. Out of interest, just before the aggression and war in 1992, Bosnia and Herzegovina ranked next to just behind Finland and Sweden for the diversity of its forests on the European scale of forest resources.

Forests are the most valuable natural ecosystems, hence there is a need for their protection. The fact that Bosnia and Herzegovina has the fourth biggest forest in Europe is our comparative advantage. Unfortunately, our relationship to this resource is irresponsible to say the least, which is the reason for the continued reduction of our forests as well as the quality of growing stock becoming rather weak.

Forests are disappearing due to changes in land use (opening of mines, construction of hydro-accumulations and travel communications, raising industrial capacity, fires, etc.). Careless management and often timber theft reduce existing stocks of wood, a lack of quality in legislation and different approaches to its application only benefit individuals.

A significant part of the forests in Bosnia and Herzegovina is mined or damaged by war activities, as a result the Bosnia and Herzegovina timber industry is not competitive in world markets. There is no overall forestry policy at the state level.

6 Primary Categories and Symbiotic Green Crimes in Bosnia and Herzegovina 109

Instead a forestry company is entrusted with forests and forest management work on the land is under the strong influence of local authorities and poorly managed policy. The concept of integrated forestry in Bosnia and Herzegovina is growing more unachievable day by day, yet it is the only way to ensure a stable system of forest protection (Sakić 2007).

Forests cover almost 46% of the country, or 0.57 ha per capita, of which the Republika Srpska is 40% forest, or 0.79 ha per capita, while the Federation of Bosnia and Herzegovina is 48% forest, or 0.66 ha per capita. Forests and forest lands in the Federation of Bosnia and Herzegovina are spread over an area of about 1,560,000 ha of which about 1,308,000 ha or 82% are owned by the state and about 277,000 ha or 18% is property under private and other legal person's ownership (Mujanović 2009).

The representation of Bosnia and Herzegovina before external institutions in the forestry sector has been the responsibility and competence of Ministry of Foreign Trade and Economic Relations. According to the constitutional provisions, the forest owners are the Federation of Bosnia and Herzegovina, the Republic of Srpska and District Brčko, with administrative and executive powers for management of the forests in their administrative boundaries being through the Ministries responsible for forestry. According to the Forests Act in the Federation, the Federal Minister transferred the responsibilities over management and utilization of forests to the cantonal ministers with portfolio competence for forestry. The Federal Bureau of Forestry, which has the function of planning the development of forestry, was set up within the Federal Ministry of Agriculture, Forestry and Water Management, while cantonal administration for forestry, which has the function of planning and administrative control over the management of state and private forests, was established in the cantonal ministries.

Table 6.1 presents the structure of registered and sentenced crime related to the illegal cutting of forests in Bosnia and Herzegovina. It shows that in 2004, only 1,042 cases were adjudicated out of 4,415 registered cases, or 24%; in 2005, only 950 applications were adjudicated out of 4,964, or 19%, in 2006, only 306 were adjudicated out of 4,947 registrations, or 6%; while in 2007, only 1,121 were adjudicated out of 4,927 registered cases, or 23%. In 2008 the number of submitted (misdemeanours and criminal offenses) notifications was 2,960, of which 1,100 were adjudicated, or 37.20%.

For the purposes of this scientific work, it is necessary to mention that the percentage of binding judgments passed in 2006 was only 6% and that this was a good indicator for measuring the politicization of Bosnia and Herzegovina society. In fact, this was the "Election year" when many are "looking through their fingers". The data can speak for itself.

Next table (Table 6.1) indicates the slowness of judicial authorities regarding criminal filings, the new Act on Offences returned these procedures to the jurisdiction of the cantons, and in 2008 the number of prosecuted misdemeanour charges increased. Of the total damage by the allegations in the amount of KM 2,148,365.37 for 2004 the charged amount was KM 65,838.20 or 3.06%; out of KM 12,236,393.4 for 2005 the charged amount was KM 40,889.57 or 3.34%; out

Table 6.1 Judicial authorities regarding criminal filings (Federal Office of Statistics 2010)

Year	2004		2005		2006		2007		2008	
	Registration	Judicial decision	Registration	Judicial decision	Registration	Judicial decision	Registration	Judicial decision	Registration	Judicial decision
Misdemeanours	2,404	517	3,093	422	3,546	161	3,631	873	1,872	895
Criminal offenses	2,011	525	1,871	528	1,401	145	1,296	248	1,088	205
Total	4,415	1,042	4,964	950	4,947	306	4,927	1,121	2,960	1,100

of KM 1,233,306.33 for 2006 the charged amount was KM 22,225.59 or 1.80%; in 2007, out of KM 1,455,955.40 the charged amount was KM 60,291.79 or 4.14%; and in 2008, the estimated damage was KM 1,057,106.00 while the charged amount was KM 36,649.00 or 3.50% (1€= 1,95 KM).

A small number of convictions and sentences are counterproductive as they encourage the theft and illegal harvesting of forests. This raises the questions of the expediency of filing and the endangering of forest keepers when other authorities do not protect state property. The non-existence of cantonal administrations to protect the forests' wealth and slow judiciary action bring into question the implementation of the government's Action Plan to Combat Illegal Activities in the Forestry Sector and Wood Industry in Bosnia and Herzegovina (2007).

6.2.3.1 Where Have the Forests Disappeared in Bosnia and Herzegovina?

In the last 15 years the state has not sufficiently taken into account the enormous amount of illegal harvesting of forests throughout the country. An estimated approximate 20% of the total pre-war (before 1992) forests have been cut down, according to environmental experts from Germany and Alaska. Poor implementation of the legal framework for the protection of natural resources has caused the current situation. The experts unanimously state that there was extremely extensive illegal logging during the war to supply the population with firewood as well as harvesting for profit. Today Bosnia and Herzegovina has fewer and fewer forests and logic suggests that the "green gold" or forests are seen as health-giving, recreational areas – the lungs of Bosnia and Herzegovina – from the perspective of citizens, while they represent a considerable source of income from the perspective of entrepreneurs. Most of the trees were cut down in western Bosnia, but also on the mountains Cincar, Prenj and Velež. The most vulnerable species were Bosnia and Herzegovina endemic species: pine, beech, mountain ash, maple, fir, spruce and munika (*Pinus heldreichii*). The forests were once the basis of development, but this is no longer so in Bosnia and Herzegovina today. Bosnia and Herzegovina forests are exploited by individuals and companies, which often illegally export raw timber to Western European countries, because Bosnia and Herzegovina is one of the few states with natural forest. But only non-governmental organizations point this out and only the occasional ambitious activist warns – at least for now to deaf ears – that 1 day it will be too late.

Bosnia and Herzegovina has over a thousand active erosion areas in forests and forest land that annually slide over 20 million m³ of forest land. Disappearing ground flora and fauna are hard to renew in the bare areas. Forest roads are inaccessible and covered with landslides. Autochthonous flora and fauna are not able to grow or be sustained and disappear. Medicinal, aromatic and edible plants, forest fruits and mushrooms are not used (Mujanović 2009).

In recent years after the war in Bosnia the amount of erosion and flooding has increased, drying up sources of drinking water and what is most worrying is that one can feel the change in the microclimate. The war has brought another great frustration for the country. The fact is that during the war Bosnia and Herzegovina lost more than 30% (320) of its graduate forestry engineers. The system of education in forestry is not in compliance with the required quality of knowledge in the new economic conditions. Little attention has been devoted to the training of staff in the fields of IT, marketing, management or foreign languages. Scientific work and professional staff development has been ignored.

As for legislation, the entity laws on forests have been enacted, which prescribes fines of €6,000–€7,500 if the user of forests or other legal entity managed forests without approved plans, if more than the allowed annual volume of trees are cut in high forest or if a biological renewal is not implemented after the allowed harvest of forest took place, if there is no plan to protect forests from fire or if there is unauthorized use of forest chemicals.

Penalties from €5,000 to €6,000 are prescribed for the forest user or other entity if they allow or perform the devastation of forests, carrying out illegal trade in forest and forest land, and if they carry out the delivery and harvesting of trees in forests without resolved property relations.

Individuals are fined from €100 to €750 if they fell the forest without an issued decision or remittance, trade forest non-wood products without a permit, cut down or illegally appropriate timber up to 2 m^3, start unauthorized fires, dispose of garbage in the forest, clear forest land without permission or commit other related offenses.

Today, unfortunately, forestry in Bosnia and Herzegovina is facing new challenges. Namely, the interference of the international community in forest policy is creating many systematic errors. The international community is ignoring the Bosnia and Herzegovina forestry sciences and professions and bringing in foreign experts whose mediocre knowledge does not offer solutions that will improve forestry and preserve our forests from further devastation. The exploitation of forests via concessions has been openly advocated, forgetting that harvest is only one of the stages in the production of economic forests. Following this road, concessionaires will obtain the forests for exploitation and we should not expect them to take a special interest in forest protection. Modern techniques and cheap labour will rationalize the production of logs solely for their own profit. Therefore, the general message to the citizens of Bosnia and Herzegovina should be to satisfy with the protection of its natural heritage before it is too late.

6.2.4 Protection of Animal Rights

In a few cases, only some animals (like cats and dogs) have legal protection in Bosnia and Herzegovina. This proves only one kind of discrimination concerning animal rights. People have made a huge distinction between different types of animals: they respect some animals merely for religious reasons, others because they can utilize them for certain activities or work, and still others because they can benefit from them in some other way. Such an understanding of animals is very degrading since they are taken into consideration only when they have a function, whether used for therapeutic, social or service purposes.

The real value of animals is constantly ignored. That is why, when people think that an animal cannot offer any of the aforementioned purposes, they use it as food. The struggle for animal rights is still in its beginnings in Bosnia and Herzegovina. It will be successful only when people realize the true value of animals. This value does not depend solely on economic values and human perception. Humans today should take into consideration all living beings, even those that at first glance are not fun, attractive or devoted. Animals were finally granted their rights in Bosnia and Herzegovina with the publication of Bosnia and Herzegovina "Official Gazette" No. 25/2009 which included the Protection and Welfare of Animals Act (Zakon o zaštiti i dobrobiti životinja 2009). Pursuant to Article IV. 4a) of the Constitution of Bosnia and Herzegovina, the Parliamentary Assembly of Bosnia and Herzegovina, at the 42nd session of the House of Representatives held on the 17th and 29th of December 2008, and on the 25th session of House of Peoples held on the 26th of February 2009, adopted the Protection and Welfare of Animals Act. This law, for the first time in the history of Bosnia and Herzegovina, formalizes certain activities connected to the protection of animals, treating issues such as: Protection of animals in keeping and breeding; Protection of animals when carrying out health and zoo-technical treatment; Killing animals; Protection of animals at slaughter; Protection of animals during transport; Protection of wildlife; Protection of animals in zoos, circuses and exhibitions; Protection of abandoned and lost animals; Protection of animals for experiments and other scientific research; Expert advice for the protection and welfare of animal. Each provision of this law is applicable to certain rights of animals in Bosnia and Herzegovina. Thus listed this way they mean nothing, as they are only written articles without their use and implementation. Therefore in this work on the violation of animal rights in Bosnia and Herzegovina we will continue by linking to the present law in its full composition and see how animals are protected by law in Bosnia and Herzegovina, and whether the given law is adequately implemented (Animal Friends in BH 2010).

6.2.4.1 Violence Against Animals

People who are violent to animals rarely stop at just that violence. Violent behaviour toward animals has long been known as an indicator of psychopathology that is dangerous not only to animals. Those who consider the life of any living creature worthless are in danger of being taken by the idea that human lives are also worthless, wrote the humanist dr. Albert Schweitzer: "Killers often begin killing and torturing animals when they are children" states Robert K. Ressler, who worked on developing the profiles of serial killers in the USA. Research has convinced sociologists, lawmakers and courts that acting violently toward animals deserves our attention. It may be the first sign of pathological violence that leads to endangering human life. The fight to protect endangered species in the world has been in full swing for years, while the majority of citizens in Bosnia and Herzegovina does not even know that animals are endangered and how much effort it takes to save them. For now, the experts in Bosnia and Herzegovina have clearly defined the six currently most vulnerable members of the animal kingdom. Thus, within the Project for the development and protection of forests in Bosnia and Herzegovina, the key endangered animals that need special attention are the following: the chamois, the large grouse, the small grouse, the bear, the wolf and the lynx. Poaching in uncontrolled forest management areas and the absence of law enforcement are the reasons why the forests in the Federation are losing indigenous wildlife. Endangered animals, protected by law, are killed for meat, for trophies, and some just for pure pleasure. Poachers do not choose a particular weapon. The animals are killed by automatic rifles, snipers and mines, which is certainly not in line with the Protection and Welfare of Animals Act. Although it is a very topical issue, the area of animal rights is still in its infancy in Bosnia and Herzegovina, so for the purposes of this study these have been the primary indications for further study of the topic.

6.3 Symbiotic Green Crime

6.3.1 Disposal of Animal Waste

This is an extremely difficult problem to solve today in Bosnia and Herzegovina. The state has fully delegated powers to lower levels of authorities (the entities or the cantons). The mere fact that in Bosnia and Herzegovina there is no unique legislation treating this problem at the state level means there is a large risk of matters going astray and out of control.

The current situation with animal waste management in Bosnia and Herzegovina is such that the responsibility of removing animal waste has been entrusted to the municipalities, which continue to engage private contractors; the only exception to this rule is the city of Sarajevo, which organized the special Public Enterprise of Sarajevo Canton "KJKP Rad" (Cantonal Public Communal Company) to perform these tasks.

In the whole territory of Bosnia and Herzegovina there are only a few burial sites and landfills in operation which meet the required standards for environmental protection. This finding represents a large risk for human health, animals and the natural environment. The disappointing fact is that dead animals in Bosnia and Herzegovina are very often buried on the farms from which they originate. The situation is no better when it comes to waste from slaughterhouses, which is usually removed along with other waste; while the animal waste from slaughter performed during religious holidays (Eid al-Adha for Muslims) is often disposed in rivers.

Annual statistics in Bosnia and Herzegovina register about 3,955 t per year of animal tissue waste and 40 t per year of dead fish. There is no adequate system in place for the disposal of waste from these activities, particularly for the disposal of animal tissue, which can lead to both environmental and health problems. Animal carcasses can pollute the soil, water and air, and thus become directly harmful to

6 Primary Categories and Symbiotic Green Crimes in Bosnia and Herzegovina 115

animals and humans. It is therefore necessary that animal corpses in general, and in particular those with infectious disease, are disposed of in an appropriate way as soon as possible. The safe removal of animal carcasses and animal waste should be processed in a facility specifically for this purpose. Exceptionally, other safe methods of disposal can be used such as incinerating the waste matter in the slaughterhouse, inserting it into pit-graves or burial at the animal cemetery or other convenient place (Mujanović 2009).

In the Republic of Srpska, the safe removal of animal carcasses is carried out at the animal cemetery located near the Manjača – Banja Luka or at locations designated by the local community. In the Federation, the safe removal of animal carcasses is done through the relevant utility companies and at locations designated by the local community. The only properly constructed pit grave in the country was built by the Cantonal Public Communal Company "RAD" in Sarajevo.

As for incineration, on the 14th of January 2009 in Sarajevo the first delivery of mobile incinerators in Bosnia and Herzegovina took place. The Federal Directorate for Civil Protection purchased a mobile incinerator model A850 (A) to be used by the Veterinary Faculty in Sarajevo.

6.3.1.1 Hazardous Waste Disposal

Besides waste of animal origin, in Bosnia and Herzegovina there is also the problem of the disposal of so-called "hazardous waste". Hazardous waste usually occurs in industry, but may also in households because hazardous substances can be found in many products that surround us such as used batteries, old medicine, paints and varnishes, various chemicals, waste motor oil, etc. Such waste contains substances that can be toxic, carcinogenic, mutagenic, infectious or flammable, and which enter the biological chain through land or water and can cause human illness and adverse effects on wildlife.

Hazardous waste can include: communal waste, industrial waste, packaging waste, construction waste, electrical and electronic waste and scrap vehicles, waste tires can be hazardous if they contain one of the properties of hazardous waste. Following the introduction and implementation of laws on waste, which are often supplemented with a number of under-laws and regulations, we now have only strictly controlled sanitary landfill. Bosnia and Herzegovina has laws, but their non-implementation has led, especially in the post-war period, to the creation of numerous waste dumps. Such laws are:

- Act on Radiation and Nuclear Safety in Bosnia and Herzegovina (Zakon o radijacijskoj i nuklearnoj sigurnosti u Bosni i Hercegovini 2007)
- Rulebook on the limits above which persons should not be exposed to radiation (Pravilnik o granicama iznad kojih osoba ne smije biti izložena radiaciji 2004)
- Rulebook on notification and authorization of activities related to ionizing radiation (Pravilnik o načinu prijave i autorizacije aktivnosti vezanih na ionizirajuće zračenje 2003)

- Protection of Population from Infectious Diseases Act (Zakon o zaštiti stanovništva od zaraznih bolesti 2005)
- Rulebook on the manner of conducting mandatory immunization (Pravilnik o načinu obavljanja obveznih imunizacija 2002)
- Waste Management Act (Zakon o gospodarjenju otpadom 2003)
- Rulebook on the management of medical waste (Pravilnik o upravljanju medicinskog otpada 2008)

This would be the positive legislative framework for the treatment of problems related to hazardous waste disposal in Bosnia and Herzegovina. In the continuation follows the presentation of the system of the non-governmental sector in Bosnia and Herzegovina popularly called "the green movements". Their activities are characterized by significant problems. In previous chapters we have seen that public awareness of environmental protection is very restricted. The green movement has a frustratingly limited financial ability in a political climate which is concerned with "more important issues", and although this is the fundamental problem for the movement, it has been considered secondary. Despite such difficult circumstances in Bosnia and Herzegovina, there are still an impressive number of organizations which are concerned with the protection of the human environment.

6.3.2 *The Green Movement in Bosnia and Herzegovina*

If we talk about those who unconditionally support and protect the environment and work hard without profit, then we are certainly talking about various eco-movements in Bosnia and Herzegovina which act as non-governmental organizations. We will use the term eco-movement to refer to any organized group of people such as voluntary organizations, humanitarian agencies, professional associations, human rights organizations, cooperatives and all similar forms of organization, provided that they are independent from government, i.e. from governmental organizations and institutions which are concerned with the preservation of the environment. These NGOs are basically self-governing bodies that operate on the basis of volunteer work, and which are therefore not under the management of the authorities. They can build different structures based on the principles of democracy and transparency, drawing on a number of legal and information resources, as well as using different types of media (Draganić et al. 2005).

Non-governmental organizations for the protection of the environment represent a link between citizens and the government sector (Draganić et al. 2005). It is very important that NGOs understand the role they can play in society and the way in which they can most effectively serve as a bridge between the government and the citizens. Mediation in the conflict of interests is an important task for these organizations. Local communities are often afraid of breaking the prohibitions which result from the declaration of protected areas (bans on grazing, logging, hunting, etc.). Non-governmental organizations must be utilized in the function of nature conservation, but also to protect local interests. This will be achieved through active participation in determining the boundaries of protected areas and the determination of zones with different levels of protection: from the strictly protected central zones to the peripheral with a lower degree of protection and development opportunities for local people (Variščić 2008).

Most eco-associations (NGOs) in Bosnia and Herzegovina tend to gather volunteer crews from schools, colleges and other institutions and try to engage professionals who will contribute in many ways to the preservation of the environment through voluntary work. NGOs do not include bodies that act as political parties, trade unions and religious communities. These movements are registered as citizen associations, distinctive for not being burdened with the struggle for power and institutionalization, but instead characterized by their massive, voluntary memberships, volunteerism and for being organized around involvement in matters which represent their key needs and interests.

In accordance with Article 11 of the Act on Associations and Foundations (Zakon o udrugama i zakladama 2003) the association must be established by at least three natural or legal persons who are citizens of Bosnia and Herzegovina, or foreigners who are permanently residing or staying longer than 1 year in the territory of Bosnia and Herzegovina, alone or together with the citizens of Bosnia and Herzegovina. The association is established by making a Founding Act. The association shall acquire the status of legal person by registration in the registry. Legal actions taken prior to entry in the register of associations creates obligations only for the individuals who have taken these actions.

International organizations and foreign donations play the dominant role in the financing of eco-movements. In some parts of Bosnia and Herzegovina there is support from local institutions (ministries and local government), but this type of financing is still insignificant. A smaller number of NGOs are trying to be self-sustaining, running commercial activities and collecting membership dues. Although there is an initiative to create a sustainable development strategy for the NGO sector in Bosnia and Herzegovina, it is still difficult to talk about real opportunities for the financing and sustainability of NGOs (Draganić et al. 2005). The number of NGOs in Bosnia and Herzegovina is somewhere around 700, but those that participate actively in planning and maintaining the environment is not more than 50.

Media relations with the environmental associations in Bosnia and Herzegovina seem almost non-existent. Most of the eco-movement has expressed dissatisfaction with media coverage of their implemented projects as well as of all other activities undertaken. This in some way confirms for us how underdeveloped awareness is in Bosnia and Herzegovina about these fundamental issues.

6.3.2.1 Government and Ecology in Bosnia and Herzegovina

The responsibility of Bosnia and Herzegovina as a state, is to protect its natural resources and the environment. Jurisdiction for the protection of the environment and concerns about nature has been entrusted by the state authority to the ministries and other institutions at entity level, such as:

(a) The Ministry of Agriculture, Forestry and Water Management of the RS
(b) The Ministry of Planning, Construction and Ecology of the RS
(c) The Ministry of Economy, Energy and Development of the RS
(d) The Republic Hydro meteorological Institute of the RS
(e) The Federal Ministry of Agriculture, Water and Forestry
(f) The Federal Ministry of the Environment and Tourism
(g) The Federal Ministry of Physical Planning

Although in 2002 and 2003 the Republic of Srpska and the Federation of Bosnia and Herzegovina (Federation) adopted acceptable legislation on the protection of the environment and nature, its implementation has not been satisfactory. This is mostly due to the significant lack of required by-laws. Since 2003, only 25 of the necessary 100 by-laws have been adopted, these legislative acts being needed to facilitate the practical application of the law. Given the current dynamics of the adoption of by-laws, we can conclude that the completion of this process is still some way off (Variščić 2008).

Eco-associations in Bosnia and Herzegovina have already emphasized the above as a significant problem. Considered an already alarming situation, the fact that environmental protection is not regulated at state level and stressing the example of behaviour of state institutions in the area of so-called entity lines of demarcation. According to the example, the entity authorities suggest that persons engaged in illegal logging continue to work freely but in "their area", or translated into plain English: in other entities. One thing is certain, as long as we continue to speak of Serbian, Croatian or Bosnian forests and rivers when referring to the territory of Bosnia and Herzegovina, eco-associations will continue to have difficulty achieving their goals and, in a direct way, the state will continue to endanger the environment (Adilović and Pehlić 2004).

Another issue highlighted by the Association for Plant Protection in Bosnia and Herzegovina is the violation of the Act on Phyto-Pharmaceutical Resources in 2005 (Zakon o fito-farmaceutskim resursima u 2005 2005). Since, this is also regulated at entity level in terms of the importation of these means, for example, no matter what is allowed to be imported into the Federation it is not possible to control because we cannot know how many were imported of these means at border crossings in the RS and thereafter arrived in the Federation of Bosnia and Herzegovina. In this way, Bosnia and Herzegovina is not able to establish a unique control system for the importation of phyto-pharmaceutical assets on its territory.

State authorities in Bosnia and Herzegovina are still trying to create certain programs for the conservation of nature and the environment. One of the measures taken has been to establish a new system of issuing environmental permits, which is based on modern European standards and procedures, and which pertains to all effects of industrial plants and facilities, and the activities of business entities in general, on certain segments of the environment and the environment as a whole. Therefore a new legal institute called environmental permits has been introduced,

6 Primary Categories and Symbiotic Green Crimes in Bosnia and Herzegovina

which, with its instruments in the form of provisions in the law and execution regulations, has a preventive effect on excessive pollution established by the limiting parameters on environmental pollution and which contributes to the preservation and protection of human health as well as that of the entire living world (Luebbe and Stroeker 1990).

However, this may be seen as a drop of water in the ocean. State authorities in Bosnia and Herzegovina could do much more. This, after all, is the duty of Bosnia and Herzegovina government both in accordance with international law as well as internal laws that the country has itself enacted.

6.4 Conclusion

In previous chapters we have tried to show the phenomenology of environmental protection in Bosnia and Herzegovina. Our aim has been to stimulate new ideas on the theme "Green Criminology in Bosnia and Herzegovina" because research of this kind has been very scarce. The focus has been deliberately directed at underlining highly important legislation such as Environmental Protection Act, the Nature Protection Act, the Air Protection Act, the Waste Management Act, the Waters Act, the Act on Fund for Environmental Protection, and the Act on Phyto-pharmaceutical means in Bosnia and Herzegovina in order to indicate the possibilities for research of violations thereof. The opening of "a new branch of applicable criminology" in Bosnia and Herzegovina, which will take its research focus as the etiology and phenomenology of environmental crime has been the main objective of this work. On this occasion, we have raised a great objection to the account of the Interior Ministries in Bosnia and Herzegovina (out of 14) for their absence of appropriate reaction and response to the reported violations of the aforementioned laws and for their lack of understanding and attitude towards this type of crime. They afford a traditional police response to conventional types of crime while "newer" crimes such as computer crime, art smuggling and, in last place, environmental crimes, are treated as secondary, less significant matters. We hope that with this article we will have stimulated young criminologists in Bosnia and Herzegovina to reflect critically on the existing practice of institutions which are responsible for formal social control in Bosnia-Herzegovina when it comes to environmental crime.

References

Adilović A, Pehlić O (2004) Zaštita životne sredine (Eng. Environmental protection). Univerzitet u Tuzli, Tuzla

Akcijski plan za borbu protiv nezakonitih aktivnosti u sektoru šumarstva i drvne industrije u Bosni i Hercegovini (2007) (Eng. Action plan to combat illegal activities in the forestry sector and wood industry in Bosnia and Herzegovina). Službene novine Federacije Bosne i Hercegovine, br. 88/2007

120 E. Muratbegović and H. Guso

Animal Friends in BH (2010) Resource document. http://www.prijatelji-zivotinja-B&H.com. Accessed 15 May 2010

Carabine E, Iganski P, Lee M, Plummer K, South N (2004) Criminology: a sociological introduction. Rutledge, London

Đonlagić M (2005) Energija i okolina (Eng. Energy and environment). Printcom, Tuzla

Draganić J, Šehović I, Pulić E (2005) Prirucnik za nevladine organizacije (Eng. Hanbook for non-governmental organizations). Regionalni centar za okolis/zivotnu sredinu za srednju i istocnu Evropu, (REC), Ured za Bosnu i Hercegovinu, Sarajevo

Federal Office of Statistics (2010) Resource document. http://www.fzs.ba. Accessed 7 May 2010

Gibbs C, Gore M, McGarrell E, Rivers L (2010) Introducing conservation criminology: towards interdisciplinary scholarship on environmental crimes and risks. Br J Criminol 50(1): 124–144

Golić B (2005) Ekologija i ekološko pravo (Eng. Ecology and Environmental Law). Pravni fakultet, Sarajevo

Korajlić N (2008) Kriminalistička metodika (Eng. Criminal investigation methods). Fakultet kriminalističkih nauka, Sarajevo

Luebbe H, Stroeker E (1990) Ekološki problemi u kulturalnoj mnijeni (Eng. Environmental issues in cultural changes). Veselin Masleša, Sarajevo

Lynch M, Stretsky P (2003) The meaning of green contrasting criminological perspectives. Theor Criminol 1(7):217–238

McLaughlin E, Municie J (2006) The sage dictionary of criminology. Sage, London

Modly D (1998) Priručni kriminalistički leksikon (Eng. Criminal investigation dictionary). Fakultet kriminalističkih nauka, Sarajevo

Mujanović A (2009) Kriminološki aspekti zaštite životne sredine u Bosni i Hercegovini (Eng. Criminological aspects of people's protection in Bosnia i Herzegovina). Fakultet kriminalističkih nauka, Sarajevo

Pearce F (2007) Green criminology: another "Yellow Logarithm" or a productive "New Paradigm?" Paper presented at the annual meeting of the American Society of Criminology, Atlanta. Resource document. http://www.allacademic.com/meta/p202306_index.html. Accessed 11 July 2010

Pravila o biljkama i postrojenjima, koji propisuje procjenu utjecaja na okoliš (2004) (Eng. Rules of plants and facilities which stipulates that environmental impact assessment). Službene novine Federacije Bosne i Hercegovine, br. 19/2004

Pravilnik o granicama iznad kojih osoba ne smije biti izložena radiaciji (2004) (Eng. Rulebook on the limits above which persons should not be exposed to radiation). Službene novine Federacije Bosne i Hercegovine, br. 8/2004

Pravilnik o načinu obavljanja obveznih imunizacija (2002) (Eng. Rulebook on the manner of conducting mandatory immunization). Službene novine Federacije Bosne i Hercegovine, br. 7/2002

Pravilnik o načinu prijave i autorizacije aktivnosti vezanih na ionizirajuće zračenje (2003) (Eng. Rulebook on notification and authorization of activities related to ionizing radiation). Službene novine Federacije Bosne i Hercegovine, br. 64/2003

Pravilnik o upravljanju medicinskog otpada (2008) (Eng. Rulebook on the management of medical waste). Službene novine Federacije Bosne i Hercegovine, br. 77/2008

Ramljak A (2002) Žrtve ekološkog kriminaliteta-masovna i intenzivna viktimizacija (Eng. Victims of green crimes – massive and intensive victimisation). Kriminalističke teme 1–2:35–57

Ramljak A, Petrović B (2005) Viktimološki pojmovnik (Eng. Victimological glossary). Udruženje diplomiranih kriminalista u Bosni i Hercegovini, Sarajevo

Sakić E (2007) Ekološki aspekti sigurnosti u Bosni i Hercegovini (Eng. Environmental aspects of safety in Bosnia and Herzegovina). Fakultet kriminalističkih nauka, Sarajevo

Šator S (2001) Okolina u Bosni i Hercegovini i pristupanje EU Ceteor (Eng. Surroundings in Bosnia and Herzegovina and EU ceteor approaches). Službeni list BH, Sarajevo

South N (1998) A green field for criminology? A proposal for a perspective. Theor Criminol 2(2):211–235

6 Primary Categories and Symbiotic Green Crimes in Bosnia and Herzegovina 121

Sućeska M (2008) Metodika otkrivanja ekološkog kriminaliteta (Methods of Detection of Environmental Crime). Fakultet kriminalističkih nauka, Sarajevo

Variščić A (2008) Zaštita prirode – Međunarodni standardi i stanje u Bosni i Hrecegovini (Eng. Environment protection – international standards and situation in Bosnia and Herzegovina). Udruženje za zaštitu okoline "Zeleni"– Neretva, Konjic

White R (2003) Environmental issues and the criminological imagination. Theor Criminol 7:483–506

Zakon o financiranju zaštite prirode Federacije Bosne i Hercegovine (2003) (Eng. Act on Fund for Environment Protection of Federation of Bosnia and Herzegovina). Službene novine Federacije Bosne I Hercegovine, br. 33/2003

Zakon o fito-farmaceutskim resursima u 2005 (2005) (Eng. Act on Phyto- Pharmaceutical Resources in 2005). Službene novine Federacije Bosne i Hercegovine, br. 29/2005

Zakon o gospodarjenju otpadom (2003) (Eng. Waste Management Act). Službene novine Federacije Bosne i Hercegovine, br. 33/2003

Zakon o izmjenama i dopunama Pravilnika o vrsti, načinu rada, volumen mjerenja i ispitivanja upotrijebljene i iskorištene vode i izvađenog materijala iz plovnog puta (2004) (Eng. Amendments to the regulations on the type, mode, volume measurements and tests on used water and extracted material from the waterway). Službene novine Federacije Bosne i Hercegovine, br. 56/2004

Zakon o izmjenama i dopunama Zakona o zaštiti voda (2003) (Eng. Act on Amendments to the Water Protection Act). Službene novine Federacije Bosne i Hercegovine, br. 54/2003

Zakon o radijacijskoj i nuklearnoj sigurnosti u Bosni i Hercegovini (2007) (Eng. Act on Radiation and Nuclear Safety in Bosnia and Herzegovina). Službene novine Federacije Bosne i Hercegovine, br. 88/2007

Zakon o udrugama i zakladama (2003) (Eng. Associations and Foundations Act). Službene novine Federacije Bosne i Hercegovine, br. 33/2003

Zakon o vodama (2006) (Eng. Waters Act). Službene novine Federacije Bosne i Hercegovine, br. 70/2006

Zakon o zašitit vode (2006) (Eng. Water Protection Act). Službene novine Federacije Bosne i Hercegovine, br. 70/2006

Zakon o zaštiti i dobrobiti životinja (2009) (Eng. Protection and Welfare of Animals Act). Službene novine Federacije Bosne i Hercegovine, br. 25/2009

Zakon o zaštiti okoliša (2003) (Eng. Environmental Protection Act). Službene novine Federacije Bosne i Hercegovine, br. 33/2003

Zakon o zaštiti okoliša (2009) (Eng. Environmental Protection Act). Službene novine Federacije Bosne i Hercegovine, br. 38/2009

Zakon o zaštiti prirode (2003) (Eng. Nature Protection Act). Službene novine Federacije Bosne i Hercegovine, br. 33/2003

Zakon o zaštiti stanovništva od zaraznih bolesti (2005) (Eng. Protection of Population from Infectious Diseases Act). Službene novine Federacije Bosne i Hercegovine, br. 29/2005

Zakon o zaštiti zraka (2003) (Eng. Air Protection Act). Službene novine Federacije Bosne i Hercegovine, br. 33/2003

Chapter 7
Usage of Special Investigation Measures in Detecting Environmental Crime: International and Macedonian Perspective

Marina Malis Sazdovska

Abstract The purpose of this chapter work is deepening the knowledge about environmental crime. Organized crime is a phenomena which is a part of modern society living and represents a treat for the real society developments. Environmental crime with some separate forms and shapes presents the environment crime, which is presented not only on the regional but although on the international, global level. One of the most profitable businesses in the sphere of crime is the illegal lodgment of waste which is put on the top of the scale of crime activities in present time. The goals that have to be achieved are more efficient fight against environment crime, and they belong in the part of sphere of criminology. Fighting against this type of crime is a part of the methodology of its detecting, substantiation and clarifying the environmental crimes and the police methods is enriched with special investigation measures and activities undertaken by the police authorized security bodies.

The chapter identifies the measures and the activities that authorized bodies take over to fight against this type of crime. It also takes into consideration the measures that are taken at a national level, and the measures taken by Interpol. During this work failures in police and other bodies operating have been addressed. The most important are the notes considering inadequately applying of measures that can result in not convincing the perpetrator. In the conclusion also the way of realization of the law estimated solutions is proposed, with an aim of protection of the environment. In this chapter a methodology is recommended about the fight against environmental crime, and because of that it is of important value for operational workers.

M.M. Sazdovska (✉)
Assistant Professor of Environmental Criminology, Faculty of security-Skopje,
University of Saint Kliment Ohridski, Republic of Macedonia,
e-mail: mmalis@fb.uklo.edu.mk

G. Meško et al. (eds.), *Understanding and Managing Threats to the Environment in South Eastern Europe*, NATO Science for Peace and Security Series C: Environmental Security, DOI 10.1007/978-94-007-0611-8_7, © Springer Science+Business Media B.V. 2011

7.1 Introduction

Today, ecological safety represents a goal of the international community, and that goal is currently the primary determination of the national states. This situation exists because today there are ecological local and global threats to stability at both, national and international level.

Environmental threats are part of the overall world threats that have influenced personal, national and international safety. Some local environmental problems evolve into global problems. Examples include illegal logging, ozone wasting, global warming, constant organic polluters, desertification, etc.

Taking into consideration that part of the definition of safety as a lack of danger, ecological safety can be defined as a state in which there is no threat for the living world or the environment (this is the definition by the author of the text).

Considering the definition, environmental crimes present an activity endangering the ecological safety and seriously violate environmental protection. In order to prevent, as well as eliminate, environmental crime, application of all available measures and means is required. Law enforcement agencies have a very important role fighting against perpetrators of environmental crimes. In some countries (Germany and United Kingdom), there are environmental police, but in some countries where there is not a special service, other competent authorities have this job. At an international level, there are international organizations and institutions authorized for the protection of the environment, for both serious violations as well as environmental crimes. Among them, the very important role of leader is given to the leading organization, Interpol. Interpol fights against this kind of criminality with all measures and activities for prevention, especially in the sphere of organized crime.

7.2 The Role of Interpol in the International Environmental Protection

Interpol, as an international organization fighting against crime in global and world frames, supports and helps all organizations, authorities and services whose goal it is to prevent and fight against international crime.

There are 184 country members whose primary aim is to gather, compare and exchange information. In 1993 an Environmental Crime Committee was established. Besides fighting against environmental crime, this committee organizes international conferences, helps in identifying problems related to eco-crime investigations and identified solutions for environmental crime.

In the frames of the Environmental Crime Committee, there are two working groups: (1) wildlife crime working group (flora and fauna) and, (2) pollution crime working group. Between 2005 and 2007, with external financial aid, Interpol employed officers with full time jobs. Specific projects, such as "Clean seas", establishing relations with organized crime investigations and conducting staff

7 Usage of Special Investigation Measures in Detecting Environmental Crime 125

training. Also a manual for oil pollution investigation was published, establishing qualitative values of crimes related to pollution in organized crime. For the purposes of the prosecutors memorandum concerning seriousness of eco-crime was issued. Furthermore, in the program of these groups projects for the eco-crime training are included with the purpose of training local government services to recognize eco-crime, to investigate and protect the health and the well-being of the community. Second way of Interpol's acting is so called eco message which has the goal to (Interpol 2010a):

1. Improve the communication between the eco-crime government services in different countries.
2. Develop a database for determining the trends of activities related to eco-crime.

Eco message includes all crimes related to ecology that have international consequences, such as: wildlife smuggling, illegal trans-border delivery of waste, pollution caused by ships, and smuggling of dangerous substances etc. The way of acting of the eco message is conducting investigations, control of cross-border deliveries of dangerous waste, training in the field of investigation techniques, forensics science, anticorruption, human rights and government services. Besides that, the wildlife working group also determines connections between the poaching of elephants with organized crime and terrorism. Interpol has implemented a database of such organized criminal acts with cooperation of non-governmental organizations to exchange information and criminal analyses (Interpol 2010b).

Initiatives by Interpol, acting of one complex mechanism, are steps toward prevention of eco-crimes. Also carrying out investigations, forming and using of databases, international trainings and tuitions, exchange of data, etc. A case study on how to detect eco-crime at an international level will be elaborated for further understanding and introduction to the procedure of investigation of eco-crimes.

Investigations at an international level include government services of different countries, predict taking measures and activities according to the national laws of those countries, and of course in cases of transnational organized crime then actions at an international level will be taken. In addition, the role and most important goals of international organizations, such as Interpol is protection of the environment. Through international exchange of data and information during eco-crime investigation cases, important evidence duding to criminal conduct can be gathered and acted upon by competent authorized officials to ensure effective arrest and prosecution.

7.2.1 Case Study: Hoax Case

Keng Liang "Anson" Wong was an international smuggler of reptiles from Penang, Malaysia. He successfully smuggled 300 protected animals to the United States concealing them in Federal Express packages, airline luggage and other disguised packaging. He was arrested in Mexico, tried and convicted in San Francisco, California. Wong was one of the largest criminal animal smugglers known to government law enforcement intelligence. The FBI identified Wong as "King Rat" due to his top black market of wildlife smuggling operations. In Malaysia he owned a private zoo, as a legal business investment to hide the illegal wildlife smuggling. Wong's illegal smuggling included animals near extinction such as the Komodo dragon from Indonesia (each cost US$30,000), Chinese alligators (US$17,000), the boa from Madagascar (US$2,500), etc. Wong's operation was largely revealed during a large train theft operation of reptiles. This operation included the theft of nearly 37 priceless tortoises (each valued at US$52,500), stolen from Madagascar. Today, Wong is serving his sentence in a California prison for smuggling endangering species (Environmental News Service 2010).

Between 1995 and 1998, the U.S. Fish and Wildlife Service (USFWS) conducted an undercover investigation of Wong and his Malaysian wildlife international crime business, "Sungai Rusa Wildlife" (Environment and Natural Resources Division 2010).

During the investigation, the USFWS documented 14 illegal shipments by Wong into the United States from Malaysia, Indonesia, and the Philippines. This shipment contained protected reptiles valued at nearly a half million dollars. In July 1998, a federal grand jury indicted Wong on 51 federal charges related to his alleged wildlife smuggling enterprise. In September 1998, an undercover USFWS agent, posing as a reptile trafficker, lured Wong to Mexico City, where he was arrested by the Government of Mexico on U.S. charges and incarcerated pending extradition to the U.S. Wong fought the extradition process for nearly two years until June 16 when he filed papers in the Mexican courts seeking to terminate his extradition fight (Environment and Natural Resources Division 2010).

Wong was charged in San Francisco with conspiracy, smuggling, money laundering, making false statements and violating federal wildlife statutes that prohibit trafficking animals protected by federal and/or international laws. Wong also was indicted by a federal grand jury in Florida in 1992 on wildlife smuggling charges. Wong and his associates are alleged to have employed a variety of schemes to clandestinely transport protected animals to the United States, such as using a human courier who concealed wildlife in airline baggage, sending animals concealed in fraudulently labelled Federal Express shipments, and concealing illegal animals within large commercial shipments of legal animals (Environment and Natural Resources Division 2010).

Approximately 300 animals were allegedly smuggled into the United States, of which approximately 70 Plowshare Tortoise were stolen in 1996 from a breeding project in their native country of Madagascar, worth US$52,500 each. Charged with Wong in the San Francisco case are James Michael Burroughs, Beau Lee Lewis, Jeffery Charles Miller, Robert G. Paluch, of the USA and Yuk Wah "Oscar" Shiu, of Hong Kong, China (Environmental News Service 2010).

The maximum penalty for the money laundering offences is up to 20 years imprisonment and/or a US$500,000 fine; the remaining charges each carry a maximum of up to 5 years imprisonment and/or US$250,000 fine. The arrests of Wong's operation resulted from cooperative efforts by the Justice Department, the Mexican Attorney General's Office, the U.S. Fish and Wildlife Service, the U.S. Customs Service, INTERPOL and the Royal Canadian Mounted Police of Canada.

7.2.2 Analysis of the Hoax Case

Examining the Keng Liang Wong case, law enforcement agencies from the USA, Canada, Mexico, Interpol etc. have realized the need of international cooperation in preventing the eco-crime. Cooperation between the authorized government entities and the procedures for detecting, investigating, and prosecuting eco-crimes, especially when it comes to trans-border crime is necessary.

In the cooperation between authorized bodies and organizations, there is a need of defining the methods of communication, realization of mutual operational-tactic field measures, arresting the perpetrators and imposing adequate sanctions.

In summary, the Hoax case is about an organized crime group in the pursuit of financial gain through the organisation of transnational illegal animal smuggling. Due to the nature of organised crime groups, careful and adequate steps must be taken to successfully detect and prosecute illegal organised crime groups. This can be done through clarification and evidence of criminal acts committed by groups, such as Wong and his group. Use of operational-tactic measures and activities, investigations and special investigations measures are the necessary in fighting organised crime.

Because of the fact that it comes to a group operation in a conspiratorial way in ransom, transport and sale of protected animals involving of a secret agent and concealed investigation was carried out. In the Hoax case, an undercover agent was involved in the investigation process (Dzukleski 2005). An undercover American agent was introduced as an animal dealer in Mexico, where he proceeded to cooperate with the Wong's group in Mexico, where later Wong was arrested. In order for all of this to have been successfully completed, the undercover agent had to be well trained, educated with the group's habits, characteristics, and behaviour. The agent also had to be educated on underground work and the criminal milieu of environmental crime. The agent kept secret, undercover communication with the secret service he belonged to in a private manner. The communication can be direct or indirect. Direct communication is through the personal contact, while indirect communication through a mediator (Batkovski 2008). In addition to the secrecy, the legality, material truth, planned approach, rationality and effectiveness, offensiveness, team work and flexibility should be considered and respected (Malis Sazdovska and Dujovski 2009).

Action by groups on an international level value teamwork to complete tasks and collaboration in communicating discoveries. Collaboration is important by all groups and individuals to successfully reach a goal. The phases of the evolutional development should be taken into consideration: forming, brainstorming, planning, performing and ending. In the Hoax case cooperation can be viewed on several occasions, such as undercover agent stationed in Mexico and the actual of Wong, in which the cooperation between the American secret agent and the Mexico police forces was necessary.

Besides the cooperation of the undercover agent, the agent must also be able to continue to infiltrate and blend in with the surrounding criminal environment. This might include a history and past story of involvement in other similar criminal activities. This gives the agent a more legitimate cover by explaining his presence in the criminal world. Through the undercover role, the agent acquires information, data, and later evidence against groups and their members. The evidence is later used to arrest and prosecute members.

To achieve the goal of an undercover agent, it is benefited to be educated and have knowledge in the fields of criminology, psychology, sociology, etc. A multi-disciplinary education approach is needed in the investigation of organised (environmental) crime.

Due the nature of the Wong's operation, the transportation of protected animals across international borders through various means was organised (use of a person-courier hiding animals in airline luggage, hiding animals and falsely labelled Fed-Ex packages, and concealing animals in big illegal commercial shipments and other ways of transport). This is referred to as *controlled shipment.*

With the intensification of the international cooperation in the fight against organized crime, controlled shipment methods by organised crime groups reviewed. Controlled shipment is often used by groups trafficking drugs, weapons, humans or illegal trade in protected species of animals and plants. Although the authorities had the information about Wong owning a private zoo in Malaysia with legally protected exotic animals intended for illegal trade on the black market, they had to follow the packages through their shipment process. This approach allows the detection of the sender and the receiver. Shipments that Wong was sending from Malaysia to other parts of the world were controlled and presented important sources of information. Police and authorities allow for packages to complete their shipment cycle to their final destination in efforts to uncovering all individuals involved in the process.

Transborder packages are unable to be tracked once leaving their original country. Therefore the detection of other offenders involved in the chain of illegal transport is often impossible and leads are lost. Competent bodies and organizations collaborate with agencies such as Interpol to inform and detect the shipment, allowing it to continue through the shipment process for further unveiling of others involved. For this type of communication and collaboration of the services from different countries authorizations and involvement of each member country have to be defined separately.

The control of communications should be valued during the investigation process of cases like the Hoax case. Maintaining communication regarding plans to fulfil controlled shipment measures, as a part of the organised crime in the sphere of environmental crime. Communication among the members of the canal of illegal trade must me traced and recorded.

When tracing the shipment it is necessary to follow the communications in the process of the investigation process such as where, how and in which way the illegal transport was performed. By sharing and communicating information about the manner and time of transport and information about the people involved in the process, authorities can gain more control of the shipment process internationally.

Covered observations and visual-sound recordings of people and their illegal activities are useful. Undetected observation is beneficial during investigating and uncovering all parties involved in organised crime and the transport process.

For example, observing Wong's zoo one would notice all of the changes and new events. For using this measure, abrupt and unusual changes of the zoo's business operations can alarm authorities for further investigation. Visual-audio recordings during investigation of alarming business operations investigations are also important for comparing changes over time and prosecuting such crime later.

Regarding the crimes of illegal trade in drugs, weapon, people, and in the Hoax case in animals, often the need of simulating purchase(s) of illicit goods is necessary, known as the measure of *seeming (simulated) purchase of objects*. As a combination of already explained measures in the undercover investigation, the secret agent poses as an interested buyer of the illicit goods. Thus, in the Wong case in Mexico, the American secret agent is playing a role of a dealer of protected animals, using the measures of control of communication, secret following and visual-sound recording of the activities.

This measure is especially effective because the criminals, in this case of environmental crimes, are caught *in flagranti*, and the crime cannot be denied. During the implementation of this measure the undercover agent is in the role of a buyer, in which way the crime of illegal trade in protected, rare and priceless animals on the black market, is clarified and proved. Consideration of the operative officers, their psychological predispositions, their previous experience and undercover actions, is necessary when choosing an agent for an assigned investigation project.

The operational officers engage associates from the criminal underground by gathering information and data about the illegal trade and simulating purchase. Because of this a special attention should be given to the status and the position of the associate, not to reveal his role as an associate. It is very important that others involved do not find out about the undercover role the agent or informant is playing. Precautions are turn to provide protection in case the agent's cover is blown.

Other case of engaging associate in undercover action is after the arrest of the group, where he is in a role of protected witness, measure that is allowed in cases of organised crime. The agent should be provided with a new identity for his own protection, in the case of being discovered, because maybe it can come to revenge from the members of the same organised group he was in (Naumovski 2008). If it comes to large areas of population there is less emphasis or providing an agent with a new identity. If it is a smaller population and working in a smaller country then the witness protection should be realized at a regional level, so the states should conclude bilateral and multilateral agreements with the neighbouring countries.

At the end of this analysis of the Hoax case, it can be concluded that the agent of international action of disposing, clarifying and proving of environmental crimes in the field of organised crime, it is one complex, comprehensive and multidisciplinary activity. In addition, in the action or undercover operation a large number of members from different countries and services should act extremely organised and coordinated in order of successful ending of the considered measures. It is important that in the context we should mention the role of Interpol, as an international organization in a role of organiser, coordinator and active participant in the mutual international actions.

7.3 Usage of Operational Measures and Activities for Eradication of the Environmental Crime in the Republic of Macedonia

Environmental crimes in Republic of Macedonia are systemized in a separate chapter in the Criminal Code of the Republic of Macedonia (Кривичниот законик на Република Македонија 1996). Namely in chapter XXII from the Criminal Code of the Republic of Macedonia, the following crimes are considered: pollution of the living environment, pollution of drinking water, production of harmful products for treating livestock or poultry, usurpation of real estate, forest logging, pollution of livestock fodder or water, unlawful hunt, unlawful fishing, endangering the living environment with waste materials, bringing dangerous materials into the country, torturing animals and serious crimes against the environment.

The numbers of environmental crimes in Macedonia are vast, but police discover only a small portion. Environmental crimes are on the margins of social activities. Often pursue a criminal case is forest devastation, which for example in 1997 was done 30 times out of 60 crimes. Devastation forest is linked with forest mafia, which is part of organized crime. They are connected with the state authorities and used firearms in carrying out criminal acts. In several cases, victims of their violence are forest officers who were killed in a crash with the perpetrators of environmental crimes. According to the records of the Ministry of Interior, the perpetrators of environmental crimes are not even 1% of total reported perpetrators. In 2005, environmental crimes estimated 159 people out of 23,814 perpetrators, which is 0.67%. Of the reported perpetrators, very few will be convicted. In 2004, 63 people reported of 127 persons or 49.6% were sentenced. This means that there are flaws in the investigation or has a mild penal policies. In 2004 63 penalties were fines, 13 prison sentences, 32 sentences suspended, out of 108 sentences (Malis Sazdovska 2005a).

This is a mild penal policy for environmental crimes in Macedonia. In Macedonia there are many types of environmental crimes, and the police and courts need to reveal these crimes. This requires their preventive and repressive actions.

In the Republic of Macedonia there are several state bodies authorized to take measures and activities in detecting the perpetrators of environmental crimes. Basic bodies eradicating environmental crime are the police, inspections, and authoritative bodies, etc. The police take measures within its competence on preventive and repressive plan. For prevention, the following measures are taken: tracing the appearances of pollution, removing environmental crime factors, informing and giving professional help. Members of the police maintain surveillance of critical sectors of environmental pollution known about environmental crimes (Malis Sazdovska 2009a). Constant information sharing process with other institutions is needed, as well as cooperation with investigators and interventions by other bodies. On many occasions the police intervene when the forest police have problem with the filed authorizations, then the police assist in the measures taken on the field.

During a repressive acting the police have on a disposal all police instruments that with the application of operational-tactic measures and investigation activities detect, prove and clarify environmental crimes and arrest offenders, cooperate with other competent bodies and act preventively through usage of repressive measures (Malis Sazdovska 2005b). Especially, important police activity during criminal handling is the giving an expert opinion during the investigation, in doing so the situational giving of an expert opinion is the most important. This expertise has great significance because of the fact that the polluting substance under the influence of the external factors and weather influences is quickly dispersed and with that the concentration of the polluting substance decreases. In the absence of the application of the principle of speed and efficiency, a lack of concentration regarding polluting substance would result in environmental crime.

Although, the usual measures that can be used by the members of the police, especially by the criminal investigations police, special measures and activities can be used in the organised crime. In some cases, environmental crime is closely connected to the organised crime, due to the detection of the environmental crimes, measures and activities considered to be part of the Criminal Code of the Republic of Macedonia. The following communications and entrance in a home or other premises or in vehicles require the following conditions of communication: under conditions and procedure established by the law; inspection and searching in computer system or in part of it, or from the computer database; secret observation, following and visual-sound recording of people and objects by technical means; seeming (simulating) purchase of goods; seeming (simulating) of giving bribe and seeming (simulating) taking bribe; controlled delivery and transport of persons and objects; using people with hidden identity for following and gathering information and data as well as opening a seeming (simulating) bank account, making a deposit gained from the conducted crime registered on seeming (simulating) legal entities or using of permanent legal entities due to gathering information.

Part of these special investigation measures and activities are used by authorized officials of the competent bodies in the Republic of Macedonia used in the cases of environmental crimes related to the organised crime as: devastation of forest performed by "forest mafia" "illegal trade of nuclear and radioactive materials". In addition to action of these bodies, one higher level of cooperation has to be performed, because regular one does not meet the needs of the fight against environmental crime. Technical equipment, the authorized bodies to be staffed, training of personnel and strengthening of the mutual collaboration are needed. Only in this way efficient acting of the authorized bodies fighting against this highly present type of crime at a national and international level is possible.

7.4 Environmental Crimes and Organised Crimes

Environmental crime – detection, prevention and suppression are gaining more attention. Certain authors consider that this type of crime is a part of the organised crime and as such the same emphasis should be devoted to it, as a type of crime that is increasing worldwide.

The damage caused by the perpetrators of the criminal acts from the area of the organised crime is extremely high. Perpetrators are using all the means that are on their disposal like corruption, threats, blackmail, use of violence, terrorist acts, money laundering, etc.

According to Article 2 from the Europol Convention (Malis Sazdovska 2009b), the following criminal acts fall under organised crime: terrorism, trafficking of drugs, trafficking of arms, trafficking of nuclear and radioactive materials, illegal migration and trafficking of human beings and trafficking of motor vehicles. Certain criminal acts from the environmental crime represent and appear in the form of organised crime.

The Europol Convention (Malis Sazdovska 2009b) in Article 2 determines the criminal acts that have the characteristics of organized crime and this article also lists other punishable behaviours that also contain these characteristics of criminal acts from the area of organized crime; among them are illegal trafficking and violation of the living environment. This on a very clear and unambiguous manner verifies that certain criminal acts from the area of the environmental crime are being handled as organised crime.

As an enhancement to the notion that the environmental crime is a part of the organised crime, we can list the conclusions of the Fifth International Conference for environmental crime that was held in June 2005. On this conference, the following conclusions were made (Interpol 2010b):

- Italy estimated that the transport of waste is the second largest profitable form of illegal transport, immediately after the illegal transport of narcotics.
- According to the study conducted by the US Government, the environmental crime is one of the fastest growing criminal activities. As a result of this type of crime the criminal syndicates worldwide are earning 22–31 billion dollars per year.
- Only in Brazil, the Brazilian national network for fight against animal snatching estimated that every year around 38 million birds, reptiles and mammals are being scattered every year.
- In France, Corrine Lepage former minister for protection of the living environment stated that the environmental crime can escalate and obtain the status as crimes against humanity.
- In India the prime minister confirmed that a multi agency unit was formed to fight against animal snatching. Worldwide there are 5,000 wild tigers remaining and half of them are in India.
- In the Philippines a local illegal fisherman was arrested for hunting with explosives and cyanide with which he endangered and partly destroyed the coral reef.

Interpol understanding the danger from this type of crime formed a Council for environmental crime. This council has the role of body that will build the strategy against the executors of criminal acts from the area of environmental crime.

7.5 Conclusion

Environmental crimes are frequent occurrence assuming big proportions in the sphere of organised crime. They refer to illegal dumping dangerous waste, illegal trade in nuclear and other dangerous materials, "forest mafia" illegal trade in

7 Usage of Special Investigation Measures in Detecting Environmental Crime

protected animals and plants etc. In order of successful fighting against this type of crime, law enforcement agencices should take adequate measures from their large range of police instruments. Namely, operations of special investigation measures should be carried out by attempting prevention from these serious environmental crimes.

That is why in the following period authorized bodies should make a good team, educate and technically equip the personnel and establish international cooperation related to the fight against the organised criminal groups. Only in this way the international community may prevent and eradicate environmental crime that is becoming a serious threat for the ecological safety.

Environmental crime as a part of the organised crime in Republic of Macedonia is present in everyday life, but unfortunately it is not adequately treated by the competent bodies. Namely, the crime police in the following period should make the activities more intensive in direction to stop this type of crime.

Republic of Macedonia is not devoting the needed attention to certain criminal acts that are serious criminal acts and can cause enormous consequences to the environment and life of the humans. For example, the devastation of the forests and the forest fires can cause massive negative consequences for the living environment, but the undertaken activities by the authorized institutions and bodies, including the police are minimal. In the future period a wide range of activity in the society by all the relevant authorized institutions and bodies is needed in order to suppress this type of crime and to protect the health of this and of the following generations.

References

Batkovski T (2008) Тактика на работа на разузнавачката служба и службата за безбедност и контраразузнавање (Eng. Intelligence, secure and counter-intelligence tactics). Jofi-sken, Skopje

Dzukleski G (2005) Прирачник за соработничка мрежа (Eng. Manual for cooperative network). Jofi-sken, Skopje

Environmental News Service (2010) International trafficker in world's rarest reptiles jailed. Resource document. http://ens-newswire.com/ens/jun2001/2001-06-08-02.asp. Accessed 25 Apr 2010

Environment and Natural Resources Division (2010) Resource document. http://www.justice.gov/enrd/Anniversary/1520.htm. Accessed 25 Apr 2010

Interpol (2010a) Resource document. http://www.interpol.int/Public/EnvironmentalCrime/Default.asp. Accessed 25 Apr 2010

Interpol (2010b) Resource document. http://www.interpol.int/Public/EnvironmentalCrime/Meetings/LyonJune2005/Default.asp. Accessed 25 Apr 2010

Кривичниот законик на Република Македонија (1996) Службен весник на Република Македонија (Eng. Criminal Code of the Republic of Macedonia), бр. 37/1996

Malis Sazdovska M (2005a) Криминалистичката полиција и еколошкиот криминал (ang. Crimes police and environmental crimes). Jofi-sken, Skopje

Malis Sazdovska M (2005b) Прирачник за разузнавачки циклус (Eng. Manual for intelligence cycle). Jofi-sken, Skopje

Malis Sazdovska M (2009a) Еколошка криминалистика (Eng. Green criminology). Solarisprint, Skopje

Malis Sazdovska M (2009b) Elimination of ecological crime as a part of organized crime in the Former Yugoslav Republic of Macedonia (FYROM). Rev Int Aff LX 1135:12–25

Malis Sazdovska M, Dujovski N (2009) Безбедносен менаџмент (Eng. Security management). Faculty of Security, Skopje

Naumovski D (2008) Прирачник за оперативни проверки (Eng. Manual for operational check). Solarisprint, Skopje

Chapter 8
Solving Problems Related to Environmental Crime Investigations

Bojan Dobovšek and Robert Praček

Abstract The purpose of this chapter is to offer an analysis of Slovenian state institutions actual responses to the environmental crime at crime scenes and to propose solutions for more efficient and effective future work of police and prosecutors. In-depth target interviews were conducted with police officers, prosecutors and judges in order to define appropriate crime scene investigation procedures for recovering the evidence supporting the ensuing procedures in courts. These in-depth interviews consisted of questions regarding different views on the problems of investigating environmental crime and, in particular, the collection of evidence gathered from a crime scene. Further, crime scene security and safety issues were explored. The main problem in investigating environmental crimes is that, according to our latest experience, it is generally difficult to collect proper evidence on which to substantiate relevant indictment charges. More specifically, though the circumstances, attributes, and consequences of such crimes may be readily available, the major remaining unresolved issue is usually the identity of the offender(s). In such cases, good cooperation between the competent institutions is of crucial importance. The dividing line between the competences or jurisdiction of particular institutions (e.g. the police, forensics, and the inspectorate) is very thin, and it often happens that these have to solve an issue falling under their mutual jurisdiction. Finally, solutions for the future crime scene procedures by the police and harmonisation of the procedures regarding environmental crime are proposed.

B. Dobovšek (✉)
Faculty of Criminal Justice and Institute of Security Strategies, University of Maribor,
Kotnikova 8, 1000 Ljubljana, Slovenia
e-mail: bojan.dobovsek@fvv.uni-mb.si

R. Praček
Criminal Technician Unit, Police Directorate Ljubljana, Ljubljana, Slovenia

G. Meško et al. (eds.), *Understanding and Managing Threats to the Environment in South Eastern Europe*, NATO Science for Peace and Security Series C: Environmental Security, DOI 10.1007/978-94-007-0611-8_8, © Springer Science+Business Media B.V. 2011

8.1 Introduction

Environmental pollution and natural disasters as its partial consequence have almost reached the top of the scale of safety problems of modern society, which are being confirmed by Brack's and Hayman's (2002) statement, that there are almost 250 signed contracts of environmental protection on the international level and even more are about to be signed. Protection of natural environment has climbed up the scales, which measure public concern from modern threats. Developed countries have adopted laws, initiated ecological taxes and founded police departments and other institutions with different kinds of authorization and aim to fight pollution, resource exhaustion and destruction of biological diversity (Brack 2002 : p. 143). Such institutions are fighting a new form of criminality, with which they do not have many experience and the legislation is not suitable.

A literature review in case of environmental criminality reveals that a wide variety of factors must be considered when defining environmental crime. Many academic disciplines are involved (Clifford and Edwards 1998; Edwards et al. 1996; Lynch and Stretesky 2007; White 2008; Eman 2008):

- Environmental criminality is a complex phenomenon, therefore investigation usually lasts long and is complicated.
- Environmental criminality is connected to the technical development and progress, therefore new forms of such criminality occur constantly.
- Environmental criminality is very diverse around the world, among individual countries and also among regions (every individual, economic system, environmental and biotic systems, etc. own it).
- Environmental criminality is sometimes imperceptible when committing such crimes and therefore it is hard to discover (insignificant perceivability of a perpetrator).
- Collectivity and anonymity of victims cause that comprehending the damage is rather abstract and indefinite.
- Perpetrators of environmental criminality often associate themselves with perpetrators abroad, which means that beside national, also foreign legal standards have to be taken into consideration.
- Typical for such criminal acts are frequency, difficulty in detection, measurement and social apathy, which makes identification, conviction and sentencing even harder.
- Some actions are often committed when attending to one's business or economic activity and involve the abuse of trust (gaining illegal benefit).
- In the field of ecology the boundaries between permitted and forbidden activities (legal and illegal) are often not clear (outsmarting or threatening the environmental legal order in the country).
- International environmental law is an incomplete system for protecting the environment, because at some points it is too extensive while at other times too lax, and on other it all depends on national interests.

8 Solving Problems Related to Environmental Crime Investigations

From the above, it is shown that there are numerous definitions of such criminality from which problems in the investigation follow. These very problems in the investigation and conviction of such criminality demand more research for solutions and improving investigation, which is also the purpose of this chapter. The chapter focuses on environmental pollution and its problems with investigation. Therefore, the authors evaluated the shortcomings, which appear when investigating of pollution, and based on such research were able to make a model for other forms of environmental criminality as well.

8.2 Investigating the Polluted Locations: Slovenian Context

Investigation at the scene of the crime is a penal investigation process, which is being done under Criminal Procedure Act (Zakon o kazenskem postopku [ZKP] 2009), by examining magistrate or police. Legal basis for performing the investigation is defined in the ZKP. At the same time, the investigation is a criminal process, in which the investigators consider rules of criminal tactics, techniques, and methodology and then reconstruct the past event and establish their own assumptions about the cause, offender and circumstances as realistic as possible, by collecting information and securing evidence (Maver 2004). For this reason, police investigate the important facts about the location, investigate the victim and evidence by direct observation, measuring, and logical thinking. Their duty is to reconstruct the past event, which could be a criminal act, suicide or an accident at work, directly with their own sense organs or technical instruments. The evidence, sketches, photos, facts and other objective statements about the circumstances at the investigation are documented by police pursuant to a protocol about the investigation of crime scenes. Documentation of the investigation is also done by video and audio devices.

For most cases including environmental crime material evidence is the most relevant. Also, the way of searching and documenting the evidence is important. Material evidence has to be obtained legally and professionally, but above all, investigators should be very careful to ensure that the traces, which could serve as important evidence in criminal proceedings, do not get destroyed. To avoid the destruction of evidence at the investigation, the police perform individual activities, among which "*securing the location or event*" is one of the most important. In cases when people were hurt in criminal acts occasionally rescue teams come to crime scenes before the police and help injured people by giving them urgent medical care. In such cases they could unintentionally destroy the traces at the crime scene (Weston et al. 2000).

Apart from the previous contents, it is necessary to know, that the priority of rescue teams (as well as police officers, fire fighters, etc.) is helping injured people or securing the property. Only when the urgent medical care and safety of all other people (rescue teams) who happen to be on this location, is ensured, it is necessary to make sure that no evidence is destroyed. At this point, it is necessary to stress that police organize and perform tasks in securing, rescuing, and helping in accordance with rules and regulations and directions from the competent commander of Civil defence or the leader of the intervention (Zakon o varstvu pred naravnimi in drugimi nesrečami [ZVNDN] 2006).

Investigations of polluted locations are rarely conducted by police. Due to the fact that police have limited knowledge in this field of expertise, such specialized training could in every investigation lead to well performed work and above all without any kind of danger for the investigators. For this kind of investigation it is also typical that it is a single and non-repeatable act, therefore it has to be performed professionally and precisely and, as a particularity of such investigations, fast. The basic purpose of performing such investigation is finding the facts and collecting the material evidence in the field. The Slovenian Criminal Investigation Department (in text below as "CID") with its forensic technicians is rather independent when investigating polluted locations, which are not so complicated to investigate. In other cases CID uses help from mobile laboratories of accredited institutions. There are not many such institutions available, therefore the use of forensic techniques is not frequent. At the same time, with regular procedures, CID tries to secure the material evidence, which could connect the suspected materials to the person causing the criminal act. Forensic technicians[1] do not have such qualifications, because there are not many such cases and cases are very complex. That's why they draw upon knowledge from traditional criminological techniques, employed in the investigation of other criminal acts. In this case, they use knowledge based on directions and instructions from the central laboratory – Centre for Forensic Science. Forensic technical investigation of environmental crime science is not different from any other investigation of a crime scene. However, the only difference is that the secured samples of evidence could be deadly (Drielak 1998).

Accordingly, research should be performed on environmental crime investigations to highlight the problems and propose solutions. To date, there is not any other empirical research on environmental crime investigation in Slovenia. Therefore, the goal of this article is to: (1) analyze the problems of investigating environmental crime; (2) evaluate the impact such problems have on further procedures in the courts; and (3) define safety issues.

For the purpose of this article, the analytical model (designed for economy) of relational database QFD was used (Quality Function Deployment) and within this database the so called matrix HOQ was tested (House of Quality) (QFD 2009).

---

[1]Investigative team is usually formed by one or more criminal investigators and one or more forensic technicians, who are experts for the different field, depends which type of crime is investigated. Criminal investigator is responsible for the scene of the crime and forensic technicians are for expert help on the scene of the crime. Terms are used in Slovenia and are not unique in Europe.

8.3 Empirical Part

8.3.1 Sample

For the analysis of the problems in the field of criminality in connection with environmental pollution, a preliminary research based on structured interviews was conducted during January 2010. Contact with the respondents was first established in the winter of 2009 and an agreement about the dates for the interviews achieved. Structured interviews, conducted by trained interviewers, took place at respondents' homes, offices and public places proposed by the interviewees (restaurants, libraries, etc.). Each interview took about one hour and was subsequently outlined. The preliminary structured interview was first tested on a sample of ten individuals (criminal investigators, prosecutors, investigative judges and academics), who were familiar with environmental issues in Slovenia. The sample consists of four criminal investigators who investigate the crime scenes, two investigative judges, two prosecutors and two lecturers of criminal investigation in Slovenia. Therefore, they were expected to reveal valid information on their perception of the studied problem. The studied respondents have been chosen upon assumption that they have knowledge on environmental crime and have experienced the problems regarding such crime. Answers were analyzed and combined to use for further research work.

8.3.2 Instruments

The structured interviews were created mainly in the spirit of a greater demand for material evidence in the court of law. Hearing of evidence in the court of law is more and more complex and mostly based upon material evidence, which is collected at the crime scene. By changing the ZKP, the value of evidence of criminalist's testimony, who can only be a prosecutor's witness, is also changing. Criminal investigators and prosecutors fail in the court of law because of procedural mistakes made unintentionally by the police, simply because of lack of knowledge. Of course, winning knowledge and investing into new working methods and new equipment demands a large budget, which nowadays is hard to ensure. Therefore the first step is the analysis of the previous work and investing in standardization of procedures and agreed education.

Because working methods and evaluation are transferred from economy into administration departments, the methods to increase quality of institutional activities are also being transferred there. That's why the QFD method was applied into police work and in this way improves the quality of investigations of polluted areas.

8.3.3 QFD in Investigating Polluted Areas

Because many mistakes happen when investigating criminal acts, the above-mentioned method suggest improvements and tried to find out solution for better work in the future. The method refers to increasing the quality of the product, which is the investigation. To increase the quality, it is necessary to first identify the field of expertise, where criminal technicians and other criminal investigators have not achieved sufficient knowledge, and based on that, carry out certain advanced trainings. From the description of the QFD and HOQ matrix, it is evident, that it is primarily about developing material products. However, in our case the product is service. Therefore it is also interested in how successful QFD or instruments of HOQ are when analyzing the service and at the same time it is possible to find out, which are the fields of interest.

The internet website Webducate was used for this task, where the teaching matrix HOQ is programmed, which should lead researcher to the goal (Webducate 2009). This matrix is meant for introducing the development of the product, service or program tools with the help of QFD (see Fig. 8.1).

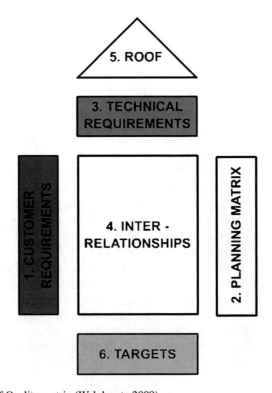

Fig. 8.1 House of Quality matrix (Webducate 2009)

8.3.3.1 Customer Requirements (WHAT)

This is the first task of the HOQ matrix, which is also the most important. It shows a suitably structured list of customer requirements for the future product, which is described in their own language. These requirements can be gained in directed interviews with the customers, where the customers should be stimulated to describe their needs about product's characteristics. This can be achieved with a suitable questionnaire, which is especially useful for unknown target group (judges, prosecutors). This part of the matrix is also called *"voice of the customer"*. All the customers' wishes should be handled systematically and structured into individual groups. In the pilot case, the customer requirements were gathered through the direct discussions with those identified as experts in environmental crime investigation. In these discussions criminalists, two investigative judges, two prosecutors, and two lecturers took part. Their requirements are defined and shown in Table 8.1 below.

To simplify the process, all customer requirements were divided into six groups. The investigation of the polluted location is divided to examination of the environment, people and animals and an evaluation of further threats to all three objects. At the suspect location criminal investigators should take into consideration the questions about searching the traces, which could lead to the eventual perpetrator. By documenting things right the differences between the consequences before and after pollution could be identified. Also the extent of the damage, and the degree to which this pollution really caused such damage to the environment, could be assessed.

Security and safety at work on such investigations requires special attention. Injured and poisoned investigators cannot do their job. It is interesting, that the

Table 8.1 Customer requirements

Documenting	Sketch of the location of pollution	Activities at the crime scene
	Photo of the location of pollution	
	Video of the location of pollution	
	Protocol of the investigation	
Threat evaluation	Threat evaluation on environment	
	Threat evaluation on people	
	Threat evaluation on animals	
Suspects	Connection to the suspects	
Consequence evaluation	Environmental situation before and after pollution	
	Situation with people before and after pollution	
	Situation with animals before and after pollution	
Health of investigators	Security and safety at work	
Report or criminal complaint	Finding signs of criminal acts	Later activities
	Finding violations of environmental and ecological regulations	

interviewed persons from judicature did not give any kind of opinion about that. However, if the investigative judge or a prosecutor wants to investigate the crime scene, this would only be possible if he/she uses suitable equipment for which he/she was trained.

8.3.3.2 Planning Matrix

This part of matrix, which is normally on the right hand side of HOQ, is used for:

1. Quantifying customers requirements and their perception to finding the characteristics of individual products.
2. Adapting these priorities regarding attributes, which are given by development team or investigator.

Values in this part of matrix are gathered with the questionnaire. The first and most important value is the so called "weight importance". It gives the relative importance of each of the customers' requirements from the customers' point of view. Because of its importance, it is usually written on the left hand side of the matrix. The questionnaire is constructed in a way, so that every customer can evaluate a certain customer requirement from 1 to 5 (1 – least important, 5 – most important), as shown in the Table 8.2 below.

In the second part of planning matrix the evaluation about how satisfied the customers are with the present product were obtained. The system of evaluating customers is the same as for evaluating customer requirements. In this way the evaluation for each individual attribute and for each individual product was obtained. In the case of investigating the polluted location by police there is no competition with other state institutions; however, there does exist information about similar work abroad. Foreign literature is also a big help, as was the manual Environmental Crime (Drielak 1998).

The respondents completed the questionnaire, in which individual requirements had to be evaluated from 1 to 5 (1 least important, 5 most important). At the same time, every requirement had to be evaluated from a "real time" viewpoint (what is the most important thing to investigate sequentially) and from a "priority" viewpoint (what are the most important things that have to be done at the investigation). It is possible for the evaluations to overlap somewhat; the average evaluation of an individual requirement was calculated by taking all the evaluations from all evaluators for an individual attribute into consideration. Furthermore, the responses are going to show, whether such analysis of hazardous materials is necessary and in what extent it is necessary for further investigation. Also "safety at work" was identified for the first time by the respondents when evaluating requirements.

Table 8.2 Part of planning matrix

Threat evaluation on environment	1 2 3 4 5
Situation with people before and after pollution	1 2 3 4 5
Finding signs of criminal acts	1 2 3 4 5

8.3.3.3 Technical Requirements (HOW)

In this part of matrix a group of developers is giving solutions, which have to be measured for realization of customer requirements. The easiest way to explain this section is with the question: "HOW" is a particular "WHAT" done? This part of matrix is called "voice of the company". The voice of the company also has to be analyzed and structured, as were the customer requirements. Criminology technicians have analyzed which kind of knowledge and activities can be offered as an answer to customer requirements. These include, but are limited to, the following:

– Sketching at the crime scene
– Taking photos at the crime scene
– Video recording at the crime scene
– Writing a protocol about the investigation
– Testing chemical, physical components of air and water and soil
– Knowing the situation before pollution
– Knowing the situation after pollution
– Collecting information during and after investigation
– Cooperation with other services at the crime scene
– Knowing the materials of pollution
– Securing the traces
– Knowing the penal code
– Knowing the environmental and ecological regulations
– Cooperation with external experts (inspectorates)
– Cooperation with health institutions
– Considering safety regulations
– Information about the first steps before arrival of investigating commission

In the Fig. 8.2, the need for improvement, if any, is shown. The arrows show the direction of improvement. If the arrow is upward, it means "the more it is present, the better".

8.3.3.4 Interrelationships

This part is actually the body of matrix and is designed for transferring customers' requirements into technical characteristics of the product. At this point, the worst possible position (with our example of service) was shown and it is revealed not earlier that in the sixth part of HOQ.

The group of developers has to evaluate the value of the relation between the technical requirement and requirement of the customer. In other words, how important is a technical requirement for a certain requirement of a customer. Evaluation is assigned one of four values: high (5 points), middle (3 points), low (1 point), no relation (0 points).

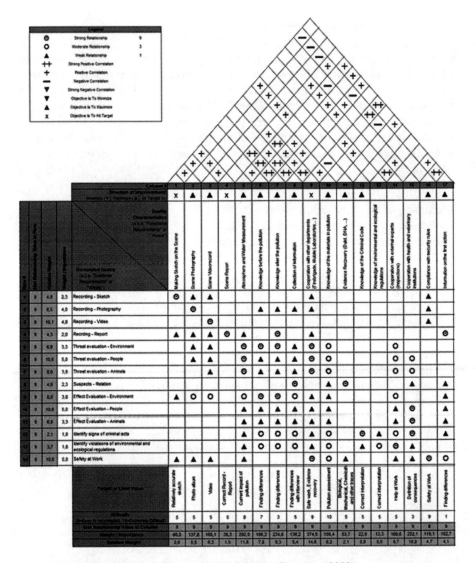

Fig. 8.2 Final matrix HOQ (QFD and House of Quality Templates 2009)

8.3.3.5 Roof

In the roof-part of the matrix individual technical characteristics and influence on each other were revealed. For each cell of the roof-part of the matrix, questions were asked, whether improvement of individual technical characteristics has a positive or a negative influence on another technical characteristic. The evaluation could be positive (extreme positive) or negative (extreme negative). They could also have no influence on one another at all. The information, which the developers get from the "roof," could be defined as one of the most important in the development. It shows, what influence a certain improvement of a product has on the final product. It also stresses the negative relations, which of course are a part of it, is crucial to find solutions, how to avoid them. This way, for example, a negative relation between securing the classic criminal technical traces and cooperation with other services occurs. With their work the services contaminate the area to a degree, that for example searching for DNA or trace evidence from tools is impossible. For example, in case of fire, if the traces were destroyed by fire brigade, therefore there would be a negative relation between them.

8.3.3.6 Targets

This is the last part of HOQ, is a summary of the whole matrix, and is placed at the end of the whole QFD. In general, it consists of three parts:

- Technical priorities
- Relative evaluation of the competition
- Aims

Technical Priorities

The relative value of each technical requirement regarding the customer requirements is simply calculated from the values given in planning matrix and relation matrix. Each value in this relation matrix is multiplied with the general value from the planning matrix, and then these values are summed up in columns of individual technical requirement.

Relative Evaluation of the Competition

Each technical requirement, set by the developers, can be set for the existing product of the analyzing company, as well as for products from the competitors. This way, target values for constructing a new product could be obtained.

Aims

The final aim of the HOQ matrix is a collection of values, which every product needs to contain. As previously mentioned, from this point the matrix can proceed into the four above-mentioned phases. This way the current technical characteristics could become customer requirements in the next matrix, for these customer requirements, the developing group should find new technical characteristics.

8.4 Statements

In the statements of HOQ matrix the activities and aims of these activities are expressed according to evaluation of importance. A higher number means higher importance and, according to results in Fig. 8.2 (bottom line) cooperation with other services is the most important among activities. Probably the results of labelling the data of several customers would create a different finding. However, even with small number of interviewed customers the results are satisfactory.

8.4.1 Cooperation with Other Services (Such as Fire Brigade, Veterinarians)

The main thing is the cooperation with services, which enables a safe fieldwork and at the same time offers specialized knowledge, which the police often do not have. The services are available around the clock, which means, that this response is almost as fast as the response of the police, if not even faster. Contacts with these services should be improved and training should be organized with police before the eventual crime occurs. This is how investigating groups could get to know their future "co-workers" in the fieldwork, and at the same time show them the most important things that need to be done at the crime scene. Such condition should lead to a solid investigation and result in improved future investigations.

8.4.2 Analyzing Air and Water

Testing air and water is necessary for a safe investigation in the contaminated area, and for the threat evaluation. Beside the report about the pollution, the police are obliged to secure such areas, ensure an undisturbed work to intervention groups, and at the same time prevent unauthorized entry into the affected area. In many cases, it happens that the local police come first to the contaminated area and are not aware of being in danger already. The police are fully dependent on cooperation with the intervention services at the crime scene and from the cooperation with the mobile laboratories.

8.4.3 Cooperation with Health and Veterinary Institutions

Cooperation with the above-mentioned institutions should not be limited only to collecting opinions about the level of noxiousness of toxic materials on humans or animals for the pre-trial proceedings, after the subject is already poisoned. Their knowledge of how individual toxic materials react on bodies of individual subjects needs to be taken into account in a proactive manner through a working database available on the central computer of the police.

8.4.4 Knowing the Situation After Pollution

Knowing the situation after pollution is one of the most important activities, which need to be noticed at the investigation of the polluted location. All three previously mentioned tasks are closely connected to the knowledge about the situation after pollution. With this knowledge the evaluation of threats to the environment, people and animals can be made. In any case, this activity is also strongly connected with the knowledge about the situation before pollution.

8.4.5 Knowing the Situation Before Pollution

With the interviews, eventual photographs and other ways of collecting information as much information as possible about what the area looked like before the pollution, were the animals ill before and last but not least was the injured person already ill before the pollution could be gained. The difference has to be established for the pre-trial proceedings as well as for the above-mentioned threat evaluation.

There are other activities that need to be performed at the crime scene, shown by HOQ matrix. However, in a detailed review these activities are directly connected to the above-mentioned ones. Documenting the location, which shows the situation after pollution, is definitely connected with knowing the situation after pollution. How this documenting needs to be done and when, maybe immediately after investigation, maybe later, is a topic of another research. However, the matrix gives the knowledge that it does not need to be done immediately at any price. The photographs from other institutions which are involved in the intervention might be of great help.

Looking for traces, which could connect criminal investigator to the eventual perpetrator, is evaluated very low. As a result it is clear that such activities do not have to be done immediately, but only when it is safe and relatively simple for the whole searching procedure as well as for the criminology technician. When removing threats, it is necessary to cooperate with intervention services to educate them on what is important for further securing of traces. Knowing the penal code and regulations, which refer to environmental problems, is necessary, however, it is not so important in the polluted area and by securing the traces.

8.5 Conclusion

Regarding the performed pilot research, the shown model offers a good basis for further research and therefore a wider research is proposed. From this research it is clear that every pollution represents a special (stressful) project, not only for criminal investigators, but also for the eventual investigating commission knowing that they might not have enough knowledge and that one mistake can lead to an unsuccessful investigation. It is necessary to ensure that such investigators are well organized, so that the investigation could be independent and complete, even in the hardest investigations of polluted areas. However, it is not just about the investigation of the polluted area. The police have to act in the field of reducing environmental criminality in some other way as well. In such cases the police should, among other things, collect the information, which could be completely technical, like for example collecting the data or measuring air pollution around a certain object, cooperation in domiciliary visits, where investigators look for dangerous waste, investigating the hazardous waste trade and a lot of other activities connected to environmental criminality. In such cases the police can, to some extent, use help from external institutions.

Organizing investigations should continue on several levels including, but not limited to, the following:

- Choosing the investigators, who have enough technical knowledge for this work
- Providing investigators' equipment for safe work on contaminated areas
- Training of investigators for this kind of work
- Offering regular health checks of investigators
- Choosing the accredited laboratories for sampling and analyzing the traces
- Establishing friendly relations between intervention services, environmental mobile laboratories and investigators
- Developing standard procedures in investigating the polluted area, which firstly assures safe work and then everything else, which is connected with the pre-trial proceedings

Analyzing the above-mentioned items would be necessary to provide a special service, following the model of most of the police departments in other countries, which would deal only with these kinds of problems. This refers not only to the technical, but also to the investigational point of view. The technical viewpoint is directly connected to the enormous financial support. Regarding the above-mentioned and the fact that the number of such events in Slovenia is very low, it would seem logical to think about forming a group of investigators, who would investigate all Slovenia, not just individual regions. These environmental investigators would do the regular work of criminalist technicians or criminal investigators in their working hours in their own region. There would not be 11 such groups any more, but only five, which would be very well trained and equipped for this kind of work. Because of a smaller number of such events, these groups would train in such investigative work and therefore the efficiency along the whole country would increase.

8 Solving Problems Related to Environmental Crime Investigations 149

It is all about first investigative activities, which have to be done by police on a contaminated area. However, further investigation would stay under supervision of an individual police administration. The environmental investigative group would decontaminate after finishing the investigation and return to their regular work safe and sound.

References

Brack D (2002) Combating international environmental crime. Resource document. Glob Environ Change 12(2): 143–147. http://www.sciencedirect.com/science?_ob=ArticleURL&_udi=B6VFV-46CRVV0-2&_user=940034&_coverDate=07%2F31%2F2002&_alid=702477340&_rdoc=1&_fmt=full&_orig=search&_cdi=6020&_sort=d&_docanchor=&view=c&_ct=1&_acct=C000048764&_version=1&_urlVersion=0&_userid=940034&md5=ecf6430ef45080dd15a01e9ffb6bfb81. Accessed 12 Sept 2007
Brack D, Hayman G (2002) International environmental crime: the nature and control of environmental black markets. Background paper for RIAA workshop. Royal Institute of International Affairs, London. http://www.chathamhouse.org.uk/files/3049_environmental_crime_ background_paper.pdf. Accessed 17 May 2010
Clifford M, Edwards TD (1998) Defining "environmental crime". In: Clifford M (ed) Environmental crime: enforcement, policy, and social responsibility. Aspen Publishers, Gaithersburg, pp 121–145
Drielak C (1998) Environmental crime, evidence gathering and investigative techniques. Charles C Thomas – Publisher LTD., Springfield
Edwards SM, Edwards TD, Fields BC (eds) (1996) Environmental crime and criminality: theoretical and practical issues. Garland Publications, New York
Eman K (2008) Uvod v fenomenološko analizo ekološke kriminalitete (Eng. Introduction to phenomenological analysis of environmental criminality). Varstvoslovje 10(1):220–239
Lynch ML, Stretesky P (2007) Green criminology in the United States. In: Beirne P, South N (eds) Issues in green criminology: confronting harms against environments, humanity and other animals. Willan Publications, Cullompton, pp 248–269
Maver D (2004) Kriminalistika: uvod, taktika, tehnika (Eng. Criminalistics: introduction, tactics, techniques). Uradni list Republike Slovenije, Ljubljana
QFD (2009) Developing the Functions of Quality or QFD (2009). Resource document. http://www.strojnistvo.com/viewtopic.php?t=1011&view=previous. Accessed 15 Nov 2009
QFD and House of Quality Templates (2009) Resource document. http://www.qfdonline.com/templates/qfd-and-house-of-quality-templates/. Accessed 15 Apr 2009
Webducate (2009) Resource document. http://www.webducate.net/qfd/qfd.html. Accessed 15 Apr 2009
Weston PB, Lushbaugh C, Kenneth M (2000) Criminal investigation: basic perspectives. Prentice Hall, Upper Saddle River
White RD (2008) Crimes against nature: environmental criminology and ecological justice. Willan Publishing, Cullompton
Zakon o kazenskem postopku [ZKP] (2009) (Eng. Criminal Procedure Act). Uradni list Republike Slovenije, št. 77/2009
Zakon o varstvu pred naravnimi in drugimi nesrečami [ZVNDN] (2006) (Eng. Act on the Protection Against Natural and Other Disasters). Official Gazette of the Republic of Slovenia, št. 51/2006

Chapter 9
The Benefits from Using Professionally Developed Models of Possible Hazardous Materials Accident Scenarios in Crime Scene Investigations

Damir Kulišić

Abstract Without the possibility of preliminary professional analysis of possible versions of scenarios of more complex accidents with hazardous materials, each crime scene investigation of causes and forensically relevant circumstances of their occurrence can be nearly impossible, with irreparable and fatal failures of securing the crime scene and in all phases of crime scene investigation process. The author analyzes some of the (possible) hazardous materials major accidents (with examples from Croatia) and explains the importance for investigation of easily accessible professionally made and updated some key documents. This includes documents of health, safety and security (HSS) hazards analysis and risk assessment, plans and programs for prevention and protection as well as plans for emergency preparedness and response, for all those companies that manufacture, store or transport hazardous materials. Police ability to locate documents and provide professional insight regarding investigations enables adequate forensic exchange of expert information among members of crime scene investigation teams and helps the quality planning of phases and procedures of crime scene investigation according to available knowledge about relevant features of the accident under investigation. This way, there will be less possibility of losing key evidence, redundant investigation procedures/costs, unnecessary dealing with less important parts of procedural, technical and operative documentation. It is necessary to update current education programs for all accident investigators. Analytical methods, techniques and software tools used for making those professional documents can be of great help in searching, finding and testing logical connections of relevant macro- and micro-traces at the scene of accident with other important clues.

D. Kulišić (✉)
Ministry of the Interior, General Police Directorate, Police Academy,
Police College, Zagreb, Croatia
e-mail: dkulisic@fkz.hr

G. Meško et al. (eds.), *Understanding and Managing Threats to the Environment in South Eastern Europe*, NATO Science for Peace and Security Series C: Environmental Security, DOI 10.1007/978-94-007-0611-8_9, © Springer Science+Business Media B.V. 2011

9.1 Introduction

People have learned, from the past and recent experiences, that accidents with hazardous materials (dangerous goods, hazardous substances),[1] can lead to potentially harmful and destructive outcomes, such as to human lives, health and environmental habitats. Also, they can cause enormous direct and indirect damage, especially if the *domino effect* occurs, certain objects or whole systems of critical (regional, national and/or transnational) infrastructure can be affected. Therefore, during accidents with hazardous materials, there are, almost always,[2] *reasonable grounds for suspicion* that the accidents occurred as a consequence of committing one or more criminal offences.

Such accidents, pursuant to the Croatian Criminal Code and the Criminal Procedure Act, are *officially* investigated, immediately after being reported (before the commencement of an investigation – by conducting *inquires* and *evidentiary actions* where there is *danger in delay*, and after the commencement of an investigation – conducting other necessary *evidentiary actions*). Suspects or responsible persons who may have *deliberately* or *negligently* caused accidents with hazardous materials (by *action* or *omission to act*) are subject to public prosecution.

Even in cases of seemingly trivial or relatively small consequences of a hazardous material incident (including the theft of hazardous material), in order to prevent possible recurrence of such incidents, as soon as the incident has been reported, it is necessary and beneficial to conduct official *inquires* and *evidentiary actions*. Otherwise, later it can turn out that (in the recurrence of the incident with far-reaching consequences) the reasons were serious safety, protection and/or security failures by a company while manufacturing, processing, storing, transporting or using hazardous materials. Or it can turn out that there were internal or external attempts of sabotage, that is, of a disguised criminal (terrorist) experimental test for an attack with certain kinds of hazardous materials or the first unsuccessful attempt of a criminal attack, which could have been discovered in time and, thus, its recurrence prevented.[3]

---

[1]Classified according to the Article 3 of the Transport of Dangerous Goods Act (Zakon o prijevozu opasnih tvari 2007) and Article 2, item 9, and the Article 26 of the Chemicals Act (Zakon o kemikalijama 2005).

[2]Except, for example, in cases of unpredictable strong earthquakes, landslides or rockslides, hurricane-force winds, extremely heavy or unexpectedly heavy snow/rain (including walnut-sized hail/ice or high/heavy layers of snow), severe floods (bursting rivers or hydroaccumulation, including a sea-level rise or *tsunami*) and other natural disasters, which could not have been predicted on the basis of prescribed (obligatory) professional security analyses/studies and thus, it was not possible to undertake appropriate prescribed or recommended measures of necessary prevention and protection; or in cases of incidents such as falling aircraft parts or (explosion of) a burning meteorite, etc. which could not be predicted and/or prevented, nor is it possible to avoid or lessen, in an appropriate manner, their possible fatally dangerous and harmful impact.

[3]The reader should, for example, take into consideration circumstantial evidence of the sabotage by disgruntled workers, in the case of the Bhopal disaster (according to the forensic expertise by the engineering consulting firm Arthur D. Little) and circumstantial evidence of the media

9 The Benefits from Using Professionally Developed Models 153

Criminal investigators emphasize the importance of obtaining reliable answers to all key questions of criminal investigation[4] only if they consistently adhere to all basic principles of criminal investigation. Of course, this also applies to investigation of accidents with hazardous materials.

The analyses of causes of major accidents with hazardous materials in the process industry, regarding storing and transporting, show that the causes can be very simple (seemingly trivial), to extraordinarily complex, mutually direct or indirectly designed. The nature of the causes can also be processual or operational, technological, technical, human, organizational, or involving communication, software, environment, safety and protection, and often security (CRAIM/UNEP 2002; USCSB 1999–2010; Bragdon et al. 2008; Crowl and Louvar 2002; Grossi and Kunreuther 2005; Garrick et al. 2008; Kletz 2009; Koehler and Brown 2010; Körvers 2004; Kulišić 1998; Lancaster 2005; Margossian 2006; Marshall 1987; Medard 1989; Newton 2008; NFPA 2008; NRC 2006; Owen et al. 2009; Ramachandran et al. 2005; Rasmussen and Svendung 2000; Sanders 2005).

Regardless of possible complexity of major accident causes, in many cases, especially those caused by or resulting in physical and/or (un)confined vapour/gas cloud explosion (UVCE/CVCE), boiling liquid expanding vapour explosion (BLEVE/"fire ball"), great fire or structure collapse, investigators find at the crime scene a large number of various traces of physical, chemical, physiological/biological and other effects and consequences of such accident. A lot of evidence, interesting for the investigation, including important/key circumstantial evidence, forensic and physical evidence of root, initial and contributive causes of an ambiguous scenario incident escalating into an accident, could be found at the crime scene with all the possibly important/key evidence for the discovery of important failures in safety and protection measures which contributed to the escalation of an accident resulting in serious/fatal consequences (Fig. 9.1). Among them there are relatively few stable or short-term stable macro- and micro-traces (physical evidence) which are of key importance for determining and proving causes of such an accident. They must be found in time, lifted and packaged by forensic technicians, quickly sent to a

---

investigation (*Le Figaro* and *L'Express*) of a possible connection between the mysterious disappearance of 18.7 kg of Cr_2O_3, a month before the accident, and the explosion of ammonium nitrate waste storage unit in "Grande Paroisse S.A." company (ex "Azoté de France/AZF", Toulouse, France, 21 September 2001). Until then an unknown organized criminal terrorist group, which adopted the same name AZF, claimed responsibility for the accident. The group had attempted to extort money from the government of France by threatening to place improvised explosive devices (IEDs) along the nation's rail lines and terrorized the French public at large in late February and early March in 2004.

[4] The Five Ws [also known as the Five Ws (and one H), or Six Ws], which are regularly asked during appropriate investigative/forensic methods, and are applied in techniques and procedures in every case of harmful or fatal incident, i.e. immediately after the information about the occurrence of possible criminal offence or its attempt, are: *what* (happened)?; *where* (did it happen)?; *when* (did it happen)?; *how* (did it happen)?; *who* (was involved and especially who caused/or attempted to cause the offence/who was the perpetrator)?; *why* (did it happen/the perpetrator's motive for committing the offence)?

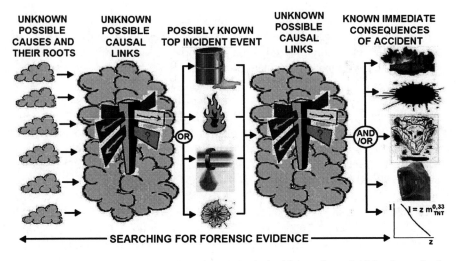

Fig. 9.1 Several key circumstantial, forensic and physical evidence of root, initial and contributive causes of an ambiguous scenario incident escalating into an accident

forensics lab, examined, and properly stored until the prescribed time; while constantly taking into account the chain of custody.

Without the possibility of investigators being preliminary familiarized with professional analysis of possible versions of scenarios of more complex accidents with hazardous materials, each crime scene investigation of causes and forensically relevant circumstances of their occurrence can be nearly impossible, with irreparable and fatal failures in the investigation process. Starting with the possible failures in timely and correct determination of the most important locations (not only scenes of effects/occurrences of accidents but also of all possible places of occurrences/ location of traces of their root causes), failures in giving the appropriate priority, scope and necessary measures to secure the locations, and thus, possible failures in conducting a criminal investigation. It must be taken into account that after the first shock, attempts can be made to destroy, eliminate and hide, even forge the evidence necessary for determination of responsibility for the accident. Also, attempts could be made to influence witnesses important for the investigation, which investigators may have overlooked. Securing consistent principles and procedures of criminal investigation (especially in extremely traumatic and traseologically complex situations) serve as most beneficial to one case and availability of maintaining and accessing official company documents ("Seveso II Directive"/Directive 96/82/EC (1996))[5] required and used in writing its "Safety Report".

---

[5] See amendments in: (a) Directive 2003/105/EC of the European Parliament and of the Council of 16 December 2003 amending Council Directive 96/82/EC on the control of major-accident hazards involving dangerous substances, and (b) Commission decision 98/433/EC of 26 June 1998 on harmonised criteria for dispensations according to Article 9 of Council Directive 96/82/EC on the control of major-accident hazards involving dangerous substances (notified under document number C (1998) 1758) (Text with EEA relevance).

9 The Benefits from Using Professionally Developed Models

9.2 The Topic and Content of the Investigation of Hazardous Materials Accidents in Industry and Transportation

The topic of the investigative interest in the investigation of any hazardous materials accident or a serious incident, which is important for security reasons, especially accidents in industry and transportation of large amounts of hazardous substances, is how to determine the manner, cause, conditions and circumstances of their occurrence (Fig. 9.2).

Mostly, the answer to the question of the occurrence of the hazardous materials accident/incident can be obtained indirectly, only after careful investigation of all physical evidence and/or circumstantial evidence related to its cause, including those related to the relevant elements of the conditions and circumstances of the accident. Sometimes, the evidence of the (most probable) cause and manner of

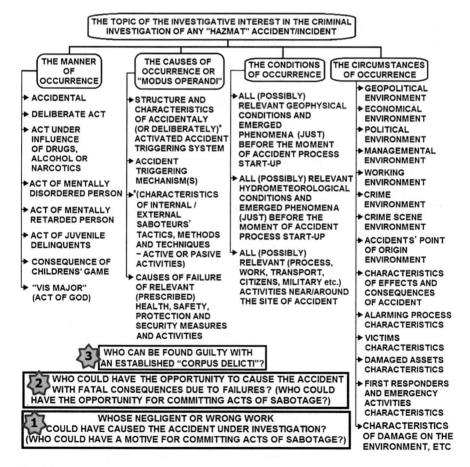

Fig. 9.2 The topic of the investigative interest in the investigation of any hazardous materials accident or a serious incident

the occurrence of the accident can be obtained only through the facts determined by the analysis of conditions and circumstances, which existed before, during and after the time of the incident/accident.

In short, after an industrial incident/accident with hazardous material has been reported and in order to preliminarily secure the incident/accident scene (because of possible risks and undertaking necessary emergency measures), a team of criminal investigators and forensic technicians should use appropriate expertise and protective equipment. Then, during the process of criminal investigation, which must be carried out in the presence of the state attorney, the team should try to:

1. Establish and document the scene and surrounding area, the exact time and other relevant geophysical and (hydro)meteorological conditions of the incident/accident occurrence as well as the kinds and features of all stable and moving things and living beings at the places where traces of hazardous materials were located, and where effects and consequences of the incident/accident were found (environment, buildings, the kinds and amounts of hazardous materials, sources of potentially hazardous energy, hardware, process ware, labour ware, etc.) before, during and after the incident/accident.

2. Establish and document:

   - All kinds and gravity of present dangers for the crime scene investigators
   - Real conditions/technical mal-functioning of systems for facility process/unit operations management and control
   - Kinds, events, a sequence of events at certain micro-locations, which preceded the occurrence of the incident (and its development into the accident), especially repairs, adding a wing/part onto a building/process unit, kinds and manners of testing of facility (units)
   - Condition/proper operation and the sequence of automatic or hand activated relevant safety, protective and alarm systems, before and during the incident/accident
   - Content of the past internal analyses concerning the results of the investigations near-miss incidents and accidents
   - Findings, requirements and recommendations of the then administrative and inspection supervisors
   - Reasons of official visits/inspection, and visits of business and private nature
   - Kinds and content of internal and external communication of employees
   - Observations about safety, and any other topic, possibly interesting for the investigation, collected from the employees of health and safety, protection and security services, managers, foremen and workers, first responders, suppliers, buyers and other business associates, neighbours, etc.

3. Establish and document all kinds of events, including the sequence and course of occurrences and events, which followed at certain micro-locations during and after the incident/accident, including measures and activities for protection, evacuation and rescue of employees, and emergency measures undertaken by internal and external first responders.

9 The Benefits from Using Professionally Developed Models

4. Establish and document all kinds, types and scopes of direct and possibly indirect consequences of the incident/accident for people, property and environment.
5. With the help of, immediately present and subsequently included, forensic technicians and acknowledged experts with necessary expertise, collect and forensically process all relevant facts which indicate to possible root, initial and contributive causes (conditions, circumstances and thus, possible manners) of incidents and their unhindered development into a possibly major accident or even disaster, including all information which can eliminate them conditionally or totally.
6. On the basis of the above-mentioned, develop (to the smallest detail important for the investigation), and, afterwards, gradually and critically analyze (according to the facts verified by forensic and criminal procedures), every possible hypothesis about likely causes (and manners) of the incident/accident and its unhindered development into the major accident or even disaster, applying common general analytical methods used in criminal investigation, helped by the appropriate analytical methods and/or techniques of health, safety, security and protection engineering (Table 9.1).
7. Identify all probable controversial factors which may be unavailable or insufficient for the investigation; in each of the considered most likely hypothesis models, forensically relevant and mutually connected factors of the cause-and-effect type should be identified. It is also necessary to try to establish and document their probable existence/nonexistence (by additional target searching of certain locations of continuously secured scene of incident/accident and/or other locations outside the area affected by the incident/accident, which are relevant for the investigation, and enlarging the number of objects or types of forensic expertise and/or undertaking other necessary evidentiary actions).

After it has been established beyond doubt that all facts confirm hypotheses and reject all null hypotheses, it can be concluded that the investigation of the incident/accident with hazardous materials is closed and completed.

From what has been discussed above it is obvious that, depending on the complexity of the industrial facility system, relevant for the investigation, its technological and working process as well as proportions and consequences of the accident (especially concerning macro- and micro-traces, relevant for the investigation, which have appeared before, during and after the accident, including the traces made by internal and external first responders), the investigators must constantly respect all principles of criminal investigation. This includes carrying out extensive, long, coordinative, technical, professional, skilled, experienced and intellectually demanding and tiring tasks, which occasionally lead to errors. Investigators' failures or mistakes are not impossible. Even the most rare, numerous and minor errors can be fatal for the success of any court proceedings, in which guilt or responsibility should be proved of any person who is directly or indirectly responsible for an incident, i.e. unhindered development of an incident into a (major) accident, or for its severe consequences.

The investigators' job can be easier if a company, which is affected by such an incident/accident (which falls into the category of those included in the Seveso II

Table 9.1 Some of the most frequently used methods and techniques of hazard analysis and risk assessment

Hazards identification and evaluation methods	Likelihood assessment methods	Consequences identification and assessment methods	Risk analysis methods
– 'Case studies' of the same or similar system with the same or similar hazardous substance – "What if?" – "Hazard and Operability Study"(HAZOP) – "Hazard Identification" (HAZID) "Failure Mode and Effect Analysis"(FMEA) – "Fault Tree Analysis"(FTA) – "Event Tree Analysis" (ETA) – Wide range of various "Human Reliability Assessment" (HRA) models or techniques (task related, time related, and context related model types)	– Historical records about malfunctioning, faults, failures, etc. of the technical components – Historical records about the most common human errors and failures – "Failure Modes Effects and Criticality Analysis"(FMECA) – "Fault Tree Analysis"(FTA) – "Event Tree Analysis"(ETA) – Wide range of various "Human Reliability Assessment" (HRA) models or techniques (task related, time related, and context related)	– "Event Tree Analysis" (ETA) – Dispersion models of hazardous substances – Fire models – Explosion models – Effect models – Mitigation models – etc.	– "Risk matrix" – "Risk histogram" – "Risk profiles" – "F-N curve" – "Risk index"

Directive),[6] maintains sufficient and detailed documentation of the facility and its technological processes, including documents of Hazard Analysis and Risk Assessment, a Safety Report, Internal Emergency Plans, and Information for External Emergency Plans. Furthermore, all data possibly stored in the computer, related to the movement and changes of all main process/operational parameters and indicators of the condition/functioning of safety, protective and security systems of critical/risk facility equipment before and during the accident (similar to the black box in an airplane). It can be even more useful if a company has an expert system for centralized surveillance, control, communication, and information collection and dissemination (intelligent processing/unit operations system). Conducting

---

[6] According to the requirements of this Directive, the Croatian government adopted the Regulation on Control of Major Industrial Accident Hazards (Uredba o sprječavanju velikih nesreća koje uključuju opasne stvari 2008) on September 30, 2008, which came into force on March 31, 2009, and which, among other things, defined the criteria for the categorization.

9 The Benefits from Using Professionally Developed Models

forensic examinations and experiments on legally *temporarily seized* up-to-date computer-simulated model of a technological process, or at least on its critical (the most sensitive/risky) process units on which the possible cause of the accident has been preliminary located, can be especially beneficial.

In contrast to the industrial facilities with hazardous substances, investigations of manners, causes, conditions and circumstances of incidents/accidents during transportation of dangerous goods are usually less complicated and technically demanding, except in cases with consequences such as a ship sinking in deep waters, a transport aircraft crashing into an industrial or residential area, or fire/ explosion of a truck/tank cars on a terminal, on a marshalling yard or in long, busy (sea, partially collapsed or flooded) tunnels. From the companies which transport dangerous goods investigators should obtain as soon as possible, all transportation and technical documentation about transport means and dangerous goods as well as documentation of the measures undertaken to prevent accidents and measures in case of accidents, within the competence of a safety consultant for transport of dangerous goods (including documents such as Internal Emergency Plan in case of marshalling yard accidents). In this case as well, computer-stored data about possibly existing and proper Intelligent Transportation System (ITS) can be very useful for the investigation.

9.3 On the Identification Procedure of Forensically Interesting Models of Possible Accident Scenarios

All experienced HSS managers are familiar with almost 40 year-old legal obligation of having professionally drawn up and up-to-date documentation of hazard analysis and risk assessment of fires, explosions, on-the-job accidents, ecological accidents, natural disasters and emergency situations caused by terrorist or war-related events. They should also have plans and programs for their prevention as well as internal and external/local community emergency plans and programs in case of any accident/ situation. All companies which produce, store, transport or sell potentially dangerous large amounts of hazardous substances and/or all companies which have been categorized into the facilities of critical local, national, transborder, regional or European infrastructure should be included in those plans.

According to the Article 9 of Seveso II Directive,[7] the operators of facilities holding large amounts of hazardous substances (or facilities in which such substances

---

[7]In Croatia, according to the Articles 13 and 15-18 of the above mentioned Regulation and obligations from the Fire Protection Act (Zakon o zaštiti od požara 2010), from the elements of the plan and evaluations based on the Rulebook on the content of fire and technological explosion protection plan (Pravilnika o sadržaju plana zaštite od požara i tehnoloških eksplozija 1994), the Rulebook on drafting fire and technological explosions risk assessment (Pravilnik o izradi procjene ugroženosti od požara i tehnološke eksplozije 1994) and the Rulebook on methodology for risk assessment and plans for the protection and rescue (Pravilnik o metodologiji za izradu procjena ugroženosti i planova zaštite i spašavanja 2008).

may appear during a major accident) are required to provide and maintain up to date Safety Reports, to demonstrate competence to authorities by providing:

1. A major-accident prevention policy and a safety management system for implementing it have been put into effect in accordance with the information set out in Annex III[8] of Seveso II Directive.
2. Major-accident hazards have been identified and that the necessary measures have been taken to prevent such accidents and to limit their consequences for man and the environment.
3. Adequate safety and reliability measures have been incorporated into the design, construction, operation and maintenance of any facility, storage facility, equipment and infrastructure connected with its operation, which are linked to major-accident hazards inside the establishment.[9]
4. Internal emergency plans have been drawn up and supplying information to enable the external plan to be drawn up in order to take the necessary measures in the event of a major accident.
5. Sufficient information to the competent authorities have been provided to enable decisions to be made in terms of the siting of new activities or developments around existing establishments.

The process of producing a Safety Report should be carried out on professional (educated, skilled and experienced), intellectual, organizational, coordinative, timely (usually long lasting and intellectually tiring) and demanding manner. This is the most risky phase in the process with the probability of making more fatal mistakes, failures or oversights and their possible consequences. This phase, called 'risk analysis', covers the hazard analysis and risk assessment of actually possible accidents and their consequences.

Its main elements are: hazard identification, accident scenario selection, scenarios' likelihood assessment, scenarios' consequence assessment, risk ranking, reliability and availability of safety systems. During this process, a team of experts in this field, on the basis of thorough examination of gathered documents and a large amount of relevant data,[10] tries to identify, thoroughly consider and illuminate all elements of the possible scenarios of each possible accidents in each process unit and each part of the facility which continuously or temporarily contains hazardous substances and/or potentially dangerous kinds of energy. Included should be all models of the worst-case scenarios. It is of extreme importance to illuminate inductively and deductively all types and kinds of different possible

---

[8]"Principles referred to in Article 7 and information referred to in Article 9 on the management system and the organization of the establishment with a view to the prevention of major accidents".

[9]"Establishment" *shall mean the whole area under the control of an operator where dangerous substances are present in one or more installations, including common or related infrastructures or activities* [Article 3, par. 1 of the Seveso II Directive 1996 (EUR 22113 2005)].

[10]For details, see Annexes of Seveso II Directive and guidance (EUR 22113 2005; OECD 2003; CRAIM/UNEP 2002) and numerous professional handbooks in this field.

9 The Benefits from Using Professionally Developed Models 161

outcomes and events. Events that may result from each actually possible kind of potential origin/root cause[11] of their occurrence or which, with the investigation in the opposite direction, can lead to the discovery of each one.

Of course, especially interesting, for investigative purposes, is unhindered occurrence of analytically identified possible scenarios of incidents which could easily/totally lead to occurrences of major or fatal accidents. Regardless of finding their probable causes inside or outside a system of an industrial complex or a technological plant, i.e. in its safety relevant narrow, wide or the widest surrounding area.

In the next phase, a team of analysts tries to assess, quantitatively and quasi-quantitatively, and sometimes, when there are no available/sufficiently reliable quantitative indicators, at least qualitatively (precisely enough), probabilities or possibilities of initiation of development of kinds and types of scenarios and all conditions and circumstances of the occurrence of undesirable events which could be the cause of initiation and development of any kind of a scenario.

As mentioned above, it is obvious that one of the key components of a Safety Report is defining appropriate scenarios of possible kinds and types of accidents with hazardous substances. The term 'scenario' in the field of safety, security and protection from accidents means the description of sequence of undesirable events, obtained by appropriate methods and techniques, which can result in a processual, technological, operational, procedural or working incident and/or accident.

According to the EUR 22113 (2005) definition, for the purpose of producing a Safety Report as required by the Seveso II Directive, a scenario is an undesirable event or a sequence of such events characterized by the loss of safe containment (LOC) or the loss of physical integrity and the immediate or delayed consequences of this occurrence. While considering possible accident scenarios for hazard analysis and risk assessment of accidents with hazardous substances, all kinds of hazards must be taken into account, including those of human and technological nature (failures, unauthorized interference, mistakes, malfunctioning, etc.) (Bragdon et al. 2008; Cavani et al. 2009; CRAIM/UNEP 2002; Grossi and Kunreuther 2005; Garrick et al. 2008; Gordon, Flin and Mearns 2005; Kletz 2009; Koehler and Brown 2010; Körvers 2004; Lancaster 2005; Maguire 2006; NRC 2006; Owen et al. 2009; Rasmussen and Svendung 2000; Sanders 2005; Spurgin 2010) as well as natural hazards (natural disasters) (Gunn 2008; McDonald 2003; Garrick et al. 2008) and malicious acts (of terrorists, employees or civil sabotage as a violent protest, thefts of safety and protective parts or equipment, vandalism by employees and other persons, etc.) (API 2005; Clements 2009; Herrmann 2002; Kulišić 2008c;

---

[11] By root cause we mean the first reason(s) which lead(s) to an unsafe event (process, act, action, condition or circumstances), which result(s) in an undesirable event or near-miss. In other words, elimination of such a cause can prevent the scenario leading to the accident. Among others, causes can be flaws in the management system, leading to the wrong design, mistakes in construction of separate parts of the facility, inappropriate material quality, assembly plant failures, maintenance or inspection failures, and even inadequate personnel policy.

162 D. Kulišić

Kulišić and Magušić 2007; Landoll 2006; Macaulay 2009; NRC 2006; Piquero et al. 2010; Reid 2005; Wortley and Mazerolle 2008),[12] which can be a "trigger" of the accident with hazardous substances.

In producing a scenario a set of different tools is used – qualitative and/or (semi) quantitative analytical methods and techniques – and appropriate necessary databases of useful information (e.g. *case studies*), which are interesting for reliable analysis of components and/or the whole unit of the risk system, within limited conditions of its harmless functioning. Many of these or similar tools are often widely used in statistics, economy, technology, etc., while planning and decision-making in the conditions of certain insecurity or threats of a danger or crisis.

The author is familiar with more than 80 methods and techniques, with various names, but which are all specific, very similar or, in a certain way, mutually combined (nowadays generally software-supported). The most often used methods and techniques for analysis of systems with hazardous substances are those presented in Table 9.1.

It is important to bear in mind some possible, especially aggravating objectively undesirable circumstances, which can significantly and profoundly influence the level of scenario reliability of causes and/or consequences of probable accidents. Moreover, the level of availability and/or reliability of the undertaken safety and protective measures and activities, as well as the substantial prolongation of the process and an increase in costs of criminal investigations (especially for the reason of necessary new specific forensic analyses) of accidents.

One of the reasons is the fact that currently over 53 million of different organic or inorganic substances or compounds (Chemical Abstract Service 2010) have been identified, i.e. isolated or synthesized. A traditional, and nowadays mounting basic health and safety problem, arises from the lack of sufficient, basic or at least any knowledge about possibly hazardous characteristics, effects and consequences, especially those which are long lasting for health and environment, as well as a knowledge about places, kinds, amounts, consumption and waste disposal of hazardous substances. Namely, in contrast to *new* substances, extensive and detailed research and assessment of all possibly hazardous characteristics has been conducted so far on relatively few potentially hazardous *existing* substances (Directive 67/548/EEC).[13]

---

[12]Thus, it is necessary (with the help of instructions and intelligence data) to carefully consider and assess all levels and possible influences of the following elements: a probability of an attack (a possibility to become a target and a possibility of its successful performance is the function of the following three variables), threat of an attack (the function of intent, motive and ability for performance of the familiar/common modus operandi of possible types of perpetrators), vulnerability to an attack (the indicator of probable successful performance of an attack, depending on the intent) and target attractiveness (which directly influences probability of an attack, depending on the perceived value and the level of interest for an attack).

[13]See amendments in: (a) Regulation (EC) NO 1272/2008 of the European Parliament and of the Council of 16 December 2008 on classification, labelling and packaging of substances and mixtures, amending and repealing Directives 67/548/EEC and 1999/45/EC, and amending Regulation (EC) No 1907/2006, and (b) Commission Directive 2009/2/EC of 15 January 2009 amending, for

9 The Benefits from Using Professionally Developed Models 163

Thus, for example, the last edition of the famous Sax's handbook about toxic, flammable, chemical and explosive properties of hazardous industrial materials includes so far only 26,000 substances (Lewis 2004). However, there are, among them, a number of substances for which no relevant data are presented or which are insufficiently known, as far as their possible long-lasting dangerous influences on health and environment are concerned.[14]

Dangerous properties of a numerous hazardous substances which have been produced, stored, transported, processed or used for decades in relatively large or even enormous amounts are meticulously examined and relatively well known to the experienced HSS managers of companies, civil employees of administrative control and inspection, emergency services experts, and especially to experts in hazardous substances working at forensic, scientific-research, educational and advising institutions.[15]

These are mainly hazardous substances whose structure is consistent with that of the international agreements or conventions on the transportation of hazardous substances, which have the status of *dangerous goods.* These include: (1) the European Agreement Concerning the International Carriage of Dangerous Goods by Road (ADR); (2) the Regulations Concerning the International Carriage of Dangerous Goods by Rail (RID); (3) the International Maritime Dangerous Goods Code Economic Commission for Europe (IMDG); (4) the European Provision Concerning the International Carriage of Dangerous Goods by Inland Waterway (ADN); and (5) the Technical Instructions for the Safe Transport of Dangerous Goods by Air (IT). Nowadays their number exceeds 2,900, including all those kinds of hazardous substances with the allocated appropriate UN identification number (UN number) (Hommel 2010).

Besides that, when the European Parliament adopted (late in 2006) the **R**egistration, **E**valuation, **A**uthorisation and Restriction of **Ch**emicals (REACH) Regulation and when it came into force in the EU on 1 June 2007, the unique,

---

the purpose of its adaptation to technical progress, for the 31st time, Council Directive 67/548/ EEC on the approximation of the laws, regulations and administrative provisions relating to the classification, packaging and labelling of dangerous substances.

[14] In this sense, a 200-page report of the U.S. President's Cancer Panel was recently published (29 April 2010), which states: *Only a few hundred of the more than 80,000 chemicals in use in the United States have been tested for safety.* It adds: *Many known or suspected carcinogens are completely unregulated* (Kristof 2010).

[15] A major breakthrough regarding HSS from potentially dangerous chemicals is expected from:

(a) further implementation of the European Regulation REACH (Regulation EC No. 1907/2006) and new risk assessments (required by the Directive 93/67/EEC for all *new* substances, Regulation EC No. 1488/94 for *existing* substances and the Directive 98/8/EC regarding *biocidal active substances* or important substances in biocidal products), full technical details for the implementation of each procedure step in hazard analysis and risk assessment of human health and environment, as described in EUR 20418 (2003).

(b) implementation of the unique "Globally Harmonized System" (GHS) of classification, labelling and packaging of dangerous substances – for details see, for example: UNECE (2009) or UNITAR (2008).

legally prescribed approach was introduced to treat enormous, large and small amounts of all possibly hazardous substances from the *chemicals* category in the EU and the European economy market. It will comprise all phases of a *life* cycle of a chemical, starting from producers, suppliers, downstream users to end users ("from cradle to grave").

Producers of chemicals, importers and end users must have, for these substances, updated Inventory of the Dangerous Substances, for the substances they produce, put on the market or use, and/or the Safety Data-Sheets (SDS), Chemical Safety Instructions and the Chemical Safety Report, depending on their weight (tonnage).

The Chemical Safety Report for the chemicals classified as *hazardous* or *very hazardous*, whose distribution is forbidden or restricted (*substances of very high concern – SVHC*), must contain exposure scenario. Moreover, if a certain product or preparation contains SVHC chemicals in the amount over 0.1%, which could be emitted during normal use (intentionally or unintentionally caused emissions). This potentially hazardous scenario for human health or environment by Chemical Safety Report must contain descriptions of one or more exposure scenarios.[16]

Another important and hardly predictable problem, which can question the level of scenario reliability of causes and/or consequences of possible accidents with hazardous substances, can arise from faster development and faster implementation of new construction materials, technical components and process units/facilities with insufficient practical experience about their reliability, endurance, expiration date, possible resistance or incompatibility when in contact with some substances or materials that are used as raw materials, additives, auxiliary or service materials (e.g. service fluids), or as unexpected/undesirable intermediate substances, by-products, products formed as a result of loss of control of chemical processes or as process/ environmental contaminants (sometimes) appear or can appear in the process.

There are other possible factors of negative influence, which are not known or observed, and can easily result in "unexpected" accident scenarios, such as: exposure of critical facility parts to the extreme process work conditions or their environment, for which they have not been previously tested[17]; accelerated aging of materials of a facility; high sensitivity and vulnerability to various external influences (of parts) from the systems dealing in electroenergy, water supply, computers, telecommunications, and safety and security systems to various external influences; an easy unauthorized access to the critical facility parts; suppressed dissatisfaction

---

[16]It is believed that this requirement will result in a significant number and scope of research, as well as in a series of new, forensically interesting/useful examination methods and techniques for potentially hazardous properties of substances in many commercial products.

[17]For example, an assumed cause of the probable malfunctioning of a blowout preventer, a crucial fail-safe mechanism on the pipe near the ocean floor, at the critical moment of a severe blow of compressed natural gas through the oil-rig pipe, in the conditions of very high pressure (>13,8 MPa) and very low temperatures at the Gulf of Mexico seabed, which on 20 April, 2010 resulted in the explosion, fire and sinking of the BP oil-rig "Deepwater Horizon", killing 11 people and setting off the largest oil spill in United States history and an environmental disaster.

9 The Benefits from Using Professionally Developed Models

on the part of employees or ecologically motivated hostility of the neighbourhood towards a company and their readiness for undertaking actions of passive/active sabotage of a facility, etc.

9.4 On the Most Frequent Types of Major Accidents with Hazardous Materials and Their Scenarios

It is well known that incidents and accidents with hazardous materials occur inside and outside the industrial facility. Depending on the type/movement of a source, the incidents with spilling/releasing of hazardous materials can occur from stationary or moving sources. Depending on manners, causes, conditions and circumstances of occurrences, they can be immediate, continuous (stationary or nonstationary) and temporary (discontinuous, multiple).

Total amounts of hazardous materials which are produced, stored and processed are often significantly larger than those found in some, mostly transportation units (e.g. cars and tank cars but not tankers). Therefore, in case of occurrence and unhindered development of a domino effect, i.e. cause-and-effect harmful events, their effects and consequences (e.g. because of inappropriate siting, distance and passive or active protection of the process units, facilities or several facilities with large amounts of hazardous materials), there is a possibility of fast occurrence of a major accident or unstoppable deterioration of consequences of the causes of the initial accident phase. One traditional, procedurally simplified deterministic approach to the preliminary preventive analysis and defining of types, kinds and levels of possible hazards (i.e. traceologically based forensic development of one of hypotheses about possible causes) in the probable scenario of initiation and development of the process incident on the one process unit to the level of a major accident or disaster (on the generalized example of possible random causes and consequences of the physical explosion of a vessel with compressed gas, hazardous for health, in an industrial complex with large amounts of various hazardous materials) is shown in Fig. 9.3.[18]

Most accidents in petro-(chemical) and oil industry result in spills or releasing of flammable, toxic, and explosive substances/materials. These accidents follow typical patterns. If carefully studied, these patterns can help anticipate, as a preventive measure (and for investigative/forensic purposes – hypothetically), potential types and scenarios of accidents.

A major accident scenario for the purposes of the Safety Report usually describes the form of the loss of safe containment (LOC) specified by its technical type (e.g. vessel rupture, pipe rupture, vessel leak, etc.) or by its chemical type

---

[18] Possible types and kinds of domino effect, which are interesting for the preventive development or for forensic reconstruction of a scenario of an accident with hazardous materials, see for example Gledhill and Lines (1998) and Cozzani et al. (2006).

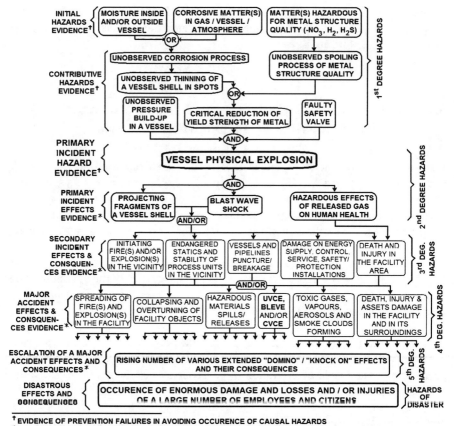

Fig. 9.3 Traditional, procedurally simplified deterministic approach to the preliminary preventive analysis and defining of types, kinds and levels of possible hazards

(e.g. exothermic *runaway* of chemical reaction(s) because of a spontaneous decomposition, autoxidation, polymerization or isomerization, etc.) and a triggered event, such as fire, explosion, release of hazardous substance(s) (EUR 22113 2005).

Decades long analyses of statistical data about types, causes and consequences of accidents with hazardous materials reveal that fires are the most common, followed by explosion and (eco)toxic release. With respect to fatalities, the order reverses, with toxic release having the greatest potential for fatalities. Material damage is consistently high for accidents involving (eco)toxic releases and explosions. The most damaging type of explosion is an unconfined vapour cloud explosion (UVCE), where a large cloud of volatile and flammable vapour is released and dispersed throughout the plant site followed by ignition and explosion of the cloud [Emergency Events Database (EM-DAT), The International Disaster Database, Center for Research on the Epidemiology of Disasters (OFDA/CRED), Catholic University of Louvain, Belgium, UNEP–APELL Disasters by Category, Major-Accident Reporting System (MARS)]

9 The Benefits from Using Professionally Developed Models 167

(Crowl and Louvar 2002; NFPA 2008). Similarly destructive can also be explosions of large amounts of stored chemically more reactive/explosive substances (e.g. ammonium nitrate in fertilizers industry, black powder in pyrotechnics production, etc.). Large-scale devastation, demolition of constructions and destructive effects of fires, which usually occur after such explosions, can extremely hinder or prevent forensic reconstruction of the actual accident scenario.

Although, traffic incidents and accidents during road transportation of dangerous goods are statistically (compared to traffic incidents and accidents happening with other transportation means), relatively rare cases – consequences of such accidents can also have fatal proportions, causing a huge number of victims among drivers, passengers and passers-by, huge material damage and long-term consequences for health and environment (Fig. 9.4). Generic elements necessary for

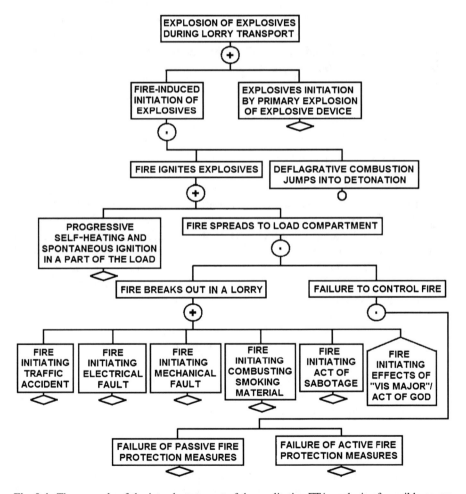

Fig. 9.4 The example of the introductory part of the qualitative FTA analysis of possible causes of explosion during road transportation of explosive substances

dangerous goods transport hazard analysis, scenarios developments, and risk assessment are described by Bragdon et al. (2008), CETU (2005), UNEP (2000), and Kulišić (2008b).

9.4.1 Scenarios of Major Accidents with Hazardous Materials, Which Are of the Greatest Professional and Public Concern

Special attention according to rigorous professional hazard analyses and risk assessments, by determined undertaking of prevention measures, appropriate administrative control and inspection and adequate readiness to react should be paid to the following most frequent scenarios of major accidents with hazardous substances and their mixtures, which can have especially hazardous effects and consequences for lives and health of a large number of people (Marshall 1987; Medard 1989; McDonald 2003; WHO 2009):

- Fire in a large building, such as numerous residential and/or official premises, large retailing shopping centres, large warehouses, large manufacturing and storing facilities with chemicals, tunnels and subways – e.g. a case of the Buncefield fire, a major conflagration after an unconfined vapour cloud explosion (UVCE) of gasoline at the British oil storage terminal in 2005.
- Physical explosion, chemical explosion of condensed hazardous matter or confined vapour cloud explosion (CVCE) can cause emission of deadly fragments of structures and building materials, demolition or serious damage to buildings, trapping victims and blocking the possibility of fast rescuing of survivors, and the wider area of the accident scene is full of thick smoke clouds, poisonous and choking gases and dust. In some cases there is a danger of sudden emission of vast amounts of hazardous materials, occurrences such as a boiling liquid expanding vapour explosion (BLEVE/"fire ball") or an unconfined vapour cloud explosion (UVCE), which can be set off at some distance from the place of emission – e.g. a case of a series of explosions, occurrences of a series of *fire balls* and a major conflagration at the offshore "Deepwater Horizon" BP oil-rig in the Gulf of Mexico or the case of disaster at the terminal of the LPG oil company "PEMEX" in Mexico in 1984.
- A sudden evident outdoor release of gas or vapour of a hazardous substance into the atmosphere (i.e., gas cloud or vapour spread fast into the surrounding area, immediately after evaporation of liquid leak – e.g. such as methyl isocyanate in Bhopal, in 1984).
- A sudden evident outdoor release of a liquid or solid aerosol of hazardous substances into the atmosphere (depending on the type of incident, e.g. in case of an explosion, and on the properties of dispersed aerosol, characteristics of the environment and hydrometeorological conditions, ranges of contaminated area around the scene of the accident can be measured in kilometres – for

9 The Benefits from Using Professionally Developed Models

example in case of a major accident with dispersed dioxin in the Italian town Seveso, in 1976).

- A prominent noticeable release of a hazardous substance into water, soil or directly into food, drink or other kinds of consuming products (during producing/ processing of food, drinks, food additives or medicine, etc. – e.g. a cases of a spill and spread liquid and suspended mineral waste containing 50–100 t of cyanide, cooper and heavy metals into the Sasar, Lapus, Somes, Tisza, and Danube Rivers from Baia Mare, Romania, in 2000).
- Sickness of a large number of people with equal or similar symptoms, when the occurrence of release/emission of dangerous substances, hazardous for health, has not been discovered on time, but belatedly by the health protection system. It usually takes days, weeks or months to discover the cause of sickness, means/ ways and the origin and genesis of the manner/mechanisms of intoxication or infection at the place of origin – especially if there is any circumstantial evidence of terrorist abuse or organized criminal illegal disposal of chemical, biological or radiological hazardous substances. See for example, Koehler and Brown (2010) and cases of mass poisoning by pesticide, endosulfan, in the Jabalpur District of India, in 2002 and with sodium bromide in Angola, in 2007.
- Imperceptible or unperceived release/emission of a hazardous substance, when the occurrence of release/emission of dangerous substances, hazardous or harmful for health, has not been discovered on time, or when, for a reason, appropriate health and safety measures have not/could not have been undertaken until the end of such occurrence.[19] It has been discovered before the appearance of symptoms, which could indicate a fatal or harmful effect of a hazardous substance.

The analyses of international and worldwide famous professional bodies which deal with the problem of modern terrorism, especially its *modus operandi*, regarding current trends of *unconventional means* abuse in attacks, show a growing trend of casualties caused by terrorist or similar criminal (organized and individual) abuse of kidnapped transportation means, as well as great vulnerability of production facilities, storage areas and transport systems with great/huge amounts of hazardous materials, which can be abused as some kinds of improvised weapons of mass destruction (WMD) or to intimidate and blackmail authorities or their owners (Kulišić 2008c, 2009).

In Croatia special preventive attention and analysis is required to address the long-term issues of economic recession, obsolete technology/facility (HSE RR509 2006; Paik and Melchers 2008; Smith 2001), chronically insufficient financial means for proper maintenance and appropriate technological and personnel renewal. Adequate attention should be paid to the analyses of levels of exposure, sensitivity, vulnerability and recovery from terrorist and other criminal attacks of all those production, storage and transport systems classified as *critical infrastructure*.

---

[19] It is possible in those incidents with hazardous substances when its effects become obvious only after a certain period of time or when the emission of a hazardous substance is greater than predicted at the beginning of its emission.

9.4.2 On Preventive and Forensically the Most Useful Way of the Presentation of Possible Scenario Models of Accidents with Hazardous Materials

Since the basic aim of producing scenario models of accidents with hazardous materials on each potentially dangerous process unit or installation is a decision-making about necessary prevention, protection and rescue measures in case of an accident, these models must be designed and graphically presented in such a way that, with expert inspection of their structure, all locations (occurrences) of possible dangers from very risky incidents/major accidents, observed by analyses, can easily be identified, including locations, purpose and function of chosen/installed undertaken measures (barriers) for their timely and efficient prevention, avoiding, eliminating or mitigating.

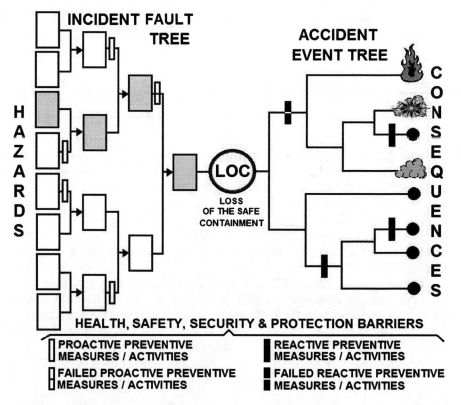

Fig. 9.5 The "Bow-tie" diagram of a scenario of a top event (a kind of an accident or attack) of its likely causes (accidental or malicious) and a scenario of a likely development of its harmful and fatal consequences (EUR 22113 2005; Kulišić 1997, 1998, 2008a, 2008b, 2008c; Kulišić and Magušić 2007)

9 The Benefits from Using Professionally Developed Models 171

One of the usual and most often recommended modern ways of model presentations of each possible scenario of accidents with hazardous materials is by using an appropriate "bow-tie" diagram. Apart from the prevention aspect, it is especially forensically interesting and useful. Diagrams, both qualitatively and quantitatively, clearly show, identify and reconsider all possible causes and consequences of a presented "top event". It shows by symbols positions of implementation/intervention of the necessary types and kinds of security, preventive, protective nature; referred to as initiation barriers or unhindered development of hazardous events. These barriers lead to the top event, presented by a scenario (the so called *proactive safety barriers*) and their harmful consequences (the so called *reactive safety barriers*) – see Fig. 9.5.

Developed and updated scenarios serve the purposes of regular, temporary and extraordinary checks, implementation of necessary improvements and confirmations that undertaken measures for safety and protection of accidents and safety, protection and defence against criminal attacks of facilities with hazardous materials are suitable for all existing, newly emerging and predictable internal and external dangers.

9.5 Advantages and Some Problems of a Quick Professional Inspection of Documents of Possible Scenarios of Accidents with Hazardous Substances and Techniques of Their Design

Of great importance for the investigation is the possibility of a quick discovery and obtaining, for preliminary professional inspection, the Safety Report and the appropriate Internal and External (local community) Emergency Plans and Programs, which are drawn up pursuant to Seveso II Directive and relevant domestic regulations and recommendations[20] (previously adopted by the advisory and inspection competent state authorities[21]). After preliminary information about all relevant characteristics, development and consequences of an accident under investigation, the

---

[20] Some of the most important Croatian legislative regulations and recommendations Nacionalna strategija kemijske sigurnosti (2008), Nacionalna strategija za prevenciju i suzbijanje terorizma (2008), Nacionalna strategija zaštite okoliša (2002), Pravilnik o registru postrojenja u kojima je utvrđena prisutnost opasnih tvari i o očevidniku prijavljenih velikih nesreća (2008), Zakon o prostornom uređenju i gradnji (2007), Zakon o zaštiti i spašavanju (2004).

[21] In Croatia, they are: the Ministry of Environmental Protection, Physical Planning and Construction, the Ministry of the Interior, the Ministry of Health, the Ministry of Economy, Labour and Entrepreneurship, the National Protection and Rescue Directorate, the Croatian Institute for Toxicology. For some issues which are not regulated by professional regulations, research, opinions and recommendations are used from the Ruđer Bošković Institute, Croatian Institute for Health Protection and Safety at Work, and other numerous Croatian and international scientific-research institutions or associations competent for certain issues in the field of health, safety, security and protection.

insight into these documents provides more proper and timely locating of exact areas, facilities/objects as well as taking all necessary measures for police securing of all areas interesting for criminal investigation (not only a preliminary crime scene). On the basis of all those information, forensic experts with appropriate professional and special knowledge necessary for the formation of the crime scene investigation team can be engaged and stages and procedures of criminal investigation thoroughly planned.

In this way, we can significantly lower: (1) the possibility of straying in a criminal investigation, the possibility of unnecessary wasting of traceologically valuable time on details and places unimportant for the investigation, and (2) the possibility of undertaking actions unnecessary for the investigation/unnecessary increase in investigation costs and unnecessary additional physical and mental wearing out of a team of investigators for the sake of irrelevant and preliminary less important investigatory and evidentiary actions (i.e. gathering and examining of irrelevant or less important information and data from usually extensive company's process, technical and operative documentation).

Investigative/forensic knowledge and skilful use of methods, techniques and software tools (refer to in Table 9.1), commonly used during hazard analyses and risk assessment and in developing the accident scenario, can be of great help and crucial importance for a successful search, discovery in time, exact identification and reliable logic checking/testing and linking macro- and micro-traces at the scene of an accident with hazardous materials, and also other important evidence and circumstantial evidence very likely relevant for the investigation (not only at the scene of an accident). Unfortunately, there are very few modern criminal investigators well informed on proper investigative applications for the purpose of process management of criminal investigation on account of searching, gathering and evaluation of necessary evidence in many types of criminal offences – not only those with natural and technical characteristics of causes/*modus operandi*, i.e. a likely *corpus delicti* the above-mentioned and similarly suitable analytical methods and techniques (or at least of the simplest FTA and ETA types) while developing and examining hypotheses. More than 50 years of undoubtedly positive experience about the usefulness of the analytical methods and their applications are shared, starting with those first-developed for analysis needs and increasing the safety and security level of nuclear ballistic missile silos (FTA). The above-mentioned analytical methods and techniques are widely used not only for technical, but for medicine diagnostics (for the comparison of general and specific symptoms). This is done to more quickly and precisely identify a real cause of an illness without wasting crucial time and means on unnecessary specialist health examinations.

While requiring an official inspection of the above-mentioned documentation, a problem can arise of determined resistance to cooperation with investigators on the part of company owners/operators affected by an accident. Motivated by an objective problem of process and business data confidentiality. However, such resistance can also be motivated by subjective reasons, i.e. fear (in all people potentially responsible for an accident) of discovery of incriminating health, safety, security, protection and/or internal emergency rescue faults or failures (i.e. the possibility of investigators' gathering key forensic evidence) related to lack of undertaking measures of

9 The Benefits from Using Professionally Developed Models 173

prevention, protection and rescue, required by regulations and designed plans (documents which are mentioned as allegedly existing/ensured or undertaken), probably well known to the responsible persons in the process, health, safety, security and/or protection management (and the company's owner himself). This problem can and must be solved as soon as possible by the court order for temporary seizure of relevant documentation with a written guarantee to the owner about the obligation of undertaking all prescribed measures for the protection of production and business data confidentiality, during and after the investigation.

Problems during investigations arise from dealing with out dated and unkept documentation. Investigators must provide additional investigative and evidentiary actions to find, check and reconstruct facts relevant to the investigation, which are missing in each of the still open hypothetical models of scenarios, of unhindered initiation, development and occurrence of accidents and their consequences.

Insufficient expertise of a company's personnel and independent development and/or application of all analytically important documents as the source for developing relevant accident scenario models and planning of prevention, protection and rescue measures in their company, is often solved by engaging eminent authorized consulting and project firms or scientific-research institutions. They work in the cooperation with: (1) the process company management; (2) experienced operators of all safety critical facility installations; (3) designers of such or similar facilities and/or manufacturers of safety critical facility installations; (4) manufacturers and service technicians of safety and protection systems; (5) experts in tactics and techniques of sheltering, evacuation and rescue of people and property; (6) experts in fire extinguishing; (7) experts in the prevention and protection from a sabotage; (8) experts in stopping the (spread) of spilling/releasing of hazardous materials, and (9) experts in prevention of further hazards and elimination of possible consequences of likely spilling/releasing of hazardous materials into the atmosphere, water and soil.

Preliminary scepticism of investigators is not always unfounded, considering availability, integrity and reliability of professional investigative quality of relevant and useful reports (especially regarding research findings) about the conducted inspections by inspectors from the competent state services who checked all obligatory safety reports before giving a positive evaluation of their professional acceptability. It is shown by the current criminal investigation of the role that likely failures (even corruption of heads or inspectors) of the competent state bodies regarding carelessly administrative "turning a blind eye" to certain conditions and circumstances of likely and existing insecurity. This could possibly "open the door" to increased occurrences of other accidents, e.g. of the BP oil-rig in the Gulf of Mexico (Cooper and Broder 2010; Urbina 2010a, b; Uhlmann 2010; Cooper and Baker 2010). The problem of accidental inspection oversights of possible key health and safety problems in companies, whose work is temporarily controlled, usually arises from the lack of professionalism/education, motivation or professional experience on the part of inspection personnel (usually low-ranking and less paid), or from superficial, formal administrative approaches to the content of inspected documents (i.e. without undertaking thorough professional analyses of the content) and formal comparisons with the actual condition within the company under the inspection (i.e. only superficial review of documentation and short "tourist" visit of the facility).

174 D. Kulišić

While engaging adequate professional and specialist experts necessary for the formation of forensic teams, investigators must undertake measures of urgent police checks of these persons to avoid the possibility of including the persons in any conflict of interest regarding the company affected by the accident.[22] Possible involvement in various frauds or ethically unacceptable/discrediting behaviours or actions can be monitored.

Besides the above-mentioned, the objective investigative problem is caused by the still relatively modest choice of available professional literature in the field of criminal investigation and forensic science, especially dealing with investigations of accidents with hazardous materials, particularly with investigations of types and causes/*modus operandi* of ecological (environmental) crime. In contrast to the relatively numerous published manuals and textbooks for forensic investigations of crime evidence, such as fire and explosion causes (of hazardous and other types of burning substances), computer crime, economic crime, burglaries, murder, sexual offences, bribery, corruption, etc., we can rarely find practical manuals, useful for the investigators, such as Interpol's "Illegal Oil Discharges from Vessels: Investigative Manual" (ICPO-OICP 2007) and some comprehensive commercially available papers (Barrs and Lynch 2004; CCPS 2003; Morrison 1999; Mudge 2008; Murphy and Morrison 2007).

9.6 Some Examples of Cases of Extensive Jeopardizing HSS of Inhabitants and Environment with Hazardous Materials, Products and Energies in Croatia Since 1982

Statistically, major accidents with hazardous materials are events that have a relatively low probability of occurrence (for disasters probability is even several hundred or a thousand times lower). Extraordinarily severe and large-scale, potentially fatal consequences are possible.

---

[22] Especially regarding their likely direct or indirect involvement in the harmful event – as: a likely responsible (associate) designer; an (associate) equipment supplier and installation contractor; a contract regular (associate) maintenance/service technician; an (associate) licenser, of a still valid attest about adequateness and functionality and/or findings and certificate about recent regular, paid by the contract, proper operation check; an inspector (or his colleagues) from the appropriate state body competent for the regular administrative control or inspection, especially if, in his/their records about the results of some last preventive reviews, there are no warnings to the management/owner of likely failures, which should be eliminated as soon as possible, and which may have caused the accident – regarding the subject and content in question, relevant to investigative/ forensic analysis, etc.

9 The Benefits from Using Professionally Developed Models 175

In the Republic of Croatia they were very rare, and chances of their occurrence, as it seems, according to the current statistics of the Ministry of the Interior,[23] are even smaller, although the above-mentioned threatening effect of the current economic recession to the delay of the facility renewal of oil and chemical industry should not be neglected. Among global factors, which have significantly influenced risk reduction from major accidents with hazardous materials in the Republic of Croatia, according to Richardson (1995) and domestic economy analysts, are modest pre-war and long-term post-war, continuously gradually decreasing the level of country's industrialization and post-war investment in other economy branches, of less risk level. The main contribution to the total safety of road traffic was a post-war construction of modern highways (on the busiest routes, important for the road transportation of dangerous goods). Not less important positive influence can be attributed to the level of information, education and safety culture of employees and citizens, a high public sensitivity to the problems of endangering the environment and implemented quality of prescribed measures and activities in the field of health, safety, defence and protection. That is why Table 9.2 shows an account of a somewhat longer period in order to present some of the accidents which had severe consequences or which, by possible consequences, caused the greatest concern and worry to the Croatian public.

9.7 Conclusions

This chapter shows that criminal investigation of the manner, causes, conditions and circumstances of incidents and accidents with hazardous materials can be a very demanding, complex, and dangerous task.

It is demanding, firstly because it requires knowledge and experience in many fields of science, many professions and skills for it to be successfully accomplished, including necessary experience in investigations of numerous potentially important traceological characteristics of the crime scene (not only the places where consequences occurred).

Since no two events are the same, even if seemingly similar, among many investigative, forensic and procedural relevant characteristics, a well-coordinated team is necessary. A team involving highly educated and highly specialized persons with appropriate professions and skills of adequate work, moral and ethical dignity, among whom at least some should be real veterans in this job.

It is complex, firstly because one should legally, safely, rationally and optimally manage, during investigations of such accidents with human and material resources, in accordance with principles of criminal investigation, regulations of police, criminal

---

[23] In 5-year-period from 2004 to 2008, because of environment pollution, only 25 criminal offences against environment [Article 250, Chapter XIX of the Criminal Code (Kazneni zakon Republike Hrvatske 1997)] were reported, 23 of which have been so far successfully solved, regarding the perpetrator and circumstances of the offence (Crime Analysis Department of the Croatian Ministry of the Interior 2004–2008).

Table 9.2 Chronological account of some especially dangerous or disturbing cases of (likely) major accidents on the process facilities, storage facilities and transport units with hazardous materials and energies, which could have resulted in severe, enormous or fatal consequences for companies, employees, local inhabitants and environment in the Republic of Croatia since 1982

Event, date and place	Process, storage or transport unit	HAZMAT (quantity)	Event cause[a]	Effects and consequences
1	2	3	4	5
Explosion (CVCE) 26 June 1982 (~08:30) Zagreb (Črnomerec) 'Pliva', a pharmaceutical company (surrounded by a densely populated residential area)	Atmospheric storage tank (R-6) for waste chlorosulfonic acid (CSA) in 8-tank storage area (40×30 m) for pharmaceutical raw substances (inorganic acids and bases)	Waste chlorosulfonic acid ($ClSO_3H$) in 100.5 m^3, partly emptied, tank volume (about 72 t of substance left before explosion)	Hot works (due to corrosion and acid leakage) on the steel shell of a tank with gradually accumulated explosive atmosphere of H_2, released in successive chemical reactions after CSA reaction with condensed H_2O[b](longer time air moisture penetration caused by poor maintenance of tank air moisture protection system)	R-6 (ejected 10–15 m) smashes steel tank holders and pipelines, flies over and falls behind a storage tank R-5. 2 dead and 2 injured workers. Acid spill in concrete retention pool, with strong toxic vapours eruption and wind dispersion to the surrounding facility buildings, and mostly to the northern part of the city. The radio Zagreb warned the citizens of that part of a city not to open windows and avoid walking down the streets. Police put blockades to prevent traffic around the company until finishing neutralization process of acid spill with $NaHCO_3$. 4 responsible persons were sentenced from 1 to 3.5 years in prison.

Explosion (UVCE) 29 July 1984 (~17:00) Zagreb (Industrial area Žitnjak) "Chromos", a chemical company (Production sector "Kutrihils")	Process facility "Etoksilacija" (production of a few types of detergent component, called "Tenzides")	Chemically unreacted ethylene oxide [$(CH_2)_2O$], a process raw material (released/ exploded quantity was unknown/forensically undetermined)	Ethylene oxide vapours released in the facility atmosphere because the chemical reactor was opened too early (before the chemical process was completed). There was any kind of explosion protection ("εx", "Ex", "S") of all outer electrical installations and devices built in that facility part (!)	1 dead and 2 injured workers (during Sunday work shift). Serious devastation of facility and surrounding buildings of neighbouring industrial companies. The explosion could produce domino effects on a few neighbouring ethylene oxide storage tanks, with possible disastrous effects to the almost whole south-eastern densely populated residential area of the town district Peščenica.
Fire 22/23 September 1988 (night) Šibenik, 'Luka Šibenik', a sea port company	The eastern part of the port storage area 'Vrulje-zapad'	NPK complex mineral fertilizers in bulk, type 15-15-15,[c] 11,300 t (delivered 3 days earlier) and type 16-16-16, abt. 4,700 t, produced in the 'INA-Petrokemija', Kutina	Possibly heat of few Hg-lamps partially buried in the top of the few, too much high, storage heaps of the mineral fertilizer (lighting was on before and during the fire)[d]	Due to possible hazard from toxic gases and aerosols (mostly NO_x, NH_3, P_2O_5 and Cl_2), which developed during the fire, about 12,000 citizens were urgently evacuated from the city, and 65 persons (including 24 police officers and a few firemen asked for a medical help). There was a fear of the explosion until the information about the types of stored fertilizers was announced. Material damage of only mineral fertilizers was estimated to about a 100–150 thousand USD.

(continued)

Table 9.2 (continued)

Event, date and place	Process, storage or transport unit	HAZMAT (quantity)	Event cause[a]	Effects and consequences
1	2	3	4	5
Explosions 7 April 1994 Zagreb (Sesvete)	Croatian Army depot " Duboki jarak"/"Gliboki jarek"	Stored ammunition, explosive devices and different kinds of military arms and equipment	Hot works in the storage area (including poor storage process control and low work safety culture)	6 dead and >20 injured. The whole depot was totally destroyed in serial successive explosions. Extensive blast, fire, or from the expelled ammunition fragments and rocket projectiles, damage in the wider town area.
War attacks on petrochemical industry 1991–1995[e] 'INA-Petrokemija' Kutina	Large production and storages complex	Unavailable information (In peace time: 1.8 million t/year of installed capacity of fertilizers and 33,000 t/year of carbon black, as finished products)[f]	See footnote e; Čavrak (1995) and Kulišić (1992)	Highly endangered was not only the factory (with its more than 3,000 employees), but also the town of Kutina with all of its population, as well as the wider area and environment. Consequences for the people and the environment that could have followed can be compared with those from chemical weapons (NH_3), or blast of tactical nuclear warhead (NH_4NO_3, see for e.g. a case of Oppau 1921, Texas City 1948, Toulouse 2001, Neyshabura/Khorasan 2004, etc.).

Fire 1 August 2002 Zagreb (Jakuševac) Hazardous waste incineration facility «PUTO» d.o.o.	A roofed-over storage area with hazardous industrial waste (42×40 m)	Different kinds of hazardous industrial waste (few hundred tons)	Keeping mutually incompatible (exothermic reactive) waste chemicals and other hazardous materials, one with another (most probably self-heating and spontaneous ignition)	The accident caused a great concern of local and state authorities and citizens for the fear of air, drink water and soil pollution with unknown kinds and quantities of spilled and, during the fire, evaporated and dispersed hazardous substances (dioxins). Material damage due to fire was estimated to about a 300,000 € (totally about 1.5 million €). 5 accused responsible company persons were released due to investigative failure to determine/refuse a possibility of eco-extremists' motivated arson.
Hazardous (choky) gas release 23 February 2007 (evening) Karlovac (town district Dubovac) Croatian brewery	Facility building with 18 beer fermentation process units	Waste CO_2 (unknown quantity)	Unauthorized connection of pipeline for releasing of waste CO_2 to the facility's underground waste water collector (instead of former proper solution for CO_2 releasing in atmosphere from the high roof pipe of the facility building, as it was regulated in facility project blueprint)	Suffocation of 1 citizen (who died several days later) and immediate death of his dog, during a walk by the stream channel (Grabiću). The incident caused a great concern of citizens for the fear of unknown hazardous gas in the atmosphere of the promenade around the brewery. The company paid the compensation to the deceased's family.

(continued)

Table 9.2 (continued)

[a]According to the forensic experts' findings/opinion, special investigative state commissions or the assumptions of the above-mentioned authors or the mass media

[b](A) $ClSO_3H + H_2O \rightarrow HCl + H_2SO_4$; (B1) $HCl + Fe$ (tank shell) $\rightarrow FeCl + H_2$; (B2) $H_2SO_4 + Fe \rightarrow FeSO_4 + H_2$

[c]The Analyses at the Forensic Science Center, Zagreb, showed, among other things, that it was a AN mixture (about 40%), KCl (about 23%), K_2SO_4, several types of phosphate salts and aromatic base oil (for preventing granules sticking together), type "Amin HOE 2497" of the German company "Hoechst", which is added in the amount of 1,22 kg/t

[d]Fire detection system, earlier installed for storing cotton, sugar and other goods, is "due to the corrosive influence of mineral fertilizers dust", 3–4 years before the fire, out of order (partly de-installed, and partly turned off)

[e]Chronology of major attacks: December 31, 1991 – Air raid (two "Orao" planes rocketing); September 12, 1993 – MRL "Orkan" – type missiles with cluster war-heads; August 6, 1995 – Air raid (two "Orao" planes rocketing); September 8, 1995 – MRL "Orkan"-type missiles with cluster war-heads; September 18, 1995 – "Luna" missile; September 26, 1995 – "Dvina" aerosol explosive missile

[f]Besides these finished products, four basic chemicals are produced as intermediates but also as possible final products: ammonia (448,000 t/year), nitric acid (415,800 t/year), 98% sulfuric acid (495,000 t/year) and phosphoric acid (165,000 t/year). There are large storages for final products as for intermediates

9 The Benefits from Using Professionally Developed Models

investigators' and forensic work, and other written or oral rules of professions, which are used for investigations. Thus, managing the team of associates, planning, organizing and coordinating investigative activities, procedures and actions should be entrusted to eminent and skilful veterans of criminal investigation in this field, a kind of experienced 'general practitioners', experienced in a quick professional observation of clues, which are important for criminal investigation (identifying key "symptoms"), setting appropriate investigative hypotheses on time (possible "diagnoses") and consistent searching for likely and precise credible answers to all open questions by forensics teams, competent independent experts and other professionals (a kind of "specialist doctors"), from other possible useful and credible sources of information.

Consequences include innocent persons wrongfully bearing the consequences of fatal mistakes and failures, which they did not commit.

Knowledge and skilful application of modern analytical methods and techniques of hazard analysis and risk assessment during team development and examination of reliability and forensic sustainability of hypothesis models about likely accident scenarios, according to circumstantial and physical evidence, available to investigators, of manners, causes, conditions and circumstances of accidents with hazardous materials, can be of great importance for the success of criminal investigation.

It should be desirable to include educational topics into the curriculum of higher education for crime investigators and lifelong in-service training of crime investigators. This will help update the traditionally used investigative and analytical safety/security applications.

Modern rigorous requirements regarding the quality of HSS management, during production, processing, storing, transportation, distribution, handling and using hazardous materials (especially those types and amounts, important for security reasons), places a burden of obligations and responsibilities on all subjects responsible for and charged with general or specific issues of public or internal HSS and/or rescue and defence. First and foremost, the operators of facilities with hazardous materials (or facilities in which such substances can emerge during major accidents), on safety advisors for transportation of dangerous goods and on employees of state administrative bodies competent for administrative control and inspection, and protection and rescue.

Efficient, socially responsible and economically acceptable risk management of major accidents with hazardous materials, and the effectiveness of undertaken emergency measures and first-step procedures (intervention) at the threatening/escalating scenes of accidents with such substances, among others, mainly depend on the thorough professional knowledge of the likely worst case and the most probable scenarios of their occurrence. Identification reliability of structure components includes all possible kinds and types of scenarios of accidents with hazardous materials. Therefore, only attentive professional observation and elucidation of a complex structure, (inter)relations and interaction of numerous likely components of root, initial, contributive and main factors of manners, causes, conditions and circumstances of their occurrence and development, as well as those of technological and technical nature and also of human and organizational nature, can all substantially contribute.

In doing so, all possibilities of unauthorized interference (including criminal attacks to the process, storage and transport units with large amounts of hazardous materials) with really possible kinds, types, *modus operandi* and assault weapons. Such items are now available to modern terrorism, organized crime and individual extremists, should not be neglected.

Therefore, besides the expected, significantly higher safety and security level of the accidents with hazardous materials, i.e. a decrease in number of such accidents, the updated and neatly documented results, easily available to investigators, of undertaken preventive activities can be useful for criminal and forensic investigation in case such an accident with hazardous materials occurs.

References

API (2005) Security guidelines for the petroleum industry. American Petroleum Institute (API), Washington, DC

Barrs FG, Lynch ML (2004) Environmental crime: a sourcebook. LFB Scholarly Publishing, New York

Bragdon CR, Bennett JCW, Bruno TA, Cook CA, James SR, Jensen T et al. (2008) In: Bragdon CR (ed) Transportation security. Butterworth-Heinemann, Burlington/Elsevier, Amsterdam

Cavani F, Centi G, Perathoner S, Trifiró S (2009) Sustainable industrial processes. WILEY-VCH, Weinheim

Čavrak B (1995) Eko-terorizam: zračni napad na Proizvodni pogon mineralnih gnojiva "INA – Petrokemija" Kutina (Hrvatska) u kolovozu 1995 (Eng. Eco-terrorism: air-raid on mineral fertilizers production plant at "INA – Petrokemija" Kutina (Croatia) in August, 1995). Safety, Security and Environmental Protection Department, Zagreb

CCPS (2003) Guidelines for investigating chemical process incidents. Center for Chemical Process Safety (CCPS), American Institute of Chemical Engineers (AIChE), New York

CETU (2005) Guide to road tunnel safety documentation: risk analyses relating to dangerous goods transport. Centre d'Etudes des Tunnels/Tunnel Study Centre, Bron Cedex

Clements BW (2009) Disasters and public health: planning and response. Elsevier/Butterworth-Heinemann, Amsterdam

Cooper H, Baker P (2010) U.S. opens criminal inquiry into oil spill. The New York Times (NYT), 1 June 2010

Cooper H, Broder JM (2010) BP's ties to agency are long and complex. The New York Times (NYT), 25 May 2010

Cozzani V, Gubinelli G, Salzano E (2006) Escalation thresholds in the assessment of domino accidental events. J Hazard Mater 129(1–3):1–21

CRAIM/UNEP (2002) Risk management guide for major industrial accidents (Intended for municipalities and industries). Council for Reducing Major Industrial Accidents/Conseil pour la réduction des accidents industriels majeurs (CRAIM), Montréal

Crowl DA, Louvar JF (2002) Chemical process safety: fundamentals with applications (Prentice Hall International Series in the Physical and Chemical Engineering Sciences), 2nd edn. Prentice Hall PTR, Upper Saddle River

EC (1996) Council Directive 96/82/EC of 9 December 1996 on the control of major- accident hazards involving dangerous substances. Off J Eur Communities L 010:13–33

EC (1998) Directive 98/8/EC of 16 February 1998 concerning the placing of biocidal products on the market. Off J Eur Communities L 123:1–63

EC (2006) Regulation No. 1907/2006 of the European Parliament and of the Council of 18 December 2006 concerning the Registration, Evaluation, Authorisation and Restriction of

9 The Benefits from Using Professionally Developed Models 183

Chemicals (REACH), establishing a European Chemicals Agency, amending Directive 1999/45/EC and repealing Council Regulation (EEC) No. 793/93 and Commission Regulation (EC) No. 1488/94 as well as Council Directive 76/769/EEC and Commission Directives 91/155/EEC, 93/67/EEC, 93/105/EC and 2000/21/EC. Off J Eur Union L 396:1–849

EEC (1967) Council Directive 67/548/EEC on the approximation of the laws, regulations and administrative provisions relating to the classification, packaging and labelling of dangerous substances. Off J Eur Union 196:1–98

EEC (1993) Commission Directive 93/67/EEC of 20 July 1993 laying down the principles for the assessment of risks to man and the environment of substances notified in accordance with Council Directive 67/548/EEC. Off J Eur Union 227:9–17

EEC (1994) Commission Regulation (EC) No. 1488/94 of 28 June 1994 laying down the principles for the assessment of risks to man and the environment of existing substances in accordance with council regulation (EEC) No. 793/93. Off J Eur Union L 161:3–11

EUR 20418 (2003) Technical Guidance Document on Risk Assessment in support of: Commission Directive 93/67/EEC on Risk Assessment for new notified substances; Commission Regulation (EC) No 1488/94 on Risk Assessment for Existing Substances; Directive 98/8/EC of the European Parliament and of the Council concerning the Placing of Biocidal Products on the Market (TGD, Part 1–4), 2nd edn. Office for Official Publications of the European Communities, Luxemburg

EUR 22113 (2005) Guidance on the Preparation of a Safety Report to Meet the Requirements of Directive 96/82/EC as Amended by Directive 2003/105/EC (Seveso II). In: Fabbri L, Struckl M, Wood M (eds) Office for Official Publications of the European Communities, Luxemburg

Garrick JB, Christie RF, Hornberger GM, Kilger M, Stetkar JW (2008) In: Garrick JB (ed) Quantifying and controlling catastrophic risks. Academic/Elsevier, London

Gledhill J, Lines I (1998) Development of methods to assess the significance of domino effects from major hazard sites (Contract Research Report 183/1998). Health and Safety Executive Books, Sudbury

Gordon R, Flin R, Mearns K (2005) Designing and evaluating a Human Factors Investigation Tool (HFIT) for accident analysis. Saf Sci 43(3):147–171

Grossi P, Kunreuther H (eds) (2005) Catastrophe modelling: a new approach to managing risk. Huebner international series on risk, insurance, and economic security. Springer, Boston

Gunn AM (2008) Encyclopedia of disasters: environmental catastrophes and human tragedies, vol 1–2. Greenwood Press, Westport

Herrmann DS (2002) A practical guide to security engineering and information assurance. Auerbach/CRC Press, Boca Raton

Hommel G (2010) Handbuch der gefährliche Güter, 24th edn. Springer, Berlin

HSE RR509 (2006) Plant ageing: management of equipment containing hazardous fluids or pressure (Research Report 509). Health and Safety Executive (HSE), Colegate (Norwich, UK)

ICPO-OICP (2007) Illegal oil discharges from vessels: investigative manual. ICPO- OICP Interpol Communication and Publications Office, Lyon

Kletz T (2009) What went wrong? – Case histories of process plant disasters and how they could have been avoided, 5th edn. Gulf Professional Publishing, Burlington

Koehler SA, Brown PA (2010) Forensic epidemiology (International Forensic Science and Investigation Series). CRC Press, Boca Raton

Körvers PMW (2004) Accident precursors: pro-active identification of safety risks in the chemical process industry. Dissertation, Technische Universiteit Eindhoven, Eindhoven

Kristof ND (2010) New alarm bells about chemicals and cancer, NYT, 5 May 2010

Kulišić D (1992) Eksplozijske i požarne opasnosti od velikih količina amonij-nitrata i nekih njegovih smjesa - I. dio (Eng. Explosion and fire risks of great quantities of the ammonium nitrate and its mixtures – part I). Policija i sigurnost 1(49):322–337

Kulišić D (1997) Sabotaže i diverzije. (Eng. Sabotages and subversions). Policija i sigurnost 6(1–2):57–131

Kulišić D (1998) Uzroci nezgoda, nesreća, požara i eksplozija: Prijedlog sustava razvrstavanja uzroka požara, eksplozija, havarija i nesreća pri radu tehnološke naravi (Eng. Causes of accidents, disasters, fires and explosions: a proposition of a classification of causes of fire, explosions, damages and disasters in technological processes). Sigurnost 40(2):95–121

Kulišić D (2008a) Mjere sigurnosti od terorističkih i inih zlonamjernih ugroza *kritične infrastrukture* – II. dio (Eng. Measures of prevention from terrorist and other malicious activities of critical infrastructure – part II). Sigurnost 50(4):343–364

Kulišić D (2008b) Požarne i eksplozijske opasnosti u prijevozu opasnih tvari (Eng. Fire and explosion hazards of dangerous goods transport). In: Javorović A, Kovačić N, Kralj R (eds) Zbornik radova stručnog savjetovanja "Novi zakon o prijevozu opasnih tvari – cestovni i zračni promet" (Eng. Proceedings of the professional conference "New Transport of Dangerous Goods Act"). Biblioteka Sigurnost Educa, Zagreb, pp 9–32

Kulišić D (2008c) Mjere sigurnosti od terorističkih i inih zlonamjernih ugroza *kritične infrastrukture* – I. dio: Kako prepoznati i ugraditi nedostajuće mjere prevencije od terorističkih i drugih zlonamjernih aktivnosti u sustavu kritične infrastrukture (Eng. Measures of prevention from terrorist and other malicious activities of *critical infrastructure* – part I: how to recognize and build in missing measures of prevention from terrorist and other malicious activities into the systems of critical infrastructure.). Sigurnost 50(3):201–226

Kulišić D (2009) Opće mjere sigurnosti i zaštite pri skladištenju i rukovanju sa *sigurnosno posebno zanimljivim* opasnim tvarima (Eng. General safety and protection measures during storing and using hazardous materials, especially interesting for security issues). In: Kralj R, Javorović A (eds) Zbornik radova Stručnog savjetovanja "Prijevoz opasnih tvari – ADR 2009." (Eng. Proceedings of the professional conference "Transport of dangerous goods – ADR 2009"). Biblioteka Sigurnost Educa, Zagreb, pp 21–50, Posebno izdanje/Special edition

Kulišić D, Magušić F (2007) O aktualnim općim pristupima i elementima za raščlambe opasnosti od zlonamjernih ugroza sigurnosti prometnih tokova (Eng. About current general approaches to and elements of danger analysis in cases of malicious endangerment of traffic routes). Policija i sigurnost 16(1–2):41–66

Lancaster JF (2005) Engineering catastrophes: causes and effects of major accidents, 3rd edn. CRC Press/Woodhead Publishing, Boca Raton/Cambridge

Landoll DJ (2006) The security risk assessment handbook: a complete guide for performing security risk assessments. Auerbach/Taylor & Francis Group, Boca Raton

Lewis RJ (ed) (2004) Sax's dangerous properties of industrial materials, vol 1–3, 11th edn. Wiley, New York

Macaulay T (2009) Critical infrastructure: understanding its component parts, vulnerabilities, operating risks, and interdependencies. CRC Press, Boca Raton

Maguire R (2006) Safety cases and safety reports: meaning, motivation and management. Ashgate Publishing, Aldershot, Hampshire

Margossian N (2006) Risques et ccidents industriels majeurs: Caractéristiques, réglementation, prévention. Dunod, Paris

Marshall VC (1987) Major chemical hazards. Ellis Horwood, Chichester

McDonald R (2003) Introduction to natural and man-made disasters and their effects on buildings. Architectural Press, Oxford

Medard LA (1989) Accidental explosions, vol 1–2. Ellis Horwood, Chichester

Morrison R (1999) Environmental forensics: principles and applications. CRC Press, Boca Raton

Mudge SM (2008) Methods in environmental forensics. CRC Press, Boca Raton

Murphy BL, Morrison RD (2007) Introduction to environmental forensics, 2nd edn. Elsevier Academic, Amsterdam

Nacionalna strategija kemijske sigurnosti (2008) (Eng. National strategy for chemical safety). Službeni list Republike Hrvatske, br. 143/2008

Nacionalna strategija za prevenciju i suzbijanje terorizma (2008) (Eng. National strategy for prevention of terrorism and for counterterrorism). Službeni list Republike Hrvatske, br. 139/2008

Nacionalna strategija zaštite okoliša (2002) (Eng. National strategy for environment protection). Službeni list Republike Hrvatske, br. 46/2002

Newton M (2008) The encyclopedia of crime scene investigation. Facts On File, New York

9 The Benefits from Using Professionally Developed Models

NFPA (2008) Fire protection handbook, vol 1–2, 12th edn. National Fire Protection Association, Quincy

NRC (2006) Terrorism and the chemical infrastructure: protecting people and reducing vulnerabilities. National Research Council (NRC), Committee on Assessing Vulnerabilities Related to the Nation's Chemical Infrastructure. The National Academies Press, Washington, DC

OECD (2003) OECD guiding principles for chemical accident prevention, preparedness and response, 2nd edn. Organisation for Economic Co-operation and Development (OECD), Paris

Owen C, Béguin P, Wackers G, Rosness R, Norros L, Nuutinen M et al (2009) In: Owen C, Béguin P, Wackers G (eds) Risky work environments: reappraising human work within fallible systems. Ashgate Publishing, Farnham, Surrey

Paik JK, Melchers RE (eds) (2008) Condition assessment of aged structures. CRC Press/ Woodhead Publishing, Boca Raton/Cambridge

Piquero AR, Weisburd D, Addington LA, Apel RJ, Ariel B, Berk R et al. (2010) In: Piquero AR, Weisburd D (eds) Handbook of Quantitative Criminology. Springer, New York

Pravilnik o izradi procjene ugroženosti od požara i tehnološke eksplozije (1994) (Eng. Rulebook on drafting fire and technological explosions risk assessment). Službeni list Republike Hrvatske, br. 35/1994, 110/2005, 28/2010

Pravilnik o metodologiji za izradu procjena ugroženosti i planova zaštite i spašavanja (2008) (Eng. Rulebook on methodology for risk assessment and plans for the protection and rescue). Službeni list Republike Hrvatske, br. 38/2008

Pravilnik o registru postrojenja u kojima je utvrđena prisutnost opasnih tvari i o očevidniku prijavljenih velikih nesreća (2008) (Eng. Rulebook on inventory of the dangerous substances present in the 'Seveso Establishment' and on information to be supplied by the operator following a major accident). Službeni list Republike Hrvatske, br. 113/2008

Pravilnika o sadržaju plana zaštite od požara i tehnoloških eksplozija (1994) (Eng. Rulebook on the content of fire and technological explosion protection plan). Službeni list Republike Hrvatske, br. 35/1994, 55/1994

Ramachandran V, Raghuram AC, Krishnan RV, Bhaumik SK (2005) Failure analysis of engineering structures: methodology and case histories. ASM International, Materials Park

Rasmussen J, Svendung I (2000) Pro-active risk management in a dynamic society. Swedish Rescue Services Agency, Karlstad

Reid RN (2005) Facility managers guide to security. Fairmont Press, Lilburn

Kazneni zakon Republike Hrtvaske (1997) (Eng. Criminal code of the Republic of Croatia). Službeni list Republike Hrvatske, br. 110/1997, 27/1998, 50/2000, 129/2000, 51/2001, 111/2003, 190/2003, 105/2004, 84/2005, 71/2006, 110/2007, 152/2008

Richardson M (1995) Effects of war on the environment: Croatia. E & FN Spon/Chapman & Hall, London

Sanders RE (2005) Chemical process safety: learning from case histories, 3rd edn. Elsevier Butterworth–Heinemann, Burlington

Smith DJ (2001) Reliability, maintainability and risk: practical methods for engineers, 6th edn. Butterworth-Heinemann, Oxford

Spurgin AJ (2010) Human reliability assessment: theory and practice. CRC Press, Boca Raton

Uhlmann DM (2010) Prosecuting crimes against the earth. The New York Times (NYT), 3 June 2010

UNECE (2009) Globally Harmonized System of classification and labelling of chemicals (GHS),Third revised edition, United Nations Economic Commission for Europe (UNECE). http://www.unece.org/trans/danger/publi/ghs/ghs_rev03/03files_e.html. Accessed 28 May 2010

UNEP (2000) TransAPELL guidance for dangerous goods transport emergency planning in local community (Technical Report 35). United Nations Environment Programme/Division of Technology Industry and Economics, Paris

UNITAR (2008) Understanding Globally Harmonized System of Classification and Labelling of Chemicals (GHS): a companion guide to the purple book. United Nations Institute for Training and Research (UNITAR), Geneva

186   D. Kulišić

Urbina I (2010a) Conflict of interest worries raised in spill tests. The New York Times (NYT), 20 May 2010

Urbina (2010b) Inspector General's inquiry faults regulators. The New York Times (NYT), 24 May 2010

Uredba o sprječavanju velikih nesreća koje uključuju opasne stvari (2008) (Eng. Regulation on control of major industrial accident hazards). Službeni list Republike Hrvatske, br. 114/2008

USCSB (1999–2010) Final Investigation Reports, U.S. Chemical Safety and Hazard Investigation Board. CSB Investigations. Resource document. http://www.csb.gov/investigations/default.aspx. Accessed 28 May 2010

WHO (2009) Manual for the public health management of chemical incidents (Inter- Organisation Programme for the Sound Management of Chemicals – IOMC). World Health Organization (WHO), Geneva

Wortley W, Mazerolle L (2008) Environmental criminology and crime analysis. Willan Publishing, Cullompton, Devon

Zakon o kemikalijama (2005) (Eng. Chemicals Act). Službeni list Republike Hrvatske, br. 150/2005, 53/2008

Zakon o prijevozu opasnih tvari (2007) (Eng. Transport of Dangerous Goods Act). Službeni list Republike Hrvatske, br. 79/2007

Zakon o prostornom uređenju i gradnji (2007) (Eng. Physical Planning and Construction Act) (2007). Službeni list Republike Hrvatske, br. 76/2007

Zakon o zaštiti i spašavanju (2004) (Eng. Disaster Protection and Rescue Act). Službeni list Republike Hrvatske, br. 174/2004, 79/2007

Zakon o zaštiti od požara (2010) (Eng. Fire Protection Act). Službeni list Republike Hrvatske, br. 92/2010

Chapter 10
Environmental-Security Aspects of Explosion in the Ammunition Storage: Opportunity for Policy Making

Zoran Keković

Abstract Since the connection between security and environmental issues has became evident, accidents in military storages as a source of environmental threats require new trend in crisis management policy and approach. This chapter presents evidence of complex of interconnected military and environmental issues based on a large accident dataset from the explosion in the ammunition storage facility located in "Paraćinske Utrine" at the Karadjordje hill. The accident examined complex situation preferable for risk assessment of storaging hazard materials and crisis management tools and methods that provide the basis for the application of methods, as well as extension new ideas for policy-making purposes. That is very important within the context of transportation and environmental issues of regional project Coridor 10. Additionally, findings stress issues of unification and harmonisation national regulation with the European Union legislation.

10.1 Introduction

The accident in "Paraćinske Utrine" storage is an example of a complex situation from the viewpoint of an explosive material stockpiling risk assessment, an accident in a hazardous material storage, after the effect of removing as well as hazard brought about by unexploded explosive weapons and ammunition. The situation which arose from the explosion at the storage facility in Paraćinske Utrine territory had the attributes of a crisis situation that called for the making of a series of important decisions in order to secure the affected population's safe living and working. Pearson and Clair (Pearson and Clair 1998) define crisis as an "event of small probability and great consequences, which imperils the life of an organization, being characterized by unclear causes, effects and means of solution, as well as a conviction that decisions must be made quickly".

Z. Keković (✉)
Faculty of Security Studies, University of Belgrade, Belgrade, Serbia
e-mail: zorankekovic@yahoo.com

G. Meško et al. (eds.), *Understanding and Managing Threats to the Environment in South Eastern Europe*, NATO Science for Peace and Security Series C: Environmental Security, DOI 10.1007/978-94-007-0611-8_10, © Springer Science+Business Media B.V. 2011

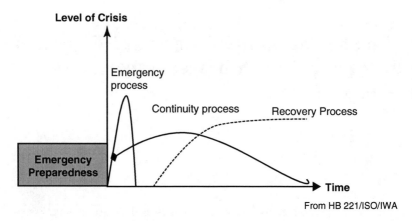

Fig. 10.1 Stages of managing the accident (Gibson et al. 2004; Raisch et al. 2006)

The situation also had the attributes of a disaster (Kešetović et al. 2009) with the challenges being numerous, complex and diverse. The multidimensional characteristics of the situation were linked with the engagement of a number of actors (public and private) depending on the stage of the situation: before, during and after the accident (Fig. 10.1). In this sense the situation was also a social construction, i.e. a cultural myth granted the status of a physical phenomenon.

During the course of the accident, the number of actors progressed. Before it occurred, the critical activity was under the jurisdiction of Ministry of Defence (MOD) and directly monitored by the appropriate ministry services. This part is functionally presented as *emergency preparedness* and includes everything the organization had done to prevent the occurrence of a harmful accident, or to minimize the negative effects in case of its occurrence. The accident is point ground zero from which three processes begin. The first one is *emergency process* which has the most powerful influence on the subsequent organization activities. On one hand, the local self-management and fire fighting units, the use of which is stipulated by the laws of Republic Serbia, and pre designed operative protocols were legally and actually limited in their response by the fact that the proprietor of the facility is the Serbian MOD. Also, there was the access restriction order to danger zones issued by the Military Police regarding the fact that fire activates the unexploded explosive weapons and ammunition left behind. This directly threatens the security of the storage facilities and objects in close proximity.

This is followed by *continuity process* as strategic and tactical capability, pre-approved by management, of an organization to plan for and respond to conditions, situations and accidents in order to continue operations at an acceptable predefined level. This primarily refers to the protection of military installations on the premises, which also falls under the jurisdiction of military security organization and services. Finally, the recovery process is supposed to facilitate the rehabilitation or reconstruction of life conditions and work activities on the premises or in the wider

10 Environmental-Security Aspects of Explosion

area that suffers the effects of the accident. In the actual case, the recovery process refers to the activities of demining performed by the Ministry in charge in collaboration with professional teams, domestic as well as from abroad and to the end of eliminating the accident after-effects.

The accident itself happened in a close proximity to a highway and the Pan European Corridor 10 with all the effects relating to the continuity and security of traffic along this important international traffic artery. Considering the proximity of urban zones, there was the danger of directly jeopardizing the people and material goods. Last but not least, the Paraćinska Utrina explosive weapon and ammunition storage is located in an area of the ecosystem which contains woods, water, arable land, and others values, all potentially endangered in the aftermath.

In the absence of explosive material stockpiling security pre-requisites, non-existence of both public as well as private sector unified policy and strategy of emergency situation management in compliance with the modern international norms and standards, the hazard of stockpiling dangerous material are still present. Location, construction and use represent physical hazards that affect property (Baranoff 2004). The difference between the public and private, profit and nonprofit segment of security risk management lies in the fact that the politics within the private and profit sector may be long-term, covering several years, that it is determined by profit interests, and that it remains that way while, on the other hand, the politics of the government administration may change due to the change of governing elite. Tait (2001) distinguished between interest-based and value based responses to risk. Interest-based conflicts are restricted to specific issue, and can be resolved by providing specific information or compensation, which value-based conflicts are rooted in normative beliefs and are more difficult to resolve. Interest-based motivation also displays economic or instrumental rationality (Tait 2001).

10.2 Historical Context and the Accident Proceeding

10.2.1 The Original Status of Explosive Material Stockpiling

The accidents prior to the explosion in Paraćin paint a vivid picture of the security, political, and economic conditions which were present in the background leading to this accident.

During the NATO air strikes on the Republic of Serbia's territory in 1999, $26,500\,m^2$ explosive weapons and ammunition stockpiling facilities were destroyed and damaged. After the bombing had stopped, the explosive weapons and ammunition were brought back to the storage facilities, their capacity had been reduced. This made the process of stockpiling, guarding, and monitoring the stockpiled explosive weapons and ammunition even more difficult. The problem of stockpiling the remaining explosive weapons and ammunition, needed by the MOD, emerged. Thus, a part of the explosive weapons and ammunition lot was kept in the open.

Along with the risk of uncontrolled detonations and theft, keeping explosive weapons and ammunition in the open shortens their aspiration date as well as accelerates element degradation. Additionally this pyrotechnically loads the infrastructure of the storage facilities as well as their surroundings. The unfortunate fact is that an explosives and ammunition can never be totally secure in the sense of "risk existence" so the acceptance of risk as "tolerable" is the optimal approach to any strategy of explosive material stockpiling risk management. However, when we talk about this kind of critical infrastructure, the danger lies in the actual presence of ammunition and explosives, and this kind of risk primarily depends on the storage infrastructure and its surroundings. In the absence of infrastructure – facilities, the explosive weapons and ammunition management is exercised in compliance with adequate standards and by applying the best practice.

Due to the shortage of financial means, the stockpiling of such materials has become a problem that could not be solved in a short time; at least not by meeting the necessary security, legal, and other requirements, and other aspects of stockpiling. This area was regulated by Guidelines for Stockpiling and Handling Ammunition and Explosive Weapons TU-V3 (edition 1970) and Technical Ammunition Storage (edition 1989). The legislation based on the UN Committee of Experts guidelines applied until 2002, when a new one – TU-V, 5105, *Guidelines for Explosive Weapons and Ammunition Stockpiling* from 2002 was created upon the suggestion of UN experts. At the moment of accident in question, the mentioned Guidelines, and *the Act on Explosive Materials and Inflammable Liquids and Gases* (Zakon o eksplozivnim materijama, zapaljivim tečnostima i gasovima 1977) had been enacted.

The "Paraćinske Utrine" storage facility is situated on "Karađorđevo Brdo" location at a close proximity of E-75 highway and only several kilometres away from urban zones (app. 2.5 km North-East of Paraćin, and 5km East of Ćupria). The storage is an industrial design and of medium capacity. Prior to the explosions, the facility encompassed a group of administrative buildings (the headquarters, dormitories and other various purpose facilities) and the technical part of the installation where industrious design allowed above ground storage for stockpiling explosive weapons and ammunition. The technical buildings and platforms for explosive weapons and ammunition keeping had prescribed protective embankments around them and adequate protection from atmospheric electric discharge in compliance with international standards. When the protection of this kind of facilities is in question, military standards apply due to their military nature, as it is the case in the majority of other countries. However, during 1999, a number of storage buildings had been demolished, causing a considerable reduction of storage capacity assigned for explosive weapons and ammunition keeping. Instead of those previous buildings, platforms for open air explosive weapons and ammunition keeping were built. At the moment of the explosion, the storage harbored 4,110 tons of various explosive weapons and ammunition.

10.2.2 The Description of the Accident and the Activities Performed During and Immediately After the Explosion

The explosion at "Paraćinske Utrine" storage facility took place early in the morning on October 10, 2006 (03:47 h). Having reviewed the available prints, the MOD experts established without doubt that there were seven powerful blasts followed by intensive light, fire, and secondary explosions that caused damage to the storage facilities as well as to infrastructure objects outside the facility. The fire build-up caused a series of secondary explosions, fires, and rocket liftoffs. At 05:56 h an extremely powerful explosion ensued, causing material damage to various objects in Paraćin and Ćupria. After the explosions had stopped, 2,810 tons of explosive weapons and ammunition with a low degree of damage remained at the storage facility.

The time window for a response was very limited both to restrain the fire and to eliminate the remaining unexploded explosive weapons and ammunition on the highway and outside the storage facility. The fire at the storage facility progressed at a rapid pace outside the facility resulting in a series of secondary explosions of different power and intensity. A strong wind that was blowing at the time also facilitated the fire spreading so quickly. Due to the heavy damage that the roofs of the storage buildings had suffered, the explosive weapons and ammunitions were exposed to the immediate effects of meteorological conditions. The fire produced an enormous cloud of smoke that covered the Paraćin and Ćupria Valley and could be seen for kilometres around. Before an adequate assessment was made, it was uncertain whether the smoke contained any dangerous substance that could pose a health threat to population.

Due to thick smoke and difficult traffic, the *police* soon closed the Pan European Corridor 10 (the highway) and took measures to block a wider area around point ground zero of the explosion – the storage facility. Several hours after the explosion, the Defence Minister, together with the Minister of Police, Military, and Security-Intelligence Agency representatives arrived to be briefed at the site, and take the necessary measures. Eminent MOD and Serbian Army Joint Chiefs of Staff experts were commissioned to make an insight into the status of unexploded explosive weapons and ammunition, and decide on the further procedure regarding those weapons and ammunition remaining in the facility objects.

Immediately after the explosion at the Serbian military storage facility on Karađorđe Hill, the Municipal Informing Centre was activated to provide more efficient coordination. A permanent telephone and radio connection to all MOD structures, as well as to all authorized state organizations in the territory of Paraćin and Ćupria Communities was established by way of this Center. The citizens were briefed by means of state and local media as well as by posted warning signs about the danger from unexploded explosive weapons and ammunition. They were further told of how to act in case of coming across any of those and in the meantime, the citizens residing in the nearest settlements were evacuated. When Paraćin Civil Protection Municipal Headquarters and the Red Cross units for giving response to disaster activated at the same time, a large group of citizens gathered together in front of the Paraćin Municipal Hall early that morning, seeking information about the accident and possible effects on their health and the environment.

10.2.3 Eliminating the After-Effects

Activities to eliminate after-effects may be categorized as:

1. Determining possible causes of explosion – lab analyses
2. Locating, marking and destroying unexploded explosive weapons and ammunition
3. Enabling life and business activity continuation in endangered territory

10.2.3.1 Determining Possible Causes of Explosion: Lab Analyses

The perimeter was contaminated with a large number of scattered unexploded explosive weapons and ammunition. In order to gain insight into the status of 20 mm anti aircraft ammunition, a commission was formed consisting of eminent MOD and Joint Chiefs of Staff experts. The research conducted was aimed at determining whether there was a transfer of fire from one cartridge to any others in the container in which 20 mm antiaircraft ammunition was kept at the moment the powder caught fire in the first cartridge. It also wanted to determine whether there was a transfer of fire from one container of cartridges to any others in the stockpile. The findings of the research established for a fact that when there is a firing or self combustion of powder in one cartridge within a case of cartridges, it causes a mechanical damage to the cartridge, but the tracer does not activate and there is no impulse transfer to any other cartridge in the container, nor there is any possibility of impulse transfer from one cartridge container to any other of the 20 mm ammunition containers in the stockpile. By analyzing the findings of control testing it may be concluded that in this period of time there were no manufactured series of 20 mm ammunition that failed to meet the quality requirements set by the regulations of product quality (PQR). The testing was done on representative samples taken from explosive weapons and ammunition storage facilities at different locations, among which were the samples of ammunition from "Paraćinske Utrine" storage facility (Technical Repair Institute 2006).

10.2.3.2 Locating, Marking and Destroying Unexploded Explosive Weapons and Ammunition

In order to gain insight into the explosive weapons and ammunition status after the explosion and to make a decision about how to further proceed regarding the

10 Environmental-Security Aspects of Explosion

explosive weapons and ammunition in the facility, a commission was formed consisting of eminent MOD and Joint Chiefs of Staff experts. The commission came to the site and established that the explosive weapons and ammunition in the remaining facility were not directly exposed to the effects of the blast. The commission assessed that in the remaining facilities there were no explosive weapons and ammunition damaged to such a degree that their moving and shipping to a place of destruction would pose a great human and environmental risk. The zone contaminated with unexploded explosive weapons and ammunition was assessed to be approximately 800 hectares large, out of which some 500 hectares were farming land on which there were some summer cottages, and approximately 300 hectares of wooded terrain (The Civil Protection Municipal Headquarters 2006).

Special search and rescue teams immediately started work on repairing the damage caused by the explosion. The following critical tasks were performed in the first days after the accident:

1. The E-75 highway was cleaned from the effects of the explosion and opened for traffic.
2. Explosive Ordnance Teams were formed to defuse and remove the explosive. They did their field work upon citizens' reports.
3. Nearby settlements and local roads were swept clean from the remaining unexploded explosive weapons and ammunition.
4. High voltage long distance power lines were cleared.
5. Overhaul of remaining unexploded explosive weapons and ammunition was done as well as their relocation to other MOD storage facilities.

In accordance with the memorandum of understanding signed by Republic Serbia (RS) Government, UNDP, and UN representatives, explosive device activation and detonation equipment and accessories were received in order to clean the explosion site.

10.2.3.3 Enabling Life and Business Continuation

The explosion in Paraćin military storage facility caused damage on 4,740 different objects with the total of 13,375 damages of different kinds (The Civil Protection Municipal Headquarters 2006).

In compliance with Art. 73 of *The Natural Disaster Act* (Zakon o elementarnim nepogodama 2005), The Lord Mare of Paraćin formed *The Municipal Operative Headquarters for Helping the Citizens* after the explosion. It was done in a military warehouse in the "July 7" settlement located in Paraćin, on October 19, 2006. The tasks of this Headquarters was: to make a list of families who needed urgent help, make records of the damage to public and privately owned objects as well as to collect material help gathered through the Permanent Conference of Cities and Municipalities in Serbia. In the meantime, a commission for survey and assessment of the damage to agricultural commodities, land, objects, forests, and cattle was formed. This was done in compliance with *The Guidelines for Unified Methodology*

194 Z. Keković

of Assessing the Damage Caused by Natural Disaster (1987). Also, a commission for monitoring the works on repairing the objects after the explosion was formed. The Headquarters held a number of meetings with the Ministry of Finance, and Ministry of Capital Investment representatives, as well as with private entrepreneurs and work contractors.

10.3 Ecological Effect of the Accident

In order to understand the decision made by these crisis situation actors, they must be placed into the context of citizens' life and property protection, primarily from the aspect of danger that unexploded explosive weapons and ammunition outside a storage facility pose to individuals who are not familiar with such explosive devices.

Explosive materials in Serbian military industry are categorized according to their effects and aero-thermo-chemical and technological properties to: explosive materials; powders; solid rocket propellants; liquid rocket propellant oxidants; liquid rocket propellants, and pyrotechnical mixtures. The main property of these materials is that they are explosive, reactive, or explosive reactive. The modes of their stockpiling are different depending on the mentioned categories. The technological procedures and ecological risks are also different ranging from their effect on air pollution to harming people's health; effect on water pollution; effect on soil pollution; and effect on long term stockpiling and raw material and final product safe keeping. Explosives and explosive materials pose the main threat. The explosives used in the defense industry in Serbia are classified into so called initial explosives (mercury fulminate, led acid, led tetra rezorcinate, and the like) that are used for powder charge production, and while used in small quantities, they are the most volatile. Therefore, there exists a danger from explosion when inappropriately handled. The explosives with which shells and large caliber bullets are filled make the second group. They are as follows: TNT, hexogen, octogen, and pentrite. The quantity of which is several hundred times over larger respective the initial explosives quantity because they produce the main energy in war projectiles fired on military targets. The third type of explosive material is powders and rocket propellants. The purpose of these substances is not is to detonate, but to propel the projectile (OSCE 2008).

The procedure of explosive material destruction is an ecological risk from two aspects. The first is the explosion and combustion products launched into the atmosphere, and the solid residuals (ashes and sediments) that remain on, or in the ground. The second aspect is noise, vibrations, and tectonic effect on the surroundings, or even the unwanted mechanical effect of the gusts of air caused by TNT explosion. Unexploded explosive material destruction falls within the jurisdiction of MOD. To this end, different sites are used that are subject to security and ecological provisions. The effect of such sites on the surroundings is usually insignificant provided that any contact closer than the one stipulated is avoided.

10 Environmental-Security Aspects of Explosion

The other ecological aspect that is not the subject of explicit research in this chapter is protection from unwanted noise and vibrations. This is made possible by building protective embankments, deflectors, and other civil construction installations by which the mechanical-tectonic effect originating from the process of explosive and inflammable waist destruction is prevented.

Measuring and research work done by authorized Ministry of Environment Protection organs showed that there was no poisoning or presence of any particles that would have caused air pollution. The total of searched ground area and amount of found and destroyed unexploded explosive weapons and ammunition is as follows:

1. 500 hectares of farming land and approximately 299 hectares of wooded land (100% of farming land; 99.66% of wooded land, which makes 99.88% of the search zone surface).
2. Detailed, under surface search (using metal detectors), 95.8 hectares of farming land that previously had been searched on surface level; 141 pieces of unexploded explosive weapons and ammunition had been found it that search.
3. The total of 89,649 pieces of unexploded explosive weapons and ammunition were found and destroyed.

Regional disposition and possible ecological impact of the explosion from the ammunition storage facility in Paraćin are very important. From the aspect of a site at which explosive materials are destroyed, the "Paraćinska Utrina" area is unsuitable due to being located in a suburban environment, and in the vicinity of Corridor 10. A certain degree of risk may be affixed to storage facility unfavourable geo-topological location, but any wider, regional threatening, larger than the town parameter ought not to be considered. However, risks do not simply exist, they are taken, run, or imposed and are inextricably linked with personal relationships. Risk analysts and other experts tend to emphasize the size of risks, while the public more often questions the potential for improbable but calamitous disasters. To communicate more effectively with the public, risk analysts need to deal with social and moral issues that are on the public's agenda (Glendon et al. 2006).

On the basis of the presented above, it may be concluded that:

(a) There is no ecological threat originating from explosive materials stockpiling in "Paraćinska Utrina" storage facility because this facility is not in function anymore.
(b) Ecological risk stemming from waste material destruction is expected. Therefore, in compliance with security provisions and other measures are taken that minimize this kind of risk.

To conclude, the effects of the explosion at ammunition storage facility in Paracin do not pose any ecological danger from air pollution with particles to any other matter (in the form of sedimentation in the water or soil, or dissemination done by wind).

10.4 The Lessons Learned

The proceeding of a crisis and its outcomes depend predominantly on the measures taken by authorized organizations. The key elements of the turnover in a crisis are decisions made by the Defense Minister and the Serbian Army Commander of the Joint Chiefs of Staff on the basis of action plan predesigned by specialists, and given priorities in the process of eliminating the crisis. Positive and negative experiences gathered in dealing with a negative event are a source of knowledge significant for future activities organizing in risk management and crisis management.

1. Stockpiling explosive material *risk management* demands hazard reduce politics and procedures, nontechnical: demographic conditions and cultural background, source of income and less control in the distribution of inhabitants (tend to move to the source of hazard); as well as technical: nature of hazard mitigation is reactive rather than preventive; less comprehensive of communication system; uncertainties level of hazard.
2. The existing Serbian Army storage facilities built for stockpiling Class 1 volatile materials are significant potential accident sites. The effects of such situations may be catastrophic in view of endangering human lives as well as large material loses and human environment jeopardizing.
3. The existing storage facilities need reurbanization along with building new storage facilities for the stockpiling of explosive weapons and ammunition.
4. Since risk management is concerned with minimizing losses, it follows that peril and vulnerability must be managed. In practice, peril can only be managed to a limited degree (Banks 2005). To ensure security of every storage facility and infrastructure, it is necessary to change the stockpiled material structure and ensure larger distance between the stockpiles kept inside.
5. Mitigation activities need to be improved: understanding the impact and trigger of hazard in area at risk, define the source of information and information flow, identify source of hazards, trigger, magnitude and frequency of hazard, define alert levels and expected community response, define constraint of urban development and regulation to support the condition.
6. Building capacity of communities should include: understanding the type and characteristics of the hazard, recognizing the source and direction of the hazard as well as the area at risk, knowledge of settlement locations, understanding of alert levels and information flow and finally the strengthening of communication and coordination amongst local communities.
7. According the principle of subsidiarity, a crisis should be managed at the lowest possible level. Within the context of the complex military and environmental issues, *crisis management* approach should not be stakeholder driven which derives critical input from the public, emergency personnel such as those initiating the warning process and those responding to emergencies, industry – partners in warning dissemination; and then finally the government which plays a supporting role, facilitating and enabling from behind the scenes to ensure the needs of all stakeholders are met. Nuclear incidents and security policy are

exempt from this rule, as they are handled centrally (Emergency Management Manual 2007).

8. After the accident, pressure from the local administration and influence by the local population increased preventing them from building any objects in the protective zones close to storage facilities.

9. Lab analyses excluded technical factor as the cause of accident, and the fact that its real cause has not yet been identified arouses public interest and a debate as to human error being factor as the cause of the accident. The simplified concept that reduces security to the control of physical access to an organization and the monitoring of property movements has recently started to broaden to include issues of health and safety, and components of risk evaluation and management, paying attention to a broader array of risks (forgeries, terrorism, emergency situations).

10. The findings presented in the preliminary assessment given by responding subjects in the zone of regional corridor K-10 in the Republic of Serbia point to the fact that there are subjects, resources, and other capacities that may be considerably affected by an accident, but they are not integrated in the integral protection system. This assumes that dangerous material stockpiling risk management capacities, and the existing danger proactive response capacities in the zone of K-10 motor way most endangered parts must be dislocated.

10.5 National Legislation and Levels of Responding to Accident in Republic Serbia

The area of emergency situation prevention and management in RS is stipulated by a number of laws and acts such as the Emergency and Civil Protection Act (Zakon o vanrednim situacijama 2009), Fire Protection Act (Zakon o zaštiti od požara 2009), Environmental Protection Act (Zakon o zaštiti životne sredine 2004), Local Self-Government Act (Zakon o lokalnoj samoupravi 2007), etc.

The competent law in this area is The Emergency and Civil Protection Act (Zakon o vanrednim situacijama 2009). Acting upon, proclaiming, and managing an emergency is regulated by this law. It also regulates the system of protecting and rescuing people, material goods, and cultural artifacts from natural disasters, technical-technological disasters – accidents and catastrophes, terrorist act after effects, war and other major disasters. It also regulates state organization competences, and the competence of autonomous provinces as well as local self-management, the role and participation of police and the Serbian Army in giving protection and rescuing. It regulates the rights and obligations of citizens, industrial associations, and other legal entities as well as entrepreneurs in regards with emergency situations. It also manages the organization and activities of the Civil Protection on protecting, rescuing, and removing the effects of natural and other disasters. The law stipulates funding, inspection supervision, international collaboration, and other issues relevant to the organization and functioning of protection and rescue system.

From the aspect of human environment protection, the Human Environment Protection Act (Zakon o zaštiti životne sredine 2004) from 2004 is in power. The process of this law enactment was followed by enactments of three more laws so that the system of actual law in this area could be completed. Those three laws enacted as a package are: the Act on Effect on Human Environment Assessment (Zakon o proceni uticaja na životnu sredinu 1992), Act on Effect on Human Environment Strategic Assessment (Zakon o strateškoj proceni uticaja na životnu sredinu 2004), and the Human Environment Polluting Integrated Prevention Act (Zakon o integrisanom sprečavanju i kontroli zagađivanja životne sredine 2004).

In RS legislation, *dangerous materials* are defined as chemicals and other materials that have harmful and dangerous properties. Any legal and/or physical entity that handles dangerous materials or applies technology dangerous to the human environment is requested to take all necessary measures to reduce the risk from any danger to the smallest possible measure. These measures are stipulated by the mentioned package of laws, but also by a set of legislative acts.

In the part *Measures of Protection from Dangerous Materials*, the Environmental Protection Act (Zakon o zaštiti životne sredine 2004) stipulates the modes of handling dangerous materials, production, shipping, processing, stockpiling, and disposing of it so as not to endanger people's lives or health, pollute the environment, and to provide for taking all other measures required by the law. According to this law, any legal and/or physical entity that produces, trades, uses, processes, stockpiles, or disposes of dangerous materials is liable to: design a plan of accident protection; take preventive and other measures of risk of accident management stated in the Plan of Accident Protection; and make a report on the security status available to the public.

Test areas that are under the competence of factories, big overhaul workshops, the Military with its own management programs all fall within a special category. Ecological problems may appear in the test areas where, in most cases, incineration of waste ammunition and deactivated explosive charges is completed. Normally, the quantities of this kind of materials to be burnt are determined in such a way that the gas combustion products do not go beyond the security zone of test area parameter, but the soil pollution after the atmospheric fallouts and its melting, as well as underground effects are subject to control according to special military regulations the results of which are not known. The categories of test areas managed by the MOD are quantified and standardized by military control programs and according to the documents from which they are governed.

In the next period of time, the endorsement of the following doctrinarian-strategic documents is expected which relate to emergency situations: a strategy of emergency situation management, assessment of threats to RS territory from natural and technical-technological catastrophes and big disasters, threats to critical national infrastructures assessment, etc.

According to the current legislation, the organization of responding in emergency situation is divided into six levels according to territorial jurisdiction and zones of responsibility: (1) Citizen responds individually according to their conviction and

10 Environmental-Security Aspects of Explosion

obligation; (2) Operator, own forces responding on site; (3) Local level, local forces responding outside the site in their own zone of responsibility – endangered zone; (4) City/county level joint local forces responding; (5) Republic/national level, engaging wider – national resources in endangered zone; (6) International level of responding within the framework of international collaboration and aid.

The groups or individuals who respond to emergency situations are not integrated at present in a unified system of protection and rescue. These groups are as follows: government subjects – the ministries, economy subjects, special organizations and ministries, public service organizations, nongovernmental sector, etc.

The subjects of the *Ministry of State Administration and local self-government* ought to respond immediately after the operators in all stages of emergency situation in their zone of responsibility. Local self-management plans, adopts and enacts *the Plan of Response and Protective Measures in Emergency Situation* in the local self-management zone of responsibility. The resources and capacities of public municipal enterprises and services belong to local self-management that forms public enterprises such as municipal, water industry, trading, agricultural, etc.

Subjects in the *Ministry of Human Environment and Spatial Planning* are as follows: The Risk Management Section, Chemical Accident Response Section, Water and Soil Protection Section, Collaboration with International Subjects and Local Administration Section, Human Environment Protection Inspection, and Human Environment Protection Agency.

The subject of the *Ministry of Infrastructure, Ministry of Health, Ministry of Agriculture, Forestry and Waters, and Ministry of Justice* play key roles in emergency situations. The subjects of the *Ministry of Infrastructure* play the important role of risk prevention and accident responding. The finishing of K-10 corridor road section that runs through Serbia falls within a special competence of this Ministry.

The *Republic Serbia Ministry of Foreign Affairs* initiates and signs bilateral and multilateral agreements on collaboration in emergency situations, completes the processes of accessing international information exchange networks, and of collaboration and aiding in emergency situations (*UN ECE-IAN, UN-OCHA-OSOC, EU-CECIS-MIC, NATO-EADRCC*).

The Republic Serbia Government has registered a certain number of economy subjects into a special category – public enterprises with their resources of special national importance for emergency management. These are the following: "Serbia Roads" Public Enterprise (PE) and Corridor 10 PE within its framework; "Serbian Railways" PE; "Serbian Waters" PE; "Serbian Forests" PE, "Electric Power Industry of Serbia" PE; "Serbian Oil Industry" – SOI PC; "Telecom Serbia"; the Information and Internet Republic Institute; and RATEL – Republic Telecommunication Agency.

Resources and capacities of special ministry organizations: Seismological Survey of Serbia and Republic Hydro meteorological Service of Serbia are of special importance in all phases of emergency situation.

The nongovernmental sector also plays a very important role in emergency situations, primarily: Serbian Auto-Moto Association; Serbian Red Cross Organization, Serbian Radio-Amateur Association with its communication system as a part of

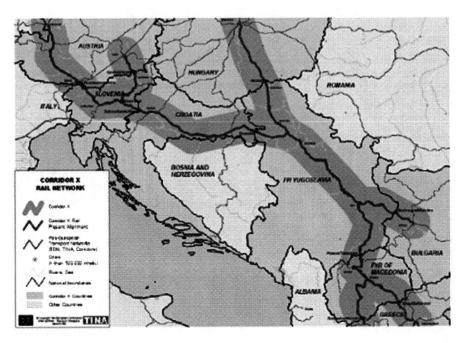

Fig. 10.2 Corridor 10

early warning system; Serbian Firefighting Association – Voluntary Firefighting Brigades; Speleologist Society; Scuba-Diving societies and clubs.

The effect of Serbian defense industry on human environment may be viewed in the context of Pan European Corridor 10 (Fig. 10.2). Namely, six out of ten pan European corridors stretch across the Balkans; the Pan European K-10 encompasses the total length of 2,360 km of roads and 2,528 km of railroads. One third of those run through Serbia in the length of 792 km.

By a human environment risk analysis in Corridor 10, a subset, or cluster of installed risks – is installed:

1. Chemical accident, explosion, fire, and pollution risk from installed operators as pollutants – very threatened sections on K-10 route.
2. Factors of risk in shipping dangerous materials – the route K-10 has been assigned EU for dangerous material shipping to the Middle East and Africa. Also, prognoses about the increase of transportation along the K-10 route through Republic Serbia point to a considerable increase of risk factors in dangerous material shipping on regional K-10.
3. Risk factors as a consequence of protest rallies and blockade of road sections. The simplified concept that reduces security to the control of physical access to an organization and the monitoring of property movements has recently started to broaden to include issues of health and safety, and components of risk evaluation and management, paying attention to a broader array of risks (forgeries, terrorism, emergency situations) (Borodzicz 2001).

4. The effect of security events must be seen in a wider sense. Security events dominate the regional image so much that they are seen as independent from the actual threat (Glaesser 2006). Gartner and Shen (1992) could empirically prove that security events bring about a change in image dimension.

10.6 Conclusion

Explosives are considered to be the most dangerous material in defense technologies. Unless handled adequately and with all necessary preventive measures of protection taken, they are extremely dangerous to the environment. Explosions at military storage facilities have multiple negative aspects to the human environment, from the primary negative effects of an explosion to the destruction of unexploded materials. The regional pollution depends on the quantity of explosive material, but also on the location of storage facility relative to geo-topological properties of the surroundings. Very often, this may be the cause of long lasting atmospheric sediments at lower altitudes.

The strategic problem of possible consequences due to inadequate storaging hazard materials in Serbia is coordination between ministries and government body, relevant local self-government bodies, city municipalities, and local communities in creation of efficient coordinating body for crisis situations preparedness and response. It includes the support in development of a plan for disaster preparedness and response. The primary goal of the future cooperation framework is strengthening capacities of local communities and civil society in identifying consequences of risks and involvement in risk reduction, as well as strengthening capacities of disaster preparedness and response.

Specifically, this chapter explains how to use set of the legal regulations on the field of environmental threats in military area, including roles of community services and first responders in decision-making process, as well as making progress in policy and procedures of storaging hazard materials.

The findings encourage the creation of conditions for prevention, preparation, and response to crises through development of an early warning system in social, political, and economic context, and development of crisis plan that includes mechanisms of cooperation and information systems on local level connected with regional, national, and international network of cooperation and information. Besides the logistic and other preconditions for crisis management, findings stress capabilities of relevant institutions and local stakeholders in taking over the responsibility for risks and crises in local environment, establishing the self-sustaining crisis management.

It is expected that the implementation of the findings would set a basis for quality risk and crises management, and contribute to the increase of overall security of local community and its relations to the complexed military and environmental surrounding in terms of prevention of crisis situations and response to them. Through implementation of the model for local communities' security condition monitoring and dissemination would be drafted, as well as methods for evaluation of applied solutions (measures).

To this end, the following recommendations *to the management of defence industry installations* may be given:

1. To adjust the existing ammunition and powder storage facilities to the modern international standard in the field of security and protection.
2. To apply more strict technological measures in the zones of certain pollution where the destruction of explosive material waste is done, and especially where the waste with the property of sedimentation in the soil is in question.
3. To better protective measures against the detonation and acoustic blast in the zones of destruction, as they are a consisting part of the surroundings security issue.
4. To raise the level of protection at work for the employees working at military installations and storage facilities.
5. To provide for a transparent system of monitoring for the degree of local environment pollution in collaboration with competent organs of local self-management as well as with human environment protection inspectors, and the MOD and Ministry of Human Environment Protection.
6. To perfect the communication between the management of defence industry installations/storage facilities and local self-management organs, ecology organizations, and state organs competent for human environment protection.

Recommendations to the local self-management organs:

1. Regarding the fact that the defence industry installations and facilities are located in urban areas, and that it is realistic to expect their further building and development, and also the fact that the degree of ecological and hazard risk increases in proportion with the increase of capacities, it is necessary to insure in the detailed surveys of the municipalities. Accordingly, the main focus turns to the management of vulnerabilities and the transfer/reduction of losses arising from the interaction of peril/vulnerability.
2. The harmonization of all economic, agro technical, community, and sanitary measures regarding the mentioned above so as to protect the health of the local population because of potential entering of dangerous materials into food chain.
3. Enhancement of community resilience through training and education to increase community respond.
4. Dissemination of information.

Recommendations to competent state organs (RS MOD and Ministry of Human Environment Protection):

1. Policy – establishment of regulation, procedures and guidelines.
2. System – monitoring, data collection and processing.
3. Establishment, deployment and examination of early warning system.
4. To pay special attention to investing in modern security equipment for storage facility protection procurement when funding the development of Serbian defence industry.

10 Environmental-Security Aspects of Explosion

5. To intensify joint ecological control of defence industry and RS Army installations and test areas.
6. To provide for civil control and transparency of informing about the human environment status monitoring on locations and in regions being used by Serbian defence industry.
7. That the Ministry of Defence in collaboration with Ministry of Human Environment Protection initiate cooperation, and access to international projects and funds directed to human environment protection by the mentioned municipalities and their competent organs as well as by the managements of defence industry installations and facilities.

References

Banks E (2005) Catastrophic risk, analysis and management. Wiley, Chichester
Baranoff E (2004) Risk management and insurance. Wiley, Danvers
Borodzicz E (2001) Security and risk: a theoretical approach to managing loss prevention. Int J Risk Secur Crime Prev 1(2):131–143
Emergency Management Manual (2007) Police emergency preparedness system, Part 1. National Police Directorate, Norway
Gartner W, Shen J (1992) The impact of Tiananmen Square on China's tourism image. J Travel Res 30(4):47–52
Gibson C, Britton N, Love G, Porter N, Fernandez E (2004) HB 221:2004. Business Continuity Management. Resource document. Standards Australia/Standards New Zealand Committee OB-007. http://www.saiglobal.com/PDFTemp/Previews/OSH/as/misc/handbook/HB221-2004.PDF+HB+221:2004.+Business+Continuity+Management. Accessed 10 May 2010
Glaesser D (2006) Crisis management in the tourism industry. Butterworth – Heinemann, Oxford, UK
Glendon AI, Clarke GS, Mackena FE (2006) Human safety and risk management. Taylor & Francis, London, New York
Guidelines for Explosive Weapons and Ammunition Stockpiling, TU-V3 (1970) Ministry of Defence, Beograd
Guidelines for Explosive Weapons and Ammunition Stockpiling, TU-V3 (1989) Ministry of Defence, Beograd
Kešetović Ž, Milašinović S, Keković Z (2009) Krizni menadžment i slični koncepti (Eng. Crisis management and related concepts). Krizovy manažment/Crisis management 8(1):55–60
OSCE (2008) Ecological risks of Republic Serbia defense industry. OSCE Mission in Serbia, Belgrade
Pearson CM, Clair JA (1998) Reafirming crisis management. Acad Manage Rev 23(1):10–32
Tait J (2001) More Faust than Frankestein: the European debate about the precautionary principle and risk generation for genetically modified crops. J Risk Res 4(2):175–189
Raisch W, Sturgeon A, Deane M (2006) ISO IWA 5:2006 Emergency Preparedness. Resource document. Emergency preparedness by International organization for standardization (ISO). American National Standards Institute – ANSI, International Center for Enterprise Preparedness (InterCEP). http://www.ansi.org/news_publications/news_story.aspx?menuid=7&articleid=1204. Accessed 10 Jun 2010
Technical Repaire Institute (2006) Laboratorijski izveštaj o vanrednoj kontoli hemijske stabilnosti baruta (Interni dokument) (Eng. Laboratory report on extraordinary control of chemical stability of gunpowder (Internal document)). Kragujevac: Tehnički remontni zavod Kragujevac (TRZ) Sektor za ispitivanje i praćenje UbS i KK – Labaratorija za ispitivanje kolekcioniranih uzoraka baruta

The Civil Protection Municipal Headquarters (2006) Report of the commission for inventory and assessment of the damage to agricultural resources, land, buildings, forests and livestock. The Civil Protection Municipal Headquarters, Paraćin

Uputstvo o jedinstvenoj metodologiji za procenu štete izazvane elementarnim nepogodama (1987) (Eng. Guidelines for unified methodology of assessing the damage caused by natural disaster). Službeni glasnik Republike Srbije, br. 27/1987

Zakon o elementarnim nepogodama (2005) (Eng. Natural Disaster Act). Službeni glasnik Republike Srbije, br. 24/2005, 61/2005

Zakon o eksplozivnim materijama, zapaljivim tečnostima i gasovima (1977) (Eng. Act on Explosive Materials and Inflammable Liquids and Gases). Službeni glasnik Socialitične Republike Srbije, br. 44/1977, 45/1984, 18/1989

Zakon o integrisanom sprečavanju i kontroli zagađivanja životne sredine (2004) (Eng. Human Environment Polluting Integrated Prevention Act). Službeni glasnik Republike Srbije, br. 135/2004

Zakon o lokalnoj samoupravi. (2007) (Eng. Local Self-Government Act). Službeni glasnik Republike Srbije, br. 129/2007

Zakon o proceni uticaja na životnu sredinu (1992) (Eng. Act on Effect on Human Environment Assessment). Službeni glasnik Republike Srbije, br. 54/1992

Zakon o strateškoj proceni uticaja na životnu sredinu (2004) (Eng. Act on Effect on Human Environment Strategic Assessment). Službeni glasnik Republike Srbije, br. 135/2004

Zakon o vanrednim situacijama (2009) (Eng. Emergency and Civil Protection Act). Službeni glasnik Republike Srbije, br. 111/2009

Zakon o zaštiti od požara (2009) (Eng. Fire Protection Act). Službeni glasnik Republike Srbije, br. 111/2009

Zakon o zaštiti životne sredine (2004) (Eng. Environmental Protection Act). Službeni glasnik Republike Srbije, br. 135/2004, 36/2009

Chapter 11
Nanotechnology: The Need for the Implementation of the Precautionary Approach beyond the EU

Dejana Dimitrijević

Abstract Nanotechnology is widely cited to be one of most exciting and important technologies for the twenty-first century. However, our knowledge of the effects nanomaterials have on human health and the environment is incomplete. The greatest potential for exposure over the next few years will be in the workplace, industry, and academic research institutions. This chapter presents the current scientific findings and understandings of the dangers nanotechnology imposes on the environment, human health, and safety with a special emphasis on current regulatory framework within the European Union (EU). This chapter analyses the overall development of this novel field by the study of two basic indicators, publications and patent applications in regard to environment, health and safety in ten South Eastern European countries. Publication data collected in worldwide databases such as those of the ISI/WOS are used to track the dynamics of science and technology development, whereas patent applications to EPO/WIPO are used to trace industry development of this novel field in South Eastern Europe (SEE). The increasing speed of nanotechnology development and commercialization has apparently run ahead of health, environmental and safety data. It is imperative for the SEE governments to develop a flexible form of the so-called adaptive management that will allow appropriate control and regulation of nanomaterials for the interim period until definitive health and safety data are obtained.

11.1 Introduction

Nanotechnology as an emerging technology inspires far-reaching expectations with respect to potential contributions of nanomaterials to welfare, health, environmental remediation, sustainable development, purification of water, waste treatment, energy

---

D. Dimitrijević (✉)
Department of Environmental Protection, Faculty of Security Studies,
University of Belgrade, Belgrade, Serbia
e-mail: ddejana@eunet.rs

G. Meško et al. (eds.), *Understanding and Managing Threats to the Environment in South Eastern Europe*, NATO Science for Peace and Security Series C: Environmental Security, DOI 10.1007/978-94-007-0611-8_11, © Springer Science+Business Media B.V. 2011

conservation, and new information and communication technologies. Frequently, nanotechnology is even regarded as the key technology of the new century, and it is compared with the industrial revolution in terms of its potential effect on society.

The Project of Emerging Nanotechnologies (PEN 2010) at the Woodrow Wilson International Center for Scholars maintains an inventory of consumer products that claim to utilize nanomaterials. The products fall into a number of different product categories such as health and fitness, home and garden, food and beverages, electronics and computers, automotive, goods for children etc. Over 60% of nanotechnology products belong to the health and fitness sector, which includes personal care products and cosmetics. In 2006 the inventory contained 212 different products, which has increased to 580 products in 2007 and 803 in 2008. As of June 20, 2010, this inventory contained 1,015 products produced by 485 companies located in 24 countries, and therefore, the inventory has grown by nearly 379% (from 212 to 1,015 products) since it was released in March 2006. According to PEN there are 154 companies in Europe, most of which are based in Germany and the UK but none from the South Eastern Europe (PEN 2010). The data need to be used with caution, as the database suffers from the problem of insufficient available information and therefore it is not an all-inclusive inventory. According to Dekkers et al. (2007) it is not possible to obtain a complete overview of all consumer products containing nanomaterials since there are products with the claim "nano" on the market that do not contain nanomaterials and have not been produced with nanotechnology either, and not all producers advertise their products as such and there is currently no legal obligation for companies to inform consumers or label products that contain nanomaterials. Recently, the European Consumers' Organization (BEUC 2009) and the European consumer voice in standardization (ANEC) have published on their website an inventory of different products claimed to contain nanomaterials applying the similar product categories and subcategories used in the Woodrow Wilson International Center for Scholars database. As of October 29, 2009 the group identified 151 products which are available in the EU market.

The International Risk Governance Council (IRGC) has identified four generations of nanotechnology products (IRGC 2007). Products already on the market are primarily passive (steady function) nanomaterials, as from 2000 what the IRGC refers to as first generation, and they are related to applications in which the nanomaterials do not change form or function. Almost all the debate about nanotechnology has focused on first-generation nanotechnology, but the second generation of the technology is now moving from science fiction to technological fact (Davies 2008).

The increasing speed of nanotechnology development and commercialization has apparently run ahead of health, environmental and safety data (Maynard et al. 2006; Oberdörster et al. 2007). The main concerns relate to those products that can lead to exposure to free nanomaterials (nanoparticles). The potential exposure is expected to be much higher if the nanomaterial consists of nanoparticles to be released in the air (as aerosols) or present in a liquid form (in an emulsion or suspension). It is broadly acknowledged that the benefits promised will fully materialize only if there is a governance system which addresses the potential risks and concerns associated with their development. It is uncertain whether the established governance systems are actually capable of adequately handling nanotechnologies and the corresponding applications within their frameworks, and as a consequence it is feared that nano-materials may cause damage to health and the environment before appropriate strategies based on quantitative risk assessment can be implemented (FramingNano 2009). This may be conceived as one of the main reasons why many stakeholders call for the implementation of a precautionary approach in order to avoid such damage, and in order to prevent vivid examples of how adverse public opinion can slow or even block the development and application of new technologies. The public backlash on nanotechnologies in general could lead to reactions similar to those observed with stem cell research, genetically modified crops, and nuclear power in recent years.

In this chapter, the overall research and development of nanomaterials in the ten SEE countries has been addressed by studying two basic indicators, publications and patent applications. The chapter provides some insight into the latest developments regarding the effects nanomaterials impose on the environment and on human health, as well as the capacity of international governance to deal with complex regulatory demands.

11.2 Publication Growth and Patent Data Generation in South Eastern Europe

The overall development of the nanotechnology field can be analyzed by studying two basic indicators – publications and patent applications. A wide range of activities has been undertaken (Heinze 2004; Igami and Okazaki 2007; Noyons et al. 2003) to allow a better understanding of the nature of nanotechnology. An analysis solely based on scientific publications, however, is not sufficient to provide an understanding of the socio-economic impacts of scientific discoveries. Relying on patent analysis is likely to be a useful way to examine the flow of continuous knowledge from science to technology. Patents are, in fact, one of the most direct and best measurable outputs of research and development (Igami and Okazaki 2007). According to the Royal Commission on Environmental Pollution (RCEP) (RCEP 2008) the number of patents registered from 1990 to 2006 for nanoparticles, nanorod, nanowire, nanocrystal, nanotube or carbon nanotubes has more than doubled every 2 years. The European Patent Office (EPO) provides a uniform application procedure for individual inventors and companies seeking patent protection in up to 40 European countries. The main task of the European Patent Office is to grant European patents and the EPO is the first choice for most European applicants. The EPO's collection of over 60 million patent documents, most of them patent applications rather than granted patents, from all over the world is available to the public via the free Esp@cenet service on the internet. In recent years, foreign applications are increasingly filed as an international application at the World Intellectual Property Organization (WIPO).

In order to estimate whether past and current nanotechnology research efforts in the ten South Eastern European countries are responding to the challenges outlined above, a review of research efforts has been undertaken, mainly in terms of peer-reviewed journal articles, proceedings papers, reviews, and patent applications as they apply to nanomaterials. This is not intended to be a complete analysis of research publications and patents within the field of nanomaterials, including environment, health and safety aspects (EHS) of nanomaterials. Instead, the intention is to provide an outline of the general direction of efforts to date.

The literature review of nanotechnology application and data generation of nanomaterials began with an analysis of the Science Citation Index (SCI) and Social Science Citation Index (SSCI) database available through the ISI Web of Science with Conference Proceedings, focusing on the years 2000 through 2010, using the 'advanced' search option. Figure 11.1a, b present the number of articles, proceedings papers and reviews published annually in Serbia, Croatia, Slovenia, Macedonia, Bosnia and Herzegovina, Greece, Hungary, Bulgaria, Romania, and Albania (RS = Serbia, HR = Croatia, SI = Slovenia, MK = Macedonia, BA = Bosnia and Herzegovina, GR = Greece, HU = Hungary, BG = Bulgaria, RO = Romania, AL = Albania). A search string for Fig. 11.1a, b is adopted from Lux Research (2007): "TS = (quantum dot OR nanostruc* OR nanopartic* OR nanotub* OR fulleren* OR nanomaterial* OR nanofib* OR nanotech* OR nanocryst* OR nanocomposit* OR nanohorn* OR nanowir* OR nanobel* OR nanopor* OR dendrimer* OR nanolith* OR nanoimp* OR nano-imp* OR dip-pen)" with document type = article, proceedings paper and review (search string taken from Lux Research 2007). The number of publications published for the first 4 months of the year 2010 was estimated by multiplying the number of papers by 3. The ISI Web of Science database was accessed and used on 1 May 2010. All data search strategies exclude obvious non relevant terms (i.e. $NaNO_3$, nanosecond, nanogram, nanosecond, nanoliter, nano-molar etc.).

A study conducted by Linkov et al. (2009a) found that on the global scale the total number of publications satisfying this search criterion increased from about 2,000 in 1995 to approximately 40,000 in 2008, and that these data are consistent with the data reported by Lux Research for the span 1990–2006 (Linkov et al. 2009a). As for SEE countries, bibliometric studies on nanoscience and nanotechnology are rather scarce. To our knowledge, there is only one publication dealing with bibliometric data concerning nanopublications in Serbia. Although the authors used different search parameters, the values they obtained closely approximate the data for Serbia presented in this study (Ševkušić and Uskoković 2009). Using a wide range of search terms, over the whole 10-year period, depicted in Fig. 11.1a, b, an average annual growth rate for Serbia is 31%, for Slovenia is 30%, followed by Romania with 28%, Croatia 35%, Greece 25%, Bulgaria 23%, and Hungary with 8% in the number of nanopublications.

Figures 11.2 and 11.3 illustrate the results in Web of Science, based on a search of the ISI Web of Knowledge on 1 May 2010, as well as EPO and WIPO databases. Figure 11.2 presents a comparison of the number of publications within different "All nano" topics in the time span 2000–2010 in SCI and SSCI (Web of Science® with Conference Proceedings) using "advanced search" options. "All Nano" corresponds to all published papers returned by the SCI and SSCI using the search

11 Nanotechnology: The Need for the Implementation of the Precautionary 209

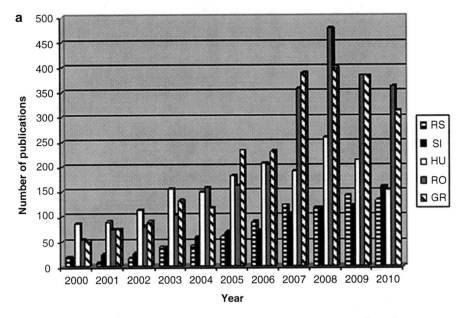

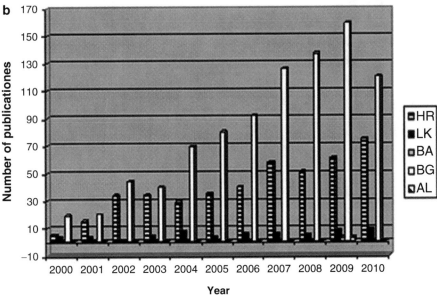

Fig. 11.1 (**a**) The number of articles, proceedings papers and reviews published annually in a field of nanotechnology in Serbia, Slovenia, Hungary, Romania, and Greece (years 2000–2010). (**b**) The number of articles, proceedings papers and reviews published annually in a field of nanotechnology in Croatia, Macedonia, Bosnia and Herzegovina, Bulgaria, Romania, and Albania (years 2000–2010)

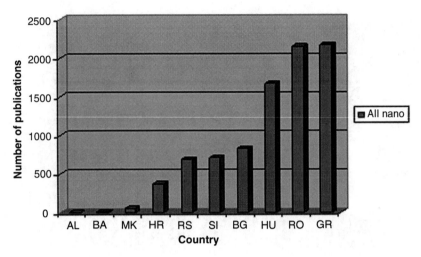

Fig. 11.2 Comparison of number of publications within different "All nano" topics for South Eastern Europe countries

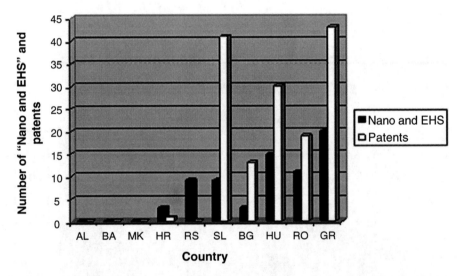

Fig. 11.3 Comparison of number of publications within different environment, health and safety (EHS) issues of nanomaterials and patent applications in the SEE countries

string as mentioned above in Fig. 11.1a, b. In Fig. 11.3, "Nano and EHS" corresponds to the combination of the original search string with the following search terms (using Boolean operator AND): "TS=(risk* OR environ* OR health OR safety OR toxi* OR occupational)" with document type=article, proceedings paper and review. Trends in nanotechnology patent applications to the EPO and WIPO in the time span 2000–2010 were presented in Fig. 11.3 with the following search string:

11 Nanotechnology: The Need for the Implementation of the Precautionary

nano*, fulleren*, quantum dot* and dendrimer* (using "advanced" search options). Searches were accessed and used on 1 May 2010.

In terms of the number of published research articles in regards to "All nano" and "Nano and EHS", Figs. 11.2 and 11.3 show that the emergence of nanomaterials has occurred much faster than the generation of corresponding EHS data. Figure 11.2 and clearly shows that only a fraction of all papers report the results for environment, health and safety characterization of nanomaterials.

Recently, Grieger et al. (2010) performed an analysis of the ISI Web of Knowledge article database, the International Council on Nanotechnology database, and the Organization for Economic Cooperation and Development (OECD 2010) research project database, which specifically focuses on EHS issues of nanomaterials. In terms of the number of published research articles and research projects within different nano-risk fields, most (60.1%) of these research efforts were within the topics of "toxicity", "ecotoxicity" and "exposure" of nanomaterials (Grieger et al. 2010).

As for patent applications, Fig. 11.3 shows SEE countries shares in nanotechnology patent applications to the EPO and WIPO from 2000 to 2010 with the following search string: nano*, fulleren*, quantum dot* and dendrimer*. It is evident, as expected, that the share of Romania, Greece, Hungary, Slovenia, and Bulgaria in patents applications is notably larger than the one of the rest of analyzed SEE countries. These countries have had an increasing number of nanotechnology patents applications especially after 2005. The biggest number of nanotechnology patent applications was recorded for Slovenia, with 18 patent applications in 2007 (priority year). This is an obvious example of the effect that the government funding of academic research has lead to an explosion in the number of patent applications in nanotechnology in Slovenia, with about 89% of the applications coming from universities and public non-university research organizations.

The expenses for an EPO application and examination are high. EPO applications represent inventions of high technological and commercial value, and the priority or first application of a patent is generally made at the domestic patent office, as this is the cheapest way of registration (Heinze 2004). According to the national patent offices in Serbia and Croatia, there are numerous patent applications made at the domestic patent offices, but for one reason or another, patent applications did not find their way to the EPO and WIPO, probably due to expenses and lack of awareness.

11.3 Environment, Health and Safety of Nanomaterials: A Brief Overview

The rapid technological development of engineered nanomaterials is expected to hold enormous potential for the development of new products and applications in a number of industrial and consumer sectors. Nevertheless, the increasing use of engineered nanomaterials makes risk assessments for human health and the environment essential. There are numerous scenarios through which humans could become exposed to engineered nanomaterials including occupational settings, environmental and consumer exposure. There is concern that nanoscale particles being exploited in certain applications might penetrate the skin and possibly even escape the immune system to reach the brain. Research has shown that the properties of manufactured nanoparticles may be different from their larger forms. This is mainly due to the larger surface area per unit mass and probable ability of nanoparticles to penetrate natural tissue barriers and cells more readily. At present, much of the concern is focused on "free" engineered nanomaterials and their effects on the environment, health and safety (EHS) during their entire life cycle. According to the Royal Society and Royal Academy of Engineering (RS and RAE) report, free manufactured nanoparticles and nanotubes (i.e. nanoparticles suspended in liquids or airborne nanoparticles) are likely to present the most immediate toxic hazard to living organisms because they might interact with organisms in the wider environment (RS and RAE 2004). There is not the same level of concern regarding fixed nanomaterials (i.e. those suspended into solid matrices or attached to surfaces), although there is clearly potential for them to become detached and enter natural ecosystems, especially when products containing them erode during use or when they are disposed of as waste or are recycled (RCEP 2008). Kaegi et al. (2008) found that titanium dioxide nanoparticles (nano-TiO_2) can be released into the water, as, was the case with paints from a house façade exposed to the external ambient environment.

Since publication of the joint landmark report of the Royal Society and Royal Academy of Engineering in 2004 related to the opportunities and uncertainties of nanotechnologies (RS and RAE 2004), more than 50 national and international reviews have considered nanoparticle risk issues (Stone et al. 2010). A comprehensive and critical scientific review of the health and environmental safety of fullerenes, carbon nanotubes (CNTs), metal and metal oxide nanomaterials has been published recently by the ENRHES, funded under the European Commission's Seventh Framework Programme, conducted by a consortium led by Edinburgh Napier University and the Institute of Occupational Medicine (IOM), UK (Stone et al. 2010). The ENRHES review study has focused mainly on nanomaterials such as airborne nanoparticles, powders and nanoparticles suspended in liquids. The state of the art review structure presents physico-chemical characterization techniques, exposure assessment, environmental fate and behaviour, human toxicity, ecotoxicity (aquatic toxicity, terrestrial toxicity, bioaccumulation, and degradability), and risk assessment of the four nanoparticle classes.

Current data on the environmental and human health risks posed by nanomaterials has often been ambiguous. It is extremely complicated to evaluate how safe or how dangerous some nanomaterials are because of our complete lack of knowledge about so many aspects of their fate and toxicology. The countless possible interactions between nanoparticles and harmful environmental chemicals may lead to unique exposures and health risks (Balbus et al. 2007). Concern was expressed about an increased risk of pulmonary damage from CNTs (Lam et al. 2004; Poland et al. 2008) and the effects of antimicrobial properties of nanosilver particles, in products such as washing machines, to the environment as well as on waste-treatment plants that clean sewage through bacterial action (Choi et al. 2008; Luoma 2008). Additionally the National Institute for Occupational Safety and Health (NIOSH) reported that nano-TiO_2 may cause carcinogenicity following pulmonary overload (NIOSH 2005).

From an environmental point of view, as more products containing nanomaterials are developed, there is a greater potential for environmental exposure. The White Paper published in 2007 by the U.S. Environmental Protection Agency (EPA) summarises what is recognized about the fate of nanomaterials in the air, in water, and in soil: (1) biodegradation, bioavailability, and bioaccumulation of nanomaterials, (2) the potential for transformation of nanomaterials to more toxic metabolites, (3) possible interactions between nanomaterials and other environmental contaminants; and (4) the applicability of current environmental fate and transport models to nanomaterials.

Nanotechnology has the potential to benefit substantially from environmental quality and sustainability through pollution prevention, wastewater treatment, and remediation. Potential nanomaterials release to the environment include direct and/or indirect releases from the manufacturing and processing of nanomaterials, releases from chemical and material manufacturing processes, and chemical cleanup activities including the remediation of contaminated sites. Releases to the environment include nanomaterials incorporated into materials used to fabricate products for consumer use including pharmaceutical products, and releases resulting from the use and disposal of consumer products containing nanoscale materials (NIOSH 2007). The possible interactions between nanomaterials and the environment are numerous and complex. The handful of studies on the toxicity of fullerenes so far suggested that they are indeed hazardous causing oxidative damage to the brain in the largemouth bass but they can be designed to be less so, by conjugating other chemicals to the surface of buckyballs, thus changing their chemical properties (Oberdörster 2004; Colvin 2003). It is plausible that soil and water organisms could take up manufactured nanoparticles escaping into the natural environment and that these particles could, depending on their surface activity interfere with vital functions, i.e. they may inhibit phagocytosis of macrophages (RS and RAE 2004). Organisms may ingest materials that have entered the water system or being deposited on vegetation. Once inhaled or ingested, nanomaterials may enter the food chain, leading to the possibility of bioaccumulation and ingestion by organisms higher up the chain. Bioaccumulation will depend on the surface activity of nanoparticles. Perhaps the greatest source of concentrated environmental exposure in the near term (less than 5 years) comes from the application of nanoparticles to soil or waters remediation. Initial studies on their potential for remediation indicate that nanoparticles of iron can travel with the groundwater over the distance of 20 m and remain reactive for 4–8 weeks (Zhang 2003).

Three modelling studies have been identified which provide useful information relating to environmental and consumer exposure (Boxall et al. 2007; Hansen et al. 2008; Mueller and Nowack 2008). The first quantitative risk assessment of nanoparticles in the environment was carried out by Mueller and Nowack (2008). The authors considered a life cycle perspectives (from production to disposal) of products containing nanosilver particles, nano-TiO_2 and CNTs in water, air and soil in Switzerland. The results of the study showed the predicted concentrations of CNTs and nano-Ag in the environment pose little risk, while nano-TiO_2 may be problematic in Swiss water bodies and therefore, indicating the need for further detailed studies. Boxall et al. (2007) in the UK developed a framework of simple models and algorithms to estimate nanomaterials concentrations in air, water and soil from direct routes of exposure. In comparison, the estimated concentrations, called predicted environmental concentrations (PECs), were calculated for nanosilver particles in water and soil and in comparison to the values obtained by Mueller and Nowack (2008), the estimate for water is lower, while that for soil is higher. Nevertheless, both values are in the same order of magnitude. On the other hand, in comparison to the values obtained by Mueller and Nowack (2008), Boxall et al. (2008) estimated the values for nano-TiO_2 in water are in the same order of magnitude while that for soil several orders of magnitude higher. Some of the difference between the PECs could be related to the fact that Boxall et al. (2007) assumed that no nanoparticles would be retained in the sewage treatment plants, whereas Mueller and Nowack (2008) assumed that 97% and 90% of the nanoparticles would be cleared in the realistic and the high exposure scenario, respectively (Hansen 2009).

Hansen et al. (2008) used the Technical Guidance Document and the best estimates available (since very few producers provide information about the content of the nanomaterials in the products), and/or worst-case assumptions for exposure assessment of nanoparticles in a number of consumer products. These products included a facial lotion, a sunscreen lotion, a fluid product for outdoor surface treatment, and a spray product for indoor surface treatment. Hansen et al. (2008) estimate consumer exposure to nano-TiO_2 to be 26, 15, and 44 µg/kg bw/year for a facial lotion, a fluid product, and a spray product containing nanoparticles, whereas, the potential uptake per kilogram body weight (bw) per day was equal to 40 mg/kg bw/day nano-TiO_2 for women and to 34 mg/kg bw/day for man respectively, if the sun lotion contains 10% nanoparticles.

11.4 Safety Measures and Regulatory Issues

The safety measures and risk assessment of manufactured nanomaterials have become the focal point of increasing consideration, mainly related to toxic hazards associated with nanoparticles and nanotubes. Indeed, many reports have been published which discuss the potential environmental and health risks associated with the manufacturing, use, distribution and disposal of nanomaterials. There are still many unanswered questions. Many scientists are raising questions regarding the manufacturing of nanomaterials and its effect on the workforce, researchers and consumers. Research has shown that the properties of manufactured nanomaterials may be different than their larger forms (RS and RAE 2004). This is mainly due to the larger surface area per unit mass and the probable ability of nanoparticles to penetrate natural tissue barriers and cells more readily. At present, the most likely place of exposure to nanoparticles and nanotubes is the workplace, including academic research institutions. Therefore, it is paramount that companies follow the usual

11 Nanotechnology: The Need for the Implementation of the Precautionary

methods of industrial hygiene. This involves the provision of personal respiratory protection, along with appropriate procedures for cleaning up accidental emissions within and outside the workplace.

The Technical Guidance Document (TGD), issued by the European Commission, supports legislation on assessment of risks of chemical substances to human health and the environment and is intended for use by the competent authorities to carry out the risk assessments on new notified substances, existing substances and on biocidal active substances or a substance of concern present in a biocidal product (European Commission 2003). The TGD's risk assessment framework developed for chemicals comprising hazard identification, dose (concentration) – response (effect) assessment, exposure assessment and risk characterization in relation to human health and the environment (European Commission 2003). In recent years, scientists, international organizations and regulatory agencies are still left with many questions as to whether nanomaterials pose a risk to human health or the environment, but have been actively working towards developing risk assessment for nanomaterials (European Commission 2003; EPA 2008; OECD 2009; SCENIHR 2007, 2009).

According to the Scientific Committee on Emerging and Newly Identified Health Risks (SCENIHR) "the TGD make very little reference to substances in particulate form" (SCENIHR 2007). With respect to human health, the existing methodologies described in the TGD, are generally likely to be able to identify the hazards associated with the use of nanoparticles. With respect to environmental exposure, the validity and appropriateness of existing technologies are not always clear. There is general consensus that current chemical risk assessment for nanomaterials, although not entirely fit for purpose, should be modified to reflect current knowledge, and with suggestions for specific improvements in methodologies and the need for new knowledge (SCENIHR 2007, 2009).

Established on 1 June 2007 and having entered into operation on 1 June 2008, Registration, Evaluation, Authorisation and Restriction of Chemicals (REACH) as a EU Regulation, "is based on the principle that manufacturers, importers and downstream users have to ensure that they manufacture, place on the market or use such substances that do not adversely affect human health or the environment" and is heavily underpinned by the precautionary principle (CEC 2008a; REACH 2008). A chemical substance is defined in Chapter 2, Article 3 as "a chemical element and its compounds in the natural state or obtained by any manufacturing process, including any additive necessary to preserve its stability and any impurity derived from the process used, but excluding any solvent which may be separated without affecting the stability of the substance or changing its composition" (EP and CEU 2006). There are no provisions in REACH referring explicitly to nanomaterials. Nevertheless, nanoscale substances are covered by the "substance" definition in REACH (CEC 2008a). REACH as a regulation applies directly in all 27 EU member states without need for domestic implementing legislation and introduces a "no data, no market" provision, prohibiting manufacture or placing on the EU market of substances for which registration risk-assessment data packages have not been submitted.

Nanomaterials are essentially treated like any other chemical substance under REACH and are subject to all of the same requirements. Substances, including substances at the nanoscale, manufactured or imported in volumes of 1 t or more per year have to be registered under REACH, while chemical safety reports are required above the 10 t/year threshold. This principle is applicable to substances in whatever size or form and for all their identified uses. The threshold levels do not reflect the fact that substances in nanoparticle form may have different health and environmental impacts per unit mass.

Frater et al. (2006) state, "given that some nanomaterials have different properties to their non-nanomaterial counterparts, it is conceivable that these thresholds are inappropriately set for the inclusion in products of nanoparticles".

For many nanoparticles, the threshold of 1 t/year of a substance per producers for registration would likely not be reached in the short term and furthermore, the usually low concentration of nanoparticles in the final article is likely to exclude many nanomaterials from the REACH legislation (Chaundry et al. 2006). No registration is required when the concentration of a substance is lower than 0.1% w/w (REACH 2008).

The European Commission is making a regulatory inventory, covering EU regulatory frameworks that are applicable to nanomaterials (chemicals, worker protection, environmental legislation, product specific legislation etc.). Recent FramingNano Report (FramingNano 2009) suggests variation in governance structures around the globe ranging from adaptation of regulatory framework, if scientific evidence indicates a need for modification, to the classification of nanomaterials as new substances being subject to nano-specific regulations. Questions regarding whether the burden to demonstrate safety should rest on the regulatory authorities, or product manufacturers, importers and downstream users of specific nanomaterials remain unsolved mainly as a result of a key difference between the USA legislation and REACH.

The European Commission's independent body, SCENIHR, has published its opinion on the most recent developments in the risk assessment of nanomaterials. SCENIHR has advised that, due to unpredictable characteristics of nanomaterials, their hazard assessment should be done on a case-by-case basis (SCENIHR 2006). SCENIHR has also reviewed the existing data on nanomaterials, data gaps, and issues that are to be considered in conducting risk assessments on nanomaterials.

Currently, there is a general consensus among regulators, scientists, policy experts and civil society organizations that there are some significant weaknesses in the application of the existing regulation to nanotechnology. According to Hansen (2010) regulatory challenges include: (1) whether to adapt existing legislation or develop a new regulatory framework, (2) whether nanomaterials should be considered as different from their bulk counterparts, (3) how to define nanotechnology and nanomaterials, and (4) how to deal with the profound limitations of risk assessment when it comes to nanomaterials. On 12th November 2008, the Royal Commission on Environmental Pollution (RCEP) published its report on *Novel Materials in the Environment: The case of nanotechnology* and concludes that despite the capacity of existing regulations to address risks posed by nanomaterials, at present they are poorly adapted to do so (Lee and Stokes 2009). The main problems of the existing legislative frameworks seem to be that metrology tools are unavailable, that thresholds are not tailored to the nanoscale, but based on bulk material, profound lack of (eco)

11 Nanotechnology: The Need for the Implementation of the Precautionary

toxicological data, and that no risk thresholds and occupational exposure limits cannot be established with existing methodologies (Hansen 2009). The different views on how to regulate nanomaterials vary considerably, ranging from the optimistic view that no regulatory attention to novel materials could be justified unless and until there were clear indications that harm is being caused, to less optimistic 'risk based' and finally, to a total moratorium that stops development of nanotechnology research and commercialization (RCEP 2008).

As mentioned previously, current regulatory system, such as those for chemical substances under REACH, are based on mass thresholds and annual production volumes. New models for risk governance have been sought as alternatives or complements to existing regulations to deal with these uncertainties and provide means on how to find practical ways to come to commonly accepted decisions in risk governance and facilitate trust among different current and potential stakeholders. Such approaches encompass voluntary measures ("soft" regulation) in nanotechnology risk governance such as codes of conduct, risk management systems or data reporting programs. A key element in self-regulating mechanisms, at least in Europe, is the Code of Conduct for responsible nanoscience and nanotechnologies research that all Member States have been recommended to use by the European Commission (CEC 2008b).

RCEP (2008) argued there is no logical reason why size of particle should in itself provide the basis for a new regulatory regime for nanomaterials, since what matters is what they do, and the implications of their properties and functionalities for environmental protection and human health. However, a highly cited report of the RS and RAE concluded that chemicals produced in the form of nanoparticles and nanotubes should be classified as new chemical substances under both the existing and proposed regulatory frameworks present in the UK and the EU (RS and RAE 2004).

The idea that voluntary measures should serve as "interim measures" until regulations based on sound science can be developed, has also been brought forward. Examples of voluntary environmental programs (VEPs) are the Voluntary Reporting Scheme for Nanomaterials established in 2006 by the Department for Environment Food and Rural Affairs (DEFRA) in the UK and the Environmental Protection Agency (US EPA) in 2008. DEFRA has received only 13 submissions in 2 years, while the US EPA in 2008 received 29 submissions on 123 different nanomaterials in 1 year (Widmer et al. 2010). Voluntary cooperation by the private sector has been established by DuPont and Environmental Defense and they formulated a detailed framework for examining the health and environmental effects of a nanomaterials (Environmental Defence and DuPont 2007). This framework urges companies to share risk assessment and management decisions with other companies within the supply chain. DuPont has made the framework mandatory in the development of nanomaterials within the company. Despite the large gaps in knowledge, it is necessary to have toxicity data on a large number of nanomaterials and products. According to Davies (2008) these data will not be generated or made known if companies are not required to do so.

However, it remains unclear to what extent such voluntary systems might fill the gap that is left open by traditional regulations, and at what price, i.e. concerning the question whether all or just few players would participate, and to what extent such voluntary efforts would be transparent and trustworthy (Widmer et al. 2010). According to Hansen (2009), the key elements of successful voluntary programs are: (1) incentives to participate for the stakeholders, (2) agency guidance and technical assistance to implement the measure or compensate for expenses, (3) signed commitments and periodical reporting in order to create some accountability despite the missing mandatory character, (4) and transparency both in design, reporting and evaluation. According to Hansen and Tickner (2007) many of these elements have not been fully addressed in VEPs on nanomaterials. The International Risk Governance Council recommends establishing systematic liaisons between government and industry to share risk information and promote socially responsible outcomes, since many problems of losing public trust derive from unnecessary secrecy (IRGC 2009). According to Hansen (2009), companies should be obligated to share information and insight into the basis of risk assessment and management decisions, and that any voluntary program on nanomaterials should be made mandatory after no more than 3 years in order to create a "threat of regulatory intervention".

Standardization has clearly been identified as one of the most important issues in the context of a responsible development of nanotechnologies since the lack of a firm definition of nanomaterials would provide a shaky basis for regulators already overwhelmed by novelty. There is no doubt that agreed definitions and a common terminology are needed in order to make progress in the ongoing debate on regulation on nanotechnologies. It has become clear that for the purpose of enhancing the comparability of scientific data, scientific definitions need to be clear and very sharp, but that on the other hand, for regulatory purposes and depending on the area regulated, regulatory definitions may need to be formulated differently (for example, one form of a nanomaterial might intentionally be explicitly considered in the regulation of cosmetics) (Widmer et al. 2010). The area of nanoscience and nanotechnology is developing new standards in Europe within the CEN/TC 352, at the international level at the ISO/TC 229. At present, it is ISO TC229 that dictates the line of the activities on nanotechnology standards. The International Organization for Standardization (ISO) TC 229 Nanotechnologies is the technical committee dealing with nanotechnologies and has 35 participating countries (and 9 observers) many liaisons to other (national) standards bodies and six international organizations contributing to its work (ISO 2008). ISO TC 229 has so far published three standards: (1) ISO TS 27687 Terminology and definitions for nano-objects – nanoparticle, nanofibre and nanoplate, (2) ISO/TR 12885 – Nanotechnologies – Health and safety practices in occupational settings relevant to nanotechnologies (ISO 2008), and (3) the latest ISO/TS 80004 Nanotechnologies – Vocabulary – Part 3: Carbon nano-objects (ISO 2010). The 33 projects have been under development with many more to come. A contribution to the standardization activities will also be made by the eight Steering Groups of the OECD Working Party on Manufactured Nanomaterials who are gathering reference data and information on characterization and safety of nanomaterials (OECD 2009). Bulgaria is the only country out of ten countries analyzed in South Eastern Europe that made its contribution to the development of standardization activities. However, the Bulgarian Institute for Standardization has so far not had standardization publications dealing with nanotechnology topics.

Currently, the Serbian Institute for Standardization is considering moving forward with standardization activities related to nanotechnology.

A nano-risk governance framework must provide structures to deal with the uncertainties and variability in available data. Recently, Linkov et al. (2007, 2009b) reported using Multi-Criteria Decision Analysis (MCDA) as a way to structure fragmented information for application in environmental management. According to Linkov et al. (2009a) the current focus should be shifted toward comparative assessment of the available possibilities. The approach suggested involves the use of the qualitative Alternative Assessment method, and quantitative MCDA methods for the interim period until definitive health and safety data are obtained. Linkov and his team suggested that nanomaterials regulatory frameworks could be built on the existing regulatory approaches with the addition of a more rigorous and transparent method for integrating technical information and expert judgment.

The precautionary principle is seen principally as a way to deal with a lack of scientific certainty due to the large knowledge gaps. In February 2000, the European Commission (EC) adopted the Communication from the Commission on the Precautionary Principle and purports, "Although the precautionary principle is not explicitly mentioned in the Treaty except in the environmental field, its scope is far wider and covers those specific circumstances where scientific evidence is insufficient, inconclusive or uncertain and there are indications through preliminary objective scientific evaluation that there are reasonable grounds for concern that the potentially dangerous effects on the environment, human, animal or plant health may be incon-sistent with chosen level of protection" (European Commission 2000). With regard to the precautionary approach to nanotechnology the EU has stated as follows: "Where the full extent of a risk is unknown, but concerns are so high that risk management measures are considered necessary, as is currently the case for nanomaterials, mea-sures must be based on the precautionary principle" (European Commission 2008). Regulation or governing the responsible development of nanotechnologies is cur-rently impeded by the prevalent uncertainties which make an efficient, science and evidence based regulatory approach practically unattainable. Therefore, the applica-tion of a precautionary approach suggests to take action before scientific evidence has completely emerged, in order to protect human health and the environment.

11.5 Conclusion

Most of the South Eastern European countries analyzed in this chapter are engaged in nanotechnology research and development, as well as in patent application. It is to be expected that these countries will be engaged in the production of raw nanomateri-als in the near future. Therefore, proper health and safety provisions are vital.

Large gaps in scientific understanding of nanomaterials are slowing down the development of a regulatory scheme. The questions whether the burden of safety should rest on the regulatory authorities or product manufacturers remain unsolved on the global scale mainly as a result of a key difference between the USA legislation and REACH. However, the 'wait and see' approach is increasingly becoming a dangerous way to overcome the present vacuum in the regulatory field. Some EU member states are very active in addressing the numerous issues related to standardization and nanoregulation. So far, no specific initiatives have been taken in any non-EU South Eastern European countries with regard to nanomaterial regulatory frameworks. Standards are not a part of the regulatory framework but can and do support regulation, and most of the SEE countries considered here have has not taken up this issue seriously. There is no doubt that transition countries should be better represented and involved in global regulatory cooperation and should be aware of the fact that a lack of action might bring significant damage to people or the environment. Responsible governments of SEE countries need to ensure that neither of these outcomes occurs.

Acknowledgments The author is grateful to Professor Edward W. Randall from The School of Biological and Chemical Sciences, Queen Mary, University of London for editorial comments.

References

Balbus JM, Maynard AD, Colvin VL, Castranova V, Daston GP, Denison RA (2007) Meeting report: hazard assessment for nanoparticles: report from an interdisciplinary workshop. Environ Health Perspect 115(11):1654–1659

Boxall ABA, Chaudhry Q, Sinclair C, Jones AD, Aitken R, Jefferson B et al (2008) Current and future predicted environmental exposure to engineered nanoparticles. Report by the Central Science Laboratory for Department of Environment, Food and Rural Affairs (DEFRA). DEFRA, Sand Hutton

CEC (2008a) Communication from the commission to the European parliament, The Council and the European Economic and Social Committee Regulatory Aspects of Nanomaterials [SEC (2008) 2036] COM (2008) 366 final. Brussels, 17.6.2008. Resource document. Commission of the European Communities. http://ec.europa.eu/nanotechnology/pdf/comm_2008_0366_en.pdf. Accessed 23 May 2010

CEC (2008b) Commission recommendation of 07/02/2008 on a code of conduct for responsible nanosciences and nanotechnologies research. Brussels, 07/02/2008 COM (2008) 424 final. Resource document. Commission of the European Communities. http://ec.europa.eu/nanotechnology/pdf/nanocode-rec_pe0894c_en.pdf. Accessed 23 May 2010

Chaundry Q, Blackburn J, Floyd P, George C, Nwaogu T, Boxall A et al (2006) A scoping study to identify gaps in environmental regulation for the products and applications of nanotechnologies. Department for Environment, Food and Rural Affairs (DEFRA), London. Resource document. DEFRA. http://www.defra.gov.uk/science/Project_Data/DocumentLibrary/CB01075/CB01075_3373_FRP.doc. Accessed 13 Apr 2010

Choi OK, Deng K, Kim NJ, Ross L, Surampalli YR, Hu ZQ (2008) The inhibitory effects of silver nanoparticles, silver ions, and silver chloride colloids on microbial growth. Water Res 42:3066–3074

Colvin VL (2003) The potential environmental impact of engineered nanomaterials. Nat Biotechnol 21(10):1166–1170

Davies JC (2008) Nanotechnology oversight: an agenda for the new administration. Woodrow Wilson International Center for Scholars Project on Emerging Nanotechnologies, Washington, DC. Resource document Woodrow Wilson International Centre for Scholars Project on Emerging Nanotechnologies. http://207.58.186.238/process/assets/files/6709/pen13.pdf. Accessed 23 May 2010

11 Nanotechnology: The Need for the Implementation of the Precautionary   221

Dekkers S, Prud'homme De Lodder LCH, de Winter R, Sips AJAM, de Jong WH (2007) Inventory of consumer products containing nanomaterials RIVM/SIR Advisory report 11124 Version July 27. Resource document. http://www.rivm.nl/bibliotheek/digitaaldepot/inventoryconsumer-products.pdf. Accessed 2 June 2010

Environmental Defence and DuPont (2007) Nano risk framework. Environmental Defense and DuPont. Nano Partnership, June 2007. Resource document. http://www.edf.org/documents/6496_Nano%20Risk%20Framework.pdf. Accessed 18 May 2010

EP and CEU (2006) Regulation (EC) No. 1907/2006 of the European Parliament and of the Council of 18 December 2006 concerning the Registration, Evaluation, Authorization and Restriction of Chemicals (REACH), establishing a European Chemicals Agency, amending Directive 99/45/EC and repealing Council Regulation (EEC) No. 793/93 and Commission Directives 91/155/EEC, 93/67/EEC, 93/105/EC and 2000/21/EC. Official Journal of the European Union. (L 396). Resource document. European Union. http://eur-lex.europa.eu/LexUriServ/LexUriServ.do?uri=oj:l:2006:396:0001:0849:en:pdf. Accessed 27 May 2010

EPA (2007) Nanotechnology white paper. Environmental Protection Agency, Washington, DC. Resource document. EPA. http://www.epa.gov/osa/pdfs/nanotech/epa-nanotechnology-white-paper-0207.pdf. Accessed 23 May 2010

EPA (2008) Nanomaterial research strategy. Environmental Protection Agency, Washington, DC. Resource document. EPA. http://www.epa.gov/nanoscience/files/nanotech_research_strategy_final.pdf. Accessed 23 May 2010

European Commission (2000) Communication from the commission on the precautionary principle. Resource document. European Commission. http://eur-lex.europa.eu/LexUriServ/LexUriServ.do?uri=COM:2000:0001:FIN:EN:PDF. Accessed 27 May 2010

European Commission (2003) Technical Guidance Document on Risk Assessment in support of Commission Directive 93/67/EEC on Risk Assessment for new notified substances, Commission Regulation (EC) No 1488/94 on Risk Assessment for existing substances and Directive 98/8/EC of the European Parliament and of the Council concerning the placing of biocidal products on the market. European Chemical Bureau, Italy. Resource document. European Commission. http://ecb.jrc.it/Documents/TECHNICAL_GUIDANCE_DOCUMENT/EDITION_2/tgdpart1_2ed.pdf. Accessed 27 May 2010

European Commission (2008) Communication from the Commission to the European Parliament, the Council and the European Economic and Social Committee: Regulatory Aspects of Nanomaterials. Brussels, 17.6.2008 COM(2008) 366 final. Resource document. European Commission. http://ec.europa.eu/nanotechnology/pdf/comm_2008_0366_en.pdf. Accessed 27 May 2010

European Consumers' Organization (BEUC) (2009) ANEC/BEUC inventory of products claiming to contain nanoparticles available on the EU market. Resource document. BEUC. http://www.beuc.eu/objects/2/files/ANEC_BEUC_nano_inventory.xls. Accessed 27 May 2010

FramingNano (2009) FramingNano mapping study on regulation and governance of nanotechnologies. FramingNano Project Consortium (January 2009). Resource document. FramingNano. http://www.framingnano.eu. Accessed 7 June 2010

Frater L, Stokes E, Lee R, Oriola T (2006) An overview of the framework of current regulation affecting the development and marketing of nanomaterials. A report for the Department for Trade and Industry (DTI) by Cardiff University. Resource document. DTI. http://www.berr.gov.uk/files/file36167.pdf. Accessed 24 May 2010

Grieger KD, Baun A, Owen R (2010) Redefining risk research priorities for nanomaterials. J Nanopart Res 12:383–392

Hansen SF (2009) Regulation and risk assessment of nanomaterials: too little, too late? Doctoral thesis. Technical University of Denmark. Resource document. http://www2.er.dtu.dk/publications/fulltext/2009/ENV2009-069.pdf. Accessed 7 June 2010

Hansen SF (2010) A global view of regulations affecting nanomaterials. Nanomed Nanobiotechnol. doi:10.1002/wnan.99

Hansen SF, Tickner JA (2007) The challenges of adopting voluntary health, safety and environment measures for manufactured nanomaterials: lessons learned from the past for more effective adoption in the future. Nanotechnol Law Bus 4(3):341–359

Hansen SF, Michelson E, Kamper A, Borling P, Stuer-Lauridsen F, Baun A (2008) Categorization framework to aid exposure assessment of nanomaterials in consumer products. Ecotoxicology 17(5):438–447

Heinze T (2004) Nanoscience and nanotechnology in Europe: analysis of publications and patent applications including comparisons with the United States. Nanotechnol Law Bus 1:427–445

Igami M, Okazaki T (2007) Capturing Nanotechnology's current state of development via analysis of patents. STI Working Paper 2007/4, OECD Directorate for Science, Technology and Industry. Resource document. OECD. http://www.oecd.org/dataoecd/6/9/38780655.pdf. Accessed 24 May 2010

International Risk Governance Council (IRGC) (2007) Policy brief: nanotechnology risk governance. Recommendations for a global, coordinated approach to the governance of potential risks. CH-1219 Geneva, Switzerland. Resource document. IRGC. www.irgc.org/IMG/pdf/PB_nanoFINAL2_2_pdf. Accessed 24 May 2010

International Risk Governance Council (IRGC) (2009) Policy brief – appropriate risk governance strategies for nanotechnology applications in food and cosmetics. CH-1219 Geneva, Switzerland. Resource document. IRGC. http://www.irgc.org/IMG/pdf/IRGC_PBnanofood_WEB.pdf. Accessed 24 May 2010

ISO (2008) ISO/TS 27687:2008. Nanotechnologies – terminology and definitions for nano-objects – nanoparticle, nanofibre and nanoplate. Resource document. ISO. http://www.iso.org/iso/catalogue_detail?csnumber=44278. Accessed 20 June 2010

ISO (2010) ISO/TS 229 Nanotechnologies. Resource document. ISO. http://www.iso.org/iso/iso_catalogue/catalogue_tc/catalogue_tc_browse.htm?commid=381983. Accessed 20 June 2010

Kaegi R, Ulrich A, Sinnet B, Vonbank R, Wichser A, Zuleeg S et al (2008) Synthetic TiO_2 nanoparticle emission from exterior facades into the aquatic environment. Environ Pollut 156:233–239

Lam CW, James JT, McCluskey R, Hunter RL (2004) Pulmonary toxicity of single- wall carbon nanotubes in mice 7 and 90 days after intratracheal instillation. J Toxicol Sci 77:126–134

Lee R, Stokes E (2009) Twenty-first century novel: regulating nanotechnologies. J Environ Law 21(3):469–482. doi:10.1093/jel/eqp028

Linkov I, Satterstrom FK, Steevens J, Ferguson E, Pleus RC (2007) Multi-criteria decision analysis and environmental risk assessment for nanomaterials. J Nanoparticles Res 9:543–554

Linkov I, Satterstrom FK, Monica JC, Hansen SF, Davis TA (2009a) Nano risk governance: current developments and future perspectives. Nanotechnol Law Bus 6:203–220

Linkov I, Steevens J, Adlakha-Hutcheon G, Bennett E, Chappell M, Colvin V et al (2009b) Emerging methods and tools for environmental risk assessment, decision-making, and policy for nanomaterials: summary of NATO advanced research workshop. J Nanoparticles Res 11:513–527

Luoma SN (2008) Silver nanotechnologies and the environment: old problems or new challenges? PEN 15. Project on Emerging Nanotechnologies, Woodrow Wilson International Centre for Scholars and PEW Charitable Trusts. Woodrow Wilson International Centre for Scholars and PEW Charitable Trusts, Washington, DC

Lux Research (2007) Small stuff in search of the big bucks: nanotechnology commercialization activities by sector. Presentation to IEEE San Francisco Bay Area Nanotechnology Council on Dec. 18, 2007. Resource document. IEEE. http://ewh.ieee.org/r6/san_francisco/nntc/events/IEEE_talk_121807_Nanotech_Commercialization_Kristin_Abkemeier_Lux_Research.pdf. Accessed 26 Jan 2010

Maynard AD, Aitken RJ, Butz T, Colvin V, Donaldson K, Oberdorster G et al (2006) Safe handling of nanotechnology. Nature 444:267–269

Mueller NC, Nowack B (2008) Exposure modelling of engineered nanoparticles in the environment. Environ Sci Technol 42:4447–4453

NIOSH (2005) Current intelligence bulletin: evaluation of health hazard and recommendations for occupational exposure to titanium dioxide. The National Institute for Occupational Safety and Health DRAFT. Resource document. NIOSH. http://www.cdc.gov/niosh/review/public/tio2/pdfs/tio2draft.pdf. Accessed 26 Jan 2010

NIOSH (2007) Progress toward safe nanotechnology in the workplace: a report from NIOSH Nanotechnology Research Centre. National Institute for Occupational Safety and Health. Resource document. NIOSH. http://www.cdc.gov/niosh/docs/2007-123/. Accessed 26 Jan 2010

Noyons ECM, Buter RK, van Raan AFJ, Schmoch U, HeinzeT, Hinze S (2003) Mapping excellence in science and technology across europe nanoscience and nanotechnology, Part 2: Nanoscience and nanotechnology. Draft report of project EC-PPN CT-2002-0001 to the European Commission. Leiden, The Netherlands: Leiden University Centre for Science and Technology Studies/Karlsruhe, Germany: Fraunhöffer Institute for Systems and Innovation Research. Resource document. European Commission. http://studies.cwts.nl/projects/ec-coe/cgi-bin/izite.pl?show=publications. Accessed 27 May 2010

Oberdörster E (2004) Manufactured nanomaterials (Fullerenes, C60) induce oxidative stress in the brain of juvenile largemouth bass. Environ Health Perspect 112(10):1058–1062

Oberdörster G, Stone V, Donaldson K (2007) Toxicology of nanoparticles: a historical perspective. Nanotoxicology 1:2–25

OECD (2009) Safety of manufactured nanomaterials. Organisation for Economic Cooperation and Development. Resource document. OECD. http://www.oecd.org/department/0,3355,en_2649_37015404_1_1_1_1_1,00.html. Accessed 27 May 2010

OECD (2010) OECD programme on the safety of manufactured nanomaterials 2009–2012: operational plans of the projects. Organisation for Economic Cooperation and Development. Resource document. OECD. http://www.olis.oecd.org/olis/2010doc.nsf/LinkTo/NT000029AE/$FILE/JT03282410.PDF. Accessed 27 May 2010

Poland CA, Duffin R, Kinloch I, Maynard A, Wallace WA, Seaton A, Stone V et al (2008) Carbon nanotubes introduced into the abdominal cavity of mice show asbestos like pathogenicity in a pilot study. Nat Nanotechnol 3(7):423–428

Project on Emerging Nanotechnologies (PEN) (2010) A nanotechnology consumer product inventory. Woodrow Wilson International Centre for Scholars. Washington, DC. Resource document. PEN. http://www.nanotechproject.org/inventories/consumer/. Accessed 20 June 2010

RCEP (2008) Novel materials in the environment: the case of nanotechnology. Royal Commission on Environmental Pollution, November 2008. Resource document. RCEP. http://www.official-documents.gov.uk/document/cm74/7468/7468.pdf. Accessed 20 June 2010

REACH (2008) Follow-up to the 6th Meeting of the REACH Competent Authorities for the Implementation of Regulation (EC) 1907/2006. Resource document. REACH. http://ec.europa.eu/environment/chemicals/reach/pdf/nanomaterials.pdf. Accessed 20 June 2010

RS and RAE (Royal Society and Royal Academy of Engineering) (2004) Nanoscience and nanotechnologies: opportunities and uncertainties. The Royal Society, London. Resource document. The Royal Society. http://www.nanotec.org.uk/report/Nano%20report%202004%20fin.pdf. Accessed 20 June 2010

SCENIHR (2006) The appropriateness of existing methodologies to assess the potential risks associated with engineered and adventitious products of nanotechnologies. European Commission Scientific Committee on Emerging and Newly Identified Health Risks. Resource document. SCENIHR. http://ec.europa.eu/health/ph_risk/committees/04_scenihr/docs/scenihr_o_003b.pdf. Accessed 24 May 2010

SCENIHR (2007) The appropriateness of the risk assessment methodology in accordance with the technical guidance documents for new and existing substances for assessing the risks of nanomaterials. European Commission, Scientific Committee on Emerging and Newly Identified Health Risks, June 21–22. Resource document. SCENIHR. http://ec.europa.eu/health/ph_risk/committees/04_scenihr/docs/scenihr_o_010.pdf. Accessed 24 May 2010

SCENIHR (2009) Risk assessment of products of nanotechnologies. European Commission: Scientific Committee on Emerging and Newly Identified Health Risks, January 19. Resource document. SCENIHR. http://ec.europa.eu/health/ph_risk/committees/04_scenihr/docs/scenihr_o_023.pdf. Accessed 24 May 2010

Ševkušić M, Uskoković DP (2009) Analiza aktivnosti u oblasti nanonauka i nanotehnologija u Srbiji na osnovu bibliometrijskih pokazatelja (Eng. State of the art in nanoscience and nanotechnology in serbia: a preliminary bibliometric analysis). Tehnika – Novi materijali 18(5):1–16

Stone V, Hankin S, Aitken R, Aschberger K, Baun A, Christensen F et al (2010) Engineered nanoparticles: review of health and environmental safety (ENRHES) 2010. Resource document. ENRHES. http://nmi.jrc.ec.europa.eu/project/ENRHES.htm. Accessed 20 June 2010

Widmer M, Meili C, Mantovani E, Porcari A (2010) The FramingNano Governance platform: a new integrated approach to the responsible development of nanotechnologies, February 2010. Resource document. FramingNano. http://www.innovationsgesellschaft.ch/media/archive2/publikationen/FramingNano_Executive_Summary_Final.pdf. Accessed 7 June 2010

Zhang W (2003) Nanoscale iron particles for environmental remediation: an overview. J Nanoparticles Res 5:323–332

Chapter 12
Management of Spring Zones of Surface Water: The Prevention of Ecological Risks on the Example of Serbia and South Eastern Europe

Miroljub Milinčić and Tijana Đorđević

Abstract Society has been living long in conditions of local and regional environmental risks caused by the growing demands, falling quality and diminishing, absolute and relative availability of water resources. Ecological risks, but also poverty of most of the world's population, are causes as well as consequences of the deficit of water resources. It has placed extremely complicated and complex tasks in front of fundamental and applied disciplines, social practice and different levels of political security decision making – how to solve environmental risks in the area of providing water, water protection, protection from water, fair distribution of water, problems of upstream and downstream interests in the basin etc. Previous practice shows that intensity adjusted interaction in due time between man and environmental risks, especially in these cases where basic problems are water resources, leads to the improvement of social organization and arrangement of space. By contrast, bad attitude towards water resources and systems for their control causes the problems in the development and decline of many societies. Although relatively rich in water resources, regions of South Eastern Europe (SEE) a high degree of ecological vulnerability. Recent growing potential of environmental risks in the area of water resources is caused by climate change and substantially inefficient transition of social system, and delay in the development of large-scale water management infrastructure.

12.1 Introduction

Prevention of environmental risks, the supporting capacity of space, but also perspective of existence and overall development of society, shows a growing dependence on those resources and environmental conditions that traditionally have

---

M. Milinčić (✉) and T. Đorđević
Faculty of Geography, University of Belgrade, Belgrade, Serbia
e-mail: mikan@gef.bg.ac.rs

G. Meško et al. (eds.), *Understanding and Managing Threats to the Environment in South Eastern Europe*, NATO Science for Peace and Security Series C: Environmental Security, DOI 10.1007/978-94-007-0611-8_12, © Springer Science+Business Media B.V. 2011

been marked as renewable (water, air, vegetation). In this context (individually and cumulatively), water resources show a very complex structure of environmental and social risks (Cutter et al. 1999; Milinčić and Videnović 2009). Regarding environmentally and economically the most important fresh water resources, potential to water deficit risk is the biggest, and level of interdependence is most complex. The basis of this "super system" consists of long-lasting, continuous, spatially more present, and consequentially more significant effects of environmental, economic and social risks. In chronological and horological terms, these risks are characterized by extreme complexity of the causes of the genesis, forms of manifestation, intensity of influence and threats to ecosystems and societies (Milinčić and Jovanović 2008).

These relations used to have a clear geospatial logic (arid and semi-arid areas), and were territorially limited. Problems caused by water deficit and environmental and economical risks have been spreading over time, mainly through the shaping of socio – economic agglomeration zones (industrialization and urbanization) to that they have reached regional, national and global scale. This development trend indicates that the environmental risks in the area of fresh water resources are not spatially and time-limited phenomenon, but a new reality of the modern world that requires an adequate theoretical and practical response within organization and society behavior (Milinčić 2009a).

Long-lasting and persistent classifying fresh water as a renewable resource has significantly contributed to the formation and channeling of the individuals' and society's negative attitude towards it. Together with misunderstanding of its nature and its negligence, it is formed public opinion that the polluted water is consequence of economic development and that it should be treated as a "price of progress". Some forms of this necessary interdependence became a part of traditional understanding – "dirty waters – higher standard"/"dirty waters – better life". Precisely, the contradiction in the relation between state and deficit of fresh water resources, on the one hand and the growing social dependence and environmental risks on the other hand, imposes an obligation of reviewing this concept. In fact, pollution and irrational (excessive and incorrect) usage of water are the main causes of its deterioration and general crisis of water resources. The situation in which humanity by its development increases its dependence of the quantitative and qualitative water resources status is one of the most significant "paradox of technology" to which modern society is exposed to (Milinčić 2009b).

Tradition of natural water regimes' transformation (construction of channels, dams, surface accumulations and spatial and temporal redistribution), was conceived by organized human societies from arid regions. By combining individual knowledge, techniques and skills of struggle for water and against it, nucleus of water management systems was made. Over time, these systems were broadening spatially and improving technically from the simplest and the cheapest ones to increasingly complex and expensive.

Serbia and South Eastern Europe in general, have been experiencing this problem for a long time. Archaeological research indicates that well digging in this region was present in the period 4,000–5,000 BC (before Christ). An important hydraulic works existed even in the period of ancient Greece. In addition, written sources, preserved artifacts and toponyms are being witness of the Roman legions and builders' exploits in the area of the Lower Srem (Progarsko, Jaračka jarčina), the North Banat (channel Zlatica-Aranka), Iron Gate (Trajan's bridge) and others. The first known surface water accumulation in Serbia is dated to the period of 15 centuries ago – reservoir near Caričin city – Zlatna and Caričin river. In its modern form, this method of surface water management and water supply has been continuously present through eight decades, and with the last quarter of the twentieth century it became one of the most promising models to provide the growing water demands (Milinčić 2001, 2009a).

12.2 About the System of Water Resources: Ecological and Social Capacities of Space

Because of its multifunctional role and importance, fresh water is a first-rate differential characteristic, latent capital and universal resource of ecumenical space. Unlike any other natural potential, at the same time water resources represent one of the major environmental, economic and social categories which are essential for fulfilling biological, productional and aspirational needs. Therefore, requirement for their security and protection is one of the most important issues of environmental risks' prevention and future's sustainability. Recent state of the problem and certainly its future territorial expansion and structural multiplication refer to that.

The lack and also poor quality of this subsistent resource and condition, which is often the case, violates environmental and social capacity of the space and quality of human life. Indirectly, that degrades and disables valorization of other available resources, and *de facto* increases overall unactivated development potentials and prospects. In fact, effects of insufficient water resources are being carried through chain reaction (domino system) across all segments of societies' life, and are rapidly generating the most serious hygienic, health, economic, political and other consequences. This is evidenced by significant positive spatial coincidence of insufficiently fresh water providence and poverty. In contrast, the activities of planning and management of fresh water resources contribute to risk prevention, capacity increasing, quality improvement and attractiveness of the entire resources database (resource and requirements) of geographical space.

Fresh water, with satisfactory quality, has long been the most reliable and the cheapest liquid available to mankind. These features, with its universal utility, are reasons why it was considered ordinary, but also why human society achieved the most intense relationship with it – existential, manufactural, aspirational. None of the material substances have ever had such a broad utilization and main significance as fresh water, and its use and consumption, by physical volume, outplays any other substance. Continuous growth of needs and society dependancy on water resources, besides absence of real alternative of significant substitution, indicate that water management becomes imperative for human societies at the expense of the great material, spatial, environmental, organizational and other constraints.

Retrospective of environmental water resources' risk prevention, should be seen through a complex mechanism of causes and consequences of human society's global development and its permanent struggle for water and against it. Relatively unfavourable or variable natural regime of water resources, and therefore one continuity of environmental risks, encourage mobilization of social groups and accelerate their progress. Capability of perception and solving environmental risks effects on human society to expand space due time where it has overcome some of the major development barriers, and by extending borders of environmental capacity area, to create conditions for a higher level of population, settlement and economic development. In fact, an interaction between human societies and ecological risks, which have a water resources at the basis, has led to improvement of social organization and arrangement of space, and to the creation of enclaves and zones of the oldest civilizations.

Development and decline of civilizations is directly related to the usage and management of water resources and systems for their control (White et al. 1964; Wittfogel 1956; Milinčić et al. 2007a; Ponting 2007; Milinčić and Šabić 2009). It is often well illustrated with numerous former developed commercial and urban centers of North Africa, Middle East and Central Asia, the Arabian Peninsula and others, which are now covered with sands. But, this geo-environmental determination does not mean that this phenomenon is characteristic only of arid and semi-arid areas. These "monuments" certify overcoming of harmful environmental risks in the field of water resources also in humid and perihumid regions from Southeast Asia (Angkor) to Central America (Yucatan).

Analysis of global water resources' availability showed continuous absolute and relative reduction in such range that deficit becomes their new dominant state, and spatial dimension of environmental risk has expansive spreading (Zeman et al. 2006). Often it is pointed to the expansion of the "geography of thirst" (Gorski 1965; Milinčić 2005). The trend of spatial spreading of water resources' availability on less than $1,000\,m^3$/inhabitants/year at the level of individual states is following (Hauchler et al. 1998) – during 1955 7 were in this situation and until 2050 will be 43 states. In the group affected by the 1950th were Djibouti, Bahrain, Jordan, Kuwait, Barbados, Malta and Singapore and since 1990 they were joined by Algeria, Burundi, Kenya, Malawi, Rwanda, Somalia, Tunisia, Israel, Yemen, Qatar, Saudi Arabia, United Arab Emirates and Cape Verde. By the year 2025 it is expected that Egypt, Ethiopia, Comoros, Lesotho, Libya, Morocco, South Africa, Oman, Syria, Haiti, Iran and Cyprus, and by 2050 Burkina Faso, Ghana, Madagascar, Nigeria, Zimbabwe, Tanzania, Togo, Uganda, Lebanon, Afghanistan and Peru are going to be affected.

12.3 Historical and Geographical Retrospective of the Problem

The first large and complex system for the water management in the function of prevention of environmental risks in agricultural production had been built 10,000 years ago (Cutter et al. 1999). Although mutual condition among agricultural economy and water management system was a gradual and substantial dispute, it is indicated to be revolutionary. This is due to difference in speed compared to the innovation diffusion at the time. Water usage in the function of the oldest agrarian communities, and agriculture as the oldest, the most permanent and for biological existence the most important section of material production, represent the beginning of intensifying relations and logarithmic growth of society's dependence on water resources. Valorization of water resources in the function of the oldest agricultural communities is the basic mechanism of space transformation, and creating artificial (eco)systems.

Overcoming of environmental risks and valuation of ecological services in whose basis are water resources is one of the driving mechanisms of the Paleolithic revolution. Its essence is in transition of tribal groups from hunting and gathering, the predominant model in 99% of the entire history of mankind, to the sedentary lifestyle and agriculture. Transition of system water resources – human society, and therefore made (agricultural) ecosystems, was not a goal for itself but a way of overcoming the environmental risks created on other foundations and an insurance mechanism of existence of human society. Since the introduction of agriculture until today, in spite of the appearance of a number of other water – dependent economic activities, it has remained the most significant cumulative abstract consumer of water resources. During the twentieth century, agriculture has achieved an absolute growth of water consumption by 9.7 times, and participation in the total consumption has dropped from 87% to 57%, so it is still its biggest consumer (Milinčić and Jovanović 2008).

Character of the interdependence of water resources and agricultural production can essentially be illustrated by saying: "Do not ask me how much soil I have, but ask me how much water I possess". FAO (Food and Agriculture Organization of the United Nations), in slightly modified form, on the occasion of World Food Day 2002 (FAO 2002), promoted this slogan as "There is no food without water". Human population growth will put in front of science and social practice, maybe existentially the most important question – how to produce more food with less water. A vicious circle is more likely: less water – less food and/or expensive water – expensive food and/or worse water quality – worse food quality.

By interacting with water resources management, human society has managed, at the micro and meso level, to increase environmental capacity of space, primarily through the creation of artificial ecosystems and perspectives for overall development. Because of its applicability and an overall importance (safety factor) such solutions have neither remained isolated nor have sunk into oblivion. On the contrary, systems for water resources control have been technically improved, territorially spread and increased in numbers. Due to its overall environmental and social importance, these systems have become the most important spatial infrastructure – basis of spreading ecumene and cultural landscape, and "civilizing" of the space. Ecumene is sometimes *a priori* identified as the territories on which, to various degrees and with the growing needs, water balance is being transformed (Shiklomanov 2000).

These spatially-oriented development processes and grouping of socio-economic components in the space, are becoming one of the main causes of new, specific and absolute, and primarily-concentrated, growth of needs and consumption, mainly of fresh water resources. When once established needs overcome possibilities of locally available sources, water resources again become a development threshold of preserving the overall quality of life and social existence. In order to ensure sufficient quantity of fresh water resources and overcoming environmental risks, society is increasing its activity in space, horizontally and vertically. This activity is usually limited on national territory area, but there are tendencies of crossing state border with different effects – from cooperation to conflict. Over time, activities in function of collecting, storing and the spatial and temporal distribution of water resources have become chronologically, strategically, spatially, ecologically, technically and by investment demands, very challenging ventures which strongly determine and transform environmental and socio-economic structure of space. Level of influence is so strong that these spatial systems are becoming a basic feature of space on the broader regional relations.

By their spatial presence, overall contribution to the function of water resources management, transformation of water regimes and prevention of environmental risks, accumulations have special significance. These hydraulic structures are known from early historic times, but confirmation for their capabilities and wide acceptance have gained in the nineteenth century. It is well presented with the following facts:

- The process of industrialization and urbanization, and overall growth, especially concentrated demand for water resources.
- Agglomerization of large consumers in cities and enlargement of water supply, and switching from municipal to a regional phase, with extreme complexity and increased system capacity.
- Development of techniques of concrete dam constructing and casting of steel pipes for water transport and others.

Full affirmation and global presence, dams have during the first half of the twentieth century, when the number of big dams increased by more than 10 times. During the 1930s in the USA (United States of America) and the USSR (Union of Soviet Socialist Republics), this form of water resources and water supply management (mega hydraulic systems) has been widely applied. Their, often general, acceptability is based on the fact that they were able to solve some of the requirements in water supply which natural regimes of local water resources (wells and springs) could not. This primarily refers to the collecting and storage, and a significant spatial and temporal (seasonal – perennial) transformation of existing regimes and to the increasing of total water resources' efficiency. Over time, despite a number of perceived deficiencies, with diverse intensity in different parts of the world, surface dams are becoming a more present, often dominant, method of improvement of water resources' state and solving the problem of prevention of environmental risks in water supply of population and economy.

During the second half of the twentieth century, problems of spatial and temporal disparities of required and available water resources were generated by the extreme growth of their accumulation needs. During the period 1950–1970 the volume of accumulated water in the USA increased by 6.5, and in the USSR by 7 times (Shiklomanov 2000). At about the same time, on a global scale the accumulation number is increased by 4, and their capacity for 8 times, which indicates a trend of the implementation of bigger facilities (Voropaev and Vendrov 1979). In the period 1950–1986 the number of dams was increased by 7 times, to about 39,000, by which reservoirs' capacity in relation to the total annual natural drainage of water reached 6% worldwide, mostly in North America – 22% and Europe – 10%. Spatial expansion of accumulation basins in the future will be governed by two thirds of the total number of Earth's rivers (Avakyan and Yakovleva 1997).

Informations on the reservoirs, particularly small ones (capacity of up to 10 million m^3) are often contradictory and unreliable. In contrast, the data on accumulation capacities greater than 0.1 km^3 are considered sufficiently reliable. Capacity of these accumulations accounts for more than 95% of the accumulated water in the world. Dynamics of the increase in the number and total capacity of reservoirs of a single volume more than 100 million m^3, by continents is presented in Table 12.1.

12.4 Cost-Benefit Analysis of Water Resources Management

Water resources are not "deployed" as people would like. Sometimes in some places there are too few, while others have too much. Spatial and temporal distribution of fresh water resources is in the inverse relation with formed social needs. Of the total annual precipitation on the Earth, 3/4 is on the territories with only 1/3 of the world population (Gleick 1993a; Gleick 1993b), which creates the impression that modern society lives under tyranny of the water cycle (Falkenmark 1990; Falkenmark 1994). However, we can not accuse nature for that, but human society and the model of disposition and development that did not take into account the natural water regime and spatial and temporal availability of water resources (Milinčić et al. 2007a; Milinčić and Šabić 2009). Therefore, the construction of dams and multifunctional reservoirs is long-lasting and widely accepted method of transforming natural water regime and its security for supply of the population and economy. The first known water accumulation for agriculture and water supply needs, probably began in southern India (Wittfogel 1956). During time, with different intensity in diverse parts of the world, there was a need for development of the inter basin water transfer system as an active way of adapting to the existing regional disparities in terms of available and needed water resources. In some

Table 12.1 Dynamics of the increase in the number and capacity of reservoirs (total capacity from 0.1 km³) by continents (Avakyan and Yakovleva 1997)

Continent		Years Before 1900	1901–1950	1951–1960	1961–1970	1971–1980	1981–1990	After 1990	Total
North America	1	25	342	178	216	113	34	7	915
	2	8.4	344.7	254.4	534	339.3	176.9	34.7	1,692
Central and South America	1	1	22	30	54	88	51	19	265
	2	0.3	8.8	28.8	96.9	251.5	349.1	236.1	971.5
Europe	1	9	104	113	172	94	76	8	576
	2	3.3	121.7	175	189.4	103.6	49.3	2.7	645
Asia	1	5	47	161	215	222	138	27	815
	2	1.7	17.9	293.6	640	484.1	321.5	221.6	1,980.4
Africa	1	1	15	21	24	27	52	6	176
	2	0.1	15	381.1	364.4	173.7	56.6	9.8	1,000.7
Australia and New Zealand	1	n.a.	10	21	18	27	12	1	89
	2	n.a.	10.6	20.1	15.5	42.4	5.9	0.3	94.8
Total for world	1	41	540	524	699	601	363	68	2,836
	2	13.8	518.7	1,153	1,840.2	1,394.3	959.3	505.2	6,384.5

(1) – number of reservoirs (2) – capacity in km³

12 Management of Spring Zones of Surface Water

countries (Poland and Pakistan), these problems are resolved by population's migration programs within the national territory, from areas deficit in water to ones which are secured of this resource (Milinčić et al. 2007b).

Long-lasting and rich experience in water resources' accumulation has not resolved the dilemma of their ability, validity and acceptability, and also on which level the spatial dispersion of man-made environmental effects is (Milinčić 2009a). In contrast, elapsed time (problem of perception, virulent effects of undesirable consequences, the synergetic effect of phenomena and processes in geographical space) revealed numerous and significant security limitations of this model of water supply. Therefore, their ability and importance in the management of water resources, prevention of environmental risks and the overall impact on environmental, natural and social components of the space are now being interpreted divergently. Construction of such large, complex and expensive hydraulic system is the most reliable way to increase total exploitable resources, and to provide necessary quantity with acceptable quality, for the supply of various categories of consumers (Markov 1986; Đorđević 1990; Milinčić and Jovanović 2008). Or such systems may only be able to resolve current imbalance between established need for water resources and natural water regime. Respectively, they have no alternative when it comes to the temporal and spatial ability of water resources' redistribution, in accordance with environmental and water management demands. Also, they are only able to optimally solve the complex problems of water use, protection of water and against one, and integrated spatial planning. In contrast, there are different views. Limited ability of preservation of reservoirs' capacity volume and quality of accumulated water, compromising water usage from these systems (Milinčić 2001). Or that such solutions for providing the necessary amount of water are only a sudden shift to much more expensive ones (Komatina 1990; Stevanović 1991; Vasileva 1977).

This and other dilemmas of the former "general acceptance" of surface water and reservoirs' resources, are being replaced with more and more frequent reexamination of their justification, positive and negative effects in short and long term. This is particularly highlighted by the fact that, due to a series of natural processes, with a smaller or greater intensity, surface reservoirs' economic and ecological value is being reduced over time. Also, there is the fact that positive effects of environmental risks' prevention emanated from the accumulated water, are carried out downstream of the reservoir, particularly in the areas of water consumers. In contrast, accumulations – which present physical occupancy of space and need for protection of accumulated water – are an obstacle to their dispersion into the water – gathering zone. This transformation of water regime, for various local communities, is a factor in producing "costs and losses", and one of the problems of regional income redistribution and social welfare (Milinčić 2004). The phenomenon of growth of accumulated water, in mid of the twentieth century, has sharpened the question of "upstream against downstream interests". This question in its initial phase was related to benefits that has a first consumer in the basin and now, it becomes a matter of interests, relationships, and collisions of spring water zones' and water consumers' zones.

234 M. Milinčić and T. Đorđević

The overall effects of existing and newly created problems considering accumulations, led to declining trend of their construction, in the end of the twentieth century. Possible answer requires activation of different solutions' variant, in accordance with natural, ecological, economic, geographical, cultural, historical and other specificities. This situation caused the change in direction of the concept of integrated water management in many countries. In addition to questioning the social and economic feasibility of already planned accumulations' construction, consuming rationalization, economic evaluation and reduction of total water consumption are also being emphasised (Davidson and Wibberley 1977). Also, in order to fit these systems into the existing natural and social structure of space, small (capacity of 200,000–2,000,000 m^3) and micro reservoirs (capacity of 20,000 m^3) are being increasingly favored. They also, besides local water supply usage, can have effectively multifunctional valorization (irrigation, protection against torrents, tourism, fisheries, etc.).

12.5 Water Resources of South Eastern European Countries and Ecological Risks

Teritory of SEE is conditionally defined and analyzed within the following countries: Albania, Greece, FYROM (the former Yugoslav Republic of Macedonia), Bulgaria, Romania, Croatia, Slovenia, B&H (Bosnia and Herzegovina), Montenegro and Serbia (autonomous province of Kosovo and Metohija is UN (United Nations) administered territory under UN Security Council Resolution 1244, of 12.06.1999). This determined region has an area of 764,949 km^2 and 66.86 million inhabitants, of which 55% is urban. Territorially the largest countries are Romania (237,500 km^2) and Greece (131,985 km^2), and the smallest are Montenegro (13,812 km^2) and Albania (28,750 km^2). The biggest share of the total urban population have Bulgaria (67%) and Greece (60%) and of rural population – B&H (57%) and Slovenia (51%).

The main feature of the SEE countries' population dynamics, during the period 1990–2000, was a descending trend of the total population of about 1.6 million (The World Bank 2003). The most significant absolute population decline was registered in Bulgaria, Romania and B&H. Trend of population growth is present in Greece, FYROM and Slovenia. Also, characteristics of a large number of dispersed population and a small average population size of settlements are registered. Almost all mountain areas in the region of SEE are underpopulated, and exposed to depopulation. Urbanization, on the one hand (growth – development), and deruralisation and depopulation on the other hand (demographic and economic stagnation), during the second half of the twentieth century produced significant spatial polarization of socio-economic and ecological space.

The region of SEE is a representative example of maxim accepted in the first UN Conference on Water held in Mar del Plata (1977) – "there would, perhaps, be enough water to satisfy all the needs, in general, if it would not always find its way to be at the wrong time at the wrong place, with the wrong quality". At the The

12 Management of Spring Zones of Surface Water 235

Third Ministerial Conference "Environment for Europe" in Sofia (1995) it was pointed out that the lack of water resources is a serious problem for 60% of industrial and urban centers of Central, Eastern and Southern Europe. These facts are of particular importance in the context of global climate change – rise of temperature and rainfall decline. The consequences of these changes, especially during vegetative period (high level of evapotranspiration) are: frequent drought, reduction of the domicile water resources and deterioration of their quality, decrease of spring yield, and often their overexploitation. Reduction of runoff threatens energetic stability of the region – Albania (83%), Montenegro (65%), B&H (61%), and Croatia (54%) are the most dependent on hydroelectric power (The World Bank 2003).

Total domicile water resources of the SEE territory are 271.4 km^3/yr., or 4,059 m^3/inhab/yr. Generally, these water resources are important, but they are unevenly distributed among individual countries (Table 12.2) and within them. Maximum availability is registered in Montenegro – 14,645 m^3/inhab/yr., and minimum in Serbia – 1,687 m^3/inhab/yr. Above average availability (per capita) have six countries (Montenegro, Slovenia, B&H, Albania, Croatia and Greece) with a total of 24.94 million inhabitants, and below average four countries (Serbia, Romania, Bulgaria and FYROM), with 41.92 million inhabitants.

Temporal (seasonal, annual and perennial) variability of water resources, and their torrent pluviometric regime make difficulties in putting periods of droughts and floods under control. Seasonal deficit of water resources, which became more often and spatially increased, cause numerous environmental, social and economic risks. This is especially characteristic of parts of Bulgaria, Romania, Serbia, FYROM, Greece and Albania. In these countries seasonal rivers' overdrying are frequent, even of these with a catchment area larger than 1,000 km^2. In Bulgaria, the average runoff in the period of 1961–1999 was reduced by 7% than that in period 1935–1984. However, it is 40% lower during the period 1985–1995 compared with

Table 12.2 Average precipitation in depth and availibility of renewable water resources in the SEE countries (FAO AQUASTAT (FAO's information system on water and agriculture) 2002; UN WWAP (The United Nations World Water Assessment Programme) 2006)

Country	Average precipitation (mm/yr)	Renewable water resources					
		Total internal		Total external		Σ	
		km^3	m^3/inhab/yr	km^3	m^3/inhab/yr	km^3	m^3/inhab/yr
Albania	1,485	26.9	8,583	14.8	4,722	41.7	13,306
B&H	1,250	36.0	9,052	2.0	503	38.0	9,555
Bulgaria[a]	650	21.0	2,642	0.3	38	21.3	2,680
Croatia	1,089	37.7	8,101	33.7	7,241	71.4	15,342
FYROM	619	5.4	2,655	1.0	492	6.4	3,147
Romania	700	42.3	1,885	169.6	7,556	211.9	9,442
Serbia	730	16.03	1,687	162	17,053	178.03	18,740
Montenegro	2,013	9.08	14,645	2.5	4,032	11.58	16,129
Greece	652	58	5,492	16.3	1,544	74.3	7,031
Slovenia	1,570	19	9,500	13	6,500	32	16,034

[a]The Danube's water resources are not taken into consideration

the period 1935–1984 (The World Bank 2003). Flood character of pluviometric regime, especially in the eastern part of territory under study, condition a greater part of the annual flow in the short-term period of high waters, which can cause major ecological and economic disasters. Also, the long period of small waters is unfavourable to the rivers as ecosystems (below biological minimum) and for all categories of consumers (below the minimum needed for water management).

Specific cause of the deficit of water resources and permanent presence of environmental risks in the wider part of the SEE territory is conditioned by its petrological and geological features. This particularly refers to the southern parts of the analyzed area, especially on islands, coastal parts of Croatia, Montenegro, Greece, B&H and Slovenia. Although it has high average rainfall amount, karstic surface cause outstanding lack of surface waters and their easy contamination. The situation is particularly unfavourable during years of drought and the tourist season (massive and highly seasonal character of tourism industry and established needs for water resources).

Such environmental risks are representative of the example of Montenegro. The average precipitation of 2,013 mm and discharge from 44 l/s/km² (phenomena present in less than 3% of world's landmass), provide an extremely high potential density of internal water resources' availability of 23,578 m³/inhab/yr. Although there are significant regional differences, hydro-geological features cause a small overall length of the river network (1,700 km) and average rivers' density of 123 m/km² of predominantly groundwater. The absence of surface water is affecting water resources in some regions of Montenegro, especially coastal, the most urbanized and the most active economic region and emerging as the most serious cause of environmental risks and constraints of development. The problem is trying to be solved by an abstraction of water from the basin of Lake Skadar, and construction of a complex regional water system that would cover the areas of most intensive water consumption in the coastal part of Montenegro.

It is increasingly stressed the problem of water allocation among different categories of consumers, local and regional communities, so as problem of shared water resources among different countries. Disputable situations will be intensified, in perspective, in trend of changing the regime of large and small water – large are becoming even bigger and small are becoming much less. Long-lasting process of transition, political tensions and conflicts had negative impact on the state of water facility management in function of water regimes, water supply, sewerage, irrigation, waste water treatment and others. In the majority of distribution networks, losses are ranging from 50% to 60%, and in the most urban zones, their capacity does not follow the consumption growth. Access to public water and sewage systems, although low in general, is especially disadvantaged in rural areas of SEE. Participation of public water supply is lowest in rural communities in Romania (34%) and FYROM (30%). In most countries (Albania, Serbia, Montenegro, B&H, FYROM) waste water level is less than 10% so that the problem of their quality is generally present. Reducing volume and adulteration of water quality, besides deterioration of parts of an irrigation system, adversely affects agricultural production in Bulgaria, FYROM, Serbia, Romania and Albania.

12 Management of Spring Zones of Surface Water   237

Due to the overall morphostructures' capacity, on the regional and mesoregional level (the Danube and the Sava faults), and formed hydrological network (convergence zone of large rivers the Danube, Sava, Drava, Tisa, Begej, Tamiš), northern parts of SEE dispose of an abundance of external water (hydrographic hub of Europe) especially in the Danube and its tributaries area. Participation of external in the total water resources is extremely large in Serbia and Bulgaria – over 90%, and in Romania – more than 80%. Some regions of these countries are forced to supplement their water needs with these water resources. An important factor of their quality is time consistency, especially in comparison to domicile water resources. However, their poor quality and inability to make an influence on it appears as the most important limiting factor of the potential utilization. In the case of Bulgaria, there are a number of limiting factors of intensive usage of these water resources – peripheral position in relation to economic, demographic and settlement spatial orientation. The least share of external in the total water resources have B&H – 5.26%, FYROM – 15.63%, Montenegro – 21.56%, Greece – 21.94% and Albania – 35%.

The increase in importance of transboundary waters should be analyzed through the effects of political fragmentation of the SEE region (disintegration of Yugoslavia). This area, with over 92% of the territory within international water basins, has become one of most internationalized in the world (international basins cover about 50% of total land area) (Table 12.2). Most major river basins are transboundary – the Danube, Sava, Drina, Drim, Krka, Neretva, Marica, Vardar, Timok, Mesta, Struma and others. More than half of the total number of transboundary river basins are shared by three or more countries. The biggest affiliation of its national territory to inter-state basins have: FYROM (97.7%), Romania (96%), B&H (93.5%), and Greece (18.8%), Albania (49.6%) and Croatia (63%) have the lowest one (The World Bank 2003) (Table 12.3).

Table 12.3 Countries of SEE and their percentage in international basins (Hydropower & Dams World Atlas and Industry Guide 1997)

Indicator		Percentage of country in international basins	
Country	Area (km²)	(km²)	(%)
Albania	28,750	14,260	49.6
B&H	51,129	47,805	93.5
Bulgaria	110,910	95,161	85.8
Croatia	56,538	35,619	63.0
FYROM	25,713	25,122	97.7
Romania	237,500	22,800	96.0
Serbia	88,361	81,292	92.0
Montenegro	13,812	9,967	72.2
Greece	131,985	24,813	18.8
Slovenia	20,251	18,205	89.9

238 M. Milinčić and T. Đorđević

In Bosnia and Herzegovina this problem is additionally complicated by entities' and cantons' borders, and various current and potential interests. On the territory of Kosovo and Metohija, which is a particularly water scarced region, a potential problem is the question of the functional valorilization of Gazivode accumulation (volume $350 \times 10^6 \, m^3$) on the Ibar river and hydrosystem Ibar–Lepenac. Functioning of the system, and economic and environmental services of accumulated water resources, can be jeopardized by the fact that different ethnic communities can control and have influence on some parts of this complex system (Milinčić 2009a; Milinčić et al 2009).

The territory of Kosovo and Metohija is by deficit of water one of the most vulnerable areas in SEE – 206 m^3/inhab/yr, and 60% of the total amount of water supplied by public water companies is mainly affecting the surface water. Only 28% of the total population is connected to the sewage system, so the numerous local water resources are contaminated and pose a threat to public health. Due to the lack of adequate sanitation and intense pollution of surface and ground water, about 80% of the rural population consume contaminated water from private wells (World Water Council 2007).

Regions of SEE are naturally predisposed with strong erosive processes – particularly hilly mountain areas of Bulgaria, Romania, Serbia, FYROM and Albania. In addition to geomorphological predisposition, significant stimulus to expressed erosion and accumulation processes comes from: the lack of areas covered with trees and the poor condition of forests, a large share of agricultural in total areas, disregard to criteria of soil fertility and erodibility in process of valorization etc. Erosion and accumulation of material are significant elements of water resources quality and quantity, possibility of their accumulation, as well as ecological and economic state of accumulation. FYROM, despite long and intensive work on erosion control and sanation, annually loses about 0.3 km^3 of useful capacity of reservoirs (The World Bank 2003).

Increased sediment production in rivers leads to backfilling of accumulation, and ultimately their consumption. Loss of the existing and small amount of suitable locations for the construction of accumulation is a particularly difficult geographical

Table 12.4 Total dam capacity in the SEE countries

Country	Capacity (km^3)	Water resources per capita (m^3/year)	Total surface water resources (%)
Albania	4.56	1,455	12
B&H	3.85	968	10
Bulgaria	5.00	629	25
Croatia	1.53	329	3
FYROM	1.70	836	27
Romania	14.00	624	11
Serbia	6.20	653	3.5
Montenegro	1.13	1,822	9.76
Greece	12.51	1,184	17
Slovenia	0.20	101	0.63

12 Management of Spring Zones of Surface Water · 239

handicap in the "fight" for the water resources management in the SEE. The fact that there exist only few locations of these kind, and that their availability in the future will be in inverse proportion to increased needs, conditioned that they are becoming a highly significant, and until now not enough appreciated, spatial resource. Their limitation is partly resulting in expendability and non-renewability, so as the necessary positive combination of a large number of morphological, hydrological and antrophogeographical elements of space.

In the order of water resources management and ones utilization, the SEE countries have a total accumulations' capacity of $51.13\,km^3$ or $764.73\,m^3$/inhab. Accumulated volume of water per capita ranges from $100\,m^3$ in Slovenia and $329\,m^3$ in Croatia, to $1,184\,m^3$ in Greece, $1,455\,m^3$ in Albania and $1,822\,m^3$ in Montenegro. With an exception of Montenegro, Albania and Greece, among all other SEE countries total capacity of reservoirs is below the world average ($1,100\,m^3$/inhab). The possibility of total renewable surface water resources accumulation on individual state level varies from 0.63% (Slovenia) to 27% (FYROM).

The tradition of construction of modern large dams and reservoirs in SEE dates from the second half of the nineteenth century, but their modern presence and the ability to accumulate water in the territories of some countries varies (Table 12.4). Trends of significant growth of reservoirs' number and ones opportunities of water accumulation began in the second half of the twentieth century. The primary reason for construction of reservoirs was need of intensive electrification, and secondary was prevention of flood waves and irrigation of agricultural areas. During that time some of the most significant accumulations were realized in the SEE region. By volume of accumulated water these are especially being distinguished – Đerdap I ($2.55\,km^3$), Đerdap II ($0.87\,km^3$), Piva Lake ($0.86\,km^3$), and by the height of dam – Mratinje (220 m) and Gazivode (107.5 m).

Over time, due to process of industrialization and urbanization, and intensive and concentrated growth of water resources consumption, there has been a shifting of focus toward the accumulation of water supply function. This applies to both the existing and the newly formed reservoirs. Therefore, an increase participation of surface, usually accumulated water resources in water supply of the population in most countries in SEE was dominant: Bulgaria – 76%, Greece – 75%, Romania – 71%, Albania – 30%, Serbia – 30%, FYROM – 20%, B&H – 11%.

12.6 General Characteristics of Water Resources in Serbia

The parameters of high average precipitation (723–737 mm), surface runoff (from 24.56 to 27.87% – 5.76–6.6 $l/s/km^2$), evapotranspiration (73.13 to 75.44%) and discharge formed on the territory of Serbia are not uniformly accepted (Milinčić 2001). This is why there are different assumptions about the average availability of internal water resources – $16\,km^3$ (Vodoprivredna osnova Republike Srbije – Water management basics of Serbia 2001), $16.7\,km^3$ (Vojnović 1995), $17.86\,km^3$ (Milinčić et al 2007b), to $18.4\,km^3$ (Gavrilović and Đorlijevski 1996). Nevertheless, this discharge ensures 1,600–1,800 m³ of internal water resources per capita per year. The areas in Serbia which are the mostly populated and economically developed (Vojvodina, Šumadija, Kolubara, Kosovo, Južno Pomoravlje) are the poorest in internal water resources.

The unfavourable situation of internal water resources' sufficiency on 1/3 of the Serbian territory is greatly facilitated with plenty of external water resources (162–173 km³/year). A positive fact is that they are spatially located mostly in the territories with low specific runoff (usually 2–4 l/s/km²) and low density of the river networks. Due to the external waters, Serbia's total renewable resources amount is over 178.03 km³, or 18,740 m³/inhab., which is 3.5 times more than Europe's average. The fact which is unfavourable is that 90–95% of external waters have an adverse balance of quality, and that Serbia, under their influence is the recipient of significant pollution.

There is a significant seasonal and annual variability of internal waters, and therefore, ones availability is being considerably less favorable. The average flow during drought and wet years is, in most rivers, in ratio 1:3–1:5, and more. The annual runoff of bigger rivers in the long period of time indicates large seasonal fluctuations: Drina 1:3, Tisa 1:4.63, Zapadna Morava 1:7.11, Južna Morava 1:7.6, Mlava 1:9.15, Timok 1:28.7, Erenik 1:39.6 (Dukić and Gavrilović 1989). Major rivers' extreme flow amplitudes, as a real representation of the flow state, are even less favorable: Ibar 1:23; Zapadna Morava 1:161; Velika Morava 1:163, Nišava 1:233, Južna Morava 1:266, Crni Timok 1:760, Beli Drim 1:920 (Dukić and Gavrilović 1989). On small rivers relations are even worse: Binačka Morava 1:7,240, Toplica 1:5,200, Sitnica 1:4,054 etc.

This pluviometric regime, as the most important element of Serbian water resources', is considered as "chaotic, almost catastrophic in its consequences" (Dukić and Gavrilović 1989:9). Also, the torrent character of Serbia's pluviometric regime (8,400 torrents) conditions that 60–75% of the total annual flow is being provided in a brief torrent. Floods, due to their long spatial presence, frequency, material damage and human casualties, were and still are one of the most destructive natural disasters and environmental risks in Serbia. The most vulnerable areas with complex geopotential structure are valleys of small and medium torrent flows and alluvial plains of large rivers.

The total length of the rivers network in Serbia is 65,980 km, or approximately 747 m/km². Due to topographical and hydro-geological conditions, distinct regional differences in terms of size and speed of runoff to the permanent watercourses are present. The differences among the regional river networks density are significant – from 123 m/km² in karstic terrains of Eastern Serbia and 225 m/km² in Vojvodina (semi water permeable sand and loess soil) to a maximum of 3,492 m/km² at the spring zone of Ribnica river, the right tributary of the river Ibar.

The quality of Serbia's water resources is unsatisfactory, with a tendency to further deterioration. Poor water quality and reduced ability for utilization are becoming more important factors of its deficit. The Serbian industry, due to its outdated technology, is a very extensive resource consumer. This is particularly valid in the domain of excessive and irrational water use and waste water emissions.

12 Management of Spring Zones of Surface Water   241

Two decades of outstanding debt and descending trend of physical volume of production and socio-economic development lags, did not significantly impact on water consumption reduction in industry and households, or reduction of the watercourses' pollution level. The average production of waste water is 3.5×10^6 m³/day, or 1.2×10^9 m³/year which represents the 7.5% of the total internal waters' runoff from the territory of Serbia.

Access to sewage system is not satisfactory, and particularly smaller towns and rural areas are faced with a much lower coverage. The situation is even worse in terms of municipal and industrial waste waters' treatment. About 87% of the total waste waters are being discharged without treatment in the recipients. This situation is characteristic for the biggest cities in Serbia: Belgrade, Novi Sad, Priština and Niš. Numerous plants were left unfinished, and the existing plants are being poorly maintained with frequent discontinuities in the work. The largest number of functional systems allows only occasional primary (mechanical) treatment of waste waters.

Long lasting and complex processes of pollution and degradation cause that water of most major rivers in Serbia has III and IV quality class. The most polluted are small and medium-sized rivers. As a result of discharges from untreated municipal and industrial waste, particularly waters of the rivers downstream of municipalities show significant decline in quality. This situation is most prominent in the following cases: Crni Timok downstream from Zaječar, Rasina downstream from Kruševac, Veliki Lug downstream from Mladenovac, Lepenica downstream from Kragujevac, Toplica downstream from Prokuplje etc. Some rivers are along the whole course out of class: Topčiderska river, Borska river, Bosut, Lugomir, Prištevka, etc. Numerous irrigation systems and small accumulations in the zones of larger urban settlements, by releasing large amounts of untreated municipal and industrial water, are being converted into the waste water collectors. The worst water quality (out of class) have many channels in Vojvodina: Galovica, Stari and Plovni Begej, channel Vrbas – Bečej, part of channel Sivac – Vrbas – Srbobran and Danube – Tisa – Danube Canal.

Southern of the Sava and the Danube rivers, groundwater and surface water courses are relatively clean at the top, but the quality deteriorates rapidly in the middle and lower parts of streams and river valleys. Their pollution is usually going below the contour line of 600 m (Milinčić 2004) and above this level water is good or acceptable quality (Class I and II).

12.7 Spring Zones of Surface Water in Serbia in the Function of Water Supply and the Prevention of Ecological Risks

On the territory of Serbia, spring zones of surface water in the function of water supply have been continuously present for eight decades. During the last quarter of the twentieth century, they are becoming one of the most promising models for providing growing water demands. In that sense, the following documents were adopted: the

Basics of long-term water supply of population and industry on the territory of Serbia without the territories of autonomous provinces (Osnove dugoročnog snabdevanja vodom stanovništva i industrije na teritoriji SR Srbije van teritorija autonomnih pokrajina, 1977) and Act on the exploitation and protection of spring zones of water sources (Zakon o iskorišćavanju i zaštiti izvorišta vodosnabdevanja 1977).

Thus, strategic decisions in the direction of: abstraction and protection of ground and surface water resources of the first rank, affirmation of surface reservoirs (as one of the key models of water resources' security) and the formation of regional and inter-regional water supply systems, were conducted. For the purpose of long-term quality water supply, in the region of central Serbia, the first rank groundwater and surface water resources (regional resources) were abstracted and protected, while the protection of the second rank waters (municipal spring zones) was left to municipalities. In the category of the first ranked spring zones of surface water, 33 territories with total area of 10,262 km² (Table 12.5), and in the category of the second ranked spring zones of surface water, 35 territories with total area of 2,341 km² are protected according to the legislation adopted in 1977. Starting solutions are being confirmed and upgraded two decades later – Spatial Plan of the Republic of Serbia (1996), the long-term strategy of development and improvement of water management in the Republic of Serbia (Strategija dugoročnog razvoja i unapređenja vodoprivrede RS 1997), Water management basics of Serbia (Vodoprivredna osnova Republike Srbije 2001) and others.

Beginning of the second half of the twentieth century was a period up to a global economic and geographical development of Serbia, with all the ancillary difficulties, by exploitation of natural water resources' regime, was possible. Small scale, disorganized and unsuitable water regime, stimulated finding of a different, temporary or permanent, local or regional solutions. However, continued expanding of municipal systems, industrialization, urbanization and the growth of standard (specific growth in consumption) made the existing bigger, and created new local and regional problems of water deficit. The growth of abstraction and water consumption had side effects in the production of waste water.

Reason for "unexpected" dealing with the problems of water resources' limitation and insufficiency was partly a consequence of current socio-economic development rate that led to doubling of consumption and producing waste waters, at intervals of eight to ten years. Growth rate of water catchment for public water supply was even faster. During the period 1950–1970 it was increased by more than eight times, from $20 \times 10^6 \, m^3$ to $170 \times 10^6 \, m^3$. In the same period, specific water consumption in urban areas rose from 100 to 285 l/inhab/day (Milinčić 2009a).

At the beginning of the twentieth century in Serbia, the first significant local and regional deficits of water resources for the supply of population and industry appeared. Kragujevac, in this situation, was particularly affected. This caused the need for conducting first surveys in the region of central Šumadija, which were related to the issues of ecological risks and water deficit prevention. In the period from 1927 to 1930 the legal acts for building and necessity of the accumulation of Grošnica have been prepared. The work on its construction lasted between 1931–1937, and during spring of 1938 water from the catchment area of 30 km² was

12 Management of Spring Zones of Surface Water 243

Table 12.5 The main characteristics of the first rank spring zones of surface water in Serbia (Milinčić 2009a; Milinčić 2001)

River	Dam profile Location	Municipality	Area of spring zone (km^2)	Usability 10^{9}m^3/year
Jablanica	Rovni	Valjevo	109	22
Kamenica	Rošci	Čačak	198	27
Čemernica	G. Gorjevnica	Čačak	138	19
Dičina	Semedraž	G. Milanovac	211	27
Gruža	Tucački Naper	Knić	359	27
Đetinja	Gradina	Užice	170	30
Sušica	Tripkova	Čajetina	163	30
Crni Rzav	Panjak	Užice	196	37
Veliki Rzav	Arilje	Arilje	554	109
Grabovačka	Cerovo	Ivanjica	72	16
Lučka	Bedina Varoš	Ivanjica	94	17
Uvac	Kokin Brod	Nova Varoš	1,100	225
Moravica	Bedina Varoš	Ivanjica	370	82
Studenica	Đakovo	Kraljevo	442	83
Ljudska	Požega	Novi Pazar	188	25
Raška	Gradina	Novi Pazar	211	27
Rasina	Ćelije	Kruševac	598	94
Toplica	Selova	Kuršumlija	350	60
Mlava	Gornjak	Žagubica	702	128
Veliki Pek	Debeli Lug	Majdanpek	182	28
Resava	Strmosten	Despotovac	121	26
Crnica	Zabrega	Paraćin	102	19
Grza	Čestobrodica	Paraćin	71	12
Crni Timok	Bogovina	Boljevac	466	100
Trg. Timok	Baranica	Knjaževac	496	64
Temštica	Temska	Pirot	755	135
Vlasina	Vlasotince	Vlasotince	918	147
Vrla	Rajkinci	Surdulica	126	23
Božička	Bosilegrad	Bosilegrad	189	43
Veternica	Barje	Leskovac	230	36
Banjska	Prvonek	Vranje	86	17
Grliška	Grlište	Zaječar	178	15
Ribnica	Paštrić	Mionica	104	16

accumulated. This was the first modern monofunctional reservoir with the purpose of population water supply in Serbia (2.13 million m^3 of total, or 1.6 million m^3 of useful volume).

At the beginning of the second half of the twentieth century, local and regional problems of water deficit in Šumadija have been renewed and new ones created, and the crisis moments of quantity and quality of local spring zones have become more distinct. These phenomena are results of adverse natural conditions and major socio-economic changes (urbanization and industrialization). Chronic water supply problems in Kragujevac were re-actualized during the decade of the 1950s. Such rapid return of water deficit and environmental risks problems caused by the growth of demands and backfilling and diminishing of Grošnica reservoir's volume. Since 1957 significant technical and biological works for protection of the reservoir were undertaken – torrent regulation and basin afforestation.

During the seventh and eighth decade of the twentieth century in order to provide fast – growing needs for water resources in Šumadija, existing, mainly monofunctional accumulations were reconstructed, and new ones built. At the same time, linking and embedding of complex hydraulic systems for transfer of water within catchments and between catchments were carried out. During the period from 1960 to 1962 the Grošnica dam was upgraded for 7.3 m, and the total volume of the reservoir was increased to $3.53 \times 10^6 \, m^3$. In order to increase water balance of Grošnica lake, during the period from 1964 to 1969, from micro-reservoir on the river Dulenska, with the pipeline, $2 \times 10^6 \, m^3$ of water was transferred annually. Problems of Aranđelovac water supply caused the requirement for establishing a complex hydropower system of three rivers catchments (Kačer, Velika and Mala Bukulja) and two reservoirs Velika Bukulja in 1965 and Garaši in 1976, whose small scale capabilities failed to serve existing and growing consumers' needs.

Lasting problems of water supply of Kragujevac and its surroundings and some settlements in the Zapadna Morava river basin, are being resolved by reservoirs constructing on the Gruža river. The first reservoir was built in 1984, according to the Basics of long-term water supply of population and industry on the territory of Serbia without the territories of autonomous provinces (1977), and became a key source (1,200 l/s) and basis of a regional Šumadija – Ibar water system.

At the same time the territory of Kosovo and Metohija has appeared as a regional critical spot considering water deficits and environmental risks. This has led to changes of focus from construction of planned multifunctional to monofunctional (for water supply) reservoirs. A striking example is the accumulation of Gračanica, built in 1965. Although planned as multifunctional (population supply, mining and irrigation of 2,300 ha of agricultural area), it had only one function – water supply of the population. After construction of accumulation Batlava, which was built in 1965 for needs of cooling thermal power plants and irrigation of 8,000 ha of agricultural land, function of the water supply for the population was added and nowadays that is its dominant feature.

On the territory of Serbia, 60 accumulations with high dams and more than 100 small and micro reservoirs have been built. In the group of high dams, there are 29 large reservoirs, with a single volume of more than 10 million m³, and the total volume of about $6 \times 10^9 \, m^3$. Other high dams' accumulations – 31, have an individual capacity of less than 10 million m³.

Of the total number of reservoirs with high dams, 22 are being used for water supply of the population (Table 12.6). Their total volume of accumulated water is $638 \times 10^6 \, m^3$, or 24% of the total accumulated internal water resources. With inclusion of these which are not yet fully constructed, and priority given monofunctional reservoirs for water supply (Stubo – Rovni, Bela stena and Selova), their volume will grow to $825 \times 10^6 \, m^3$, or on almost 1/3 of total accumulated internal waters. From 30 major reservoirs, 16 are being used for water supply of population: 10 are

12 Management of Spring Zones of Surface Water

Table 12.6 Accumulations in the function of water supply in Serbia (Milinčić 2009a)

Accumulation	River	Year of construction	Drainage basin area (km²)	Volume (10⁶ m³)	Purpose
Grošnica	Grošnica	1937	30	4	W
Vlasina	Vlasina	1949	349	176	P,W
Kokin Brod	Uvac	1962	1,057	250	P,W
Velika Bukulja	Velika Bukulja	1965	6	0.4	W
Gračanka	Gračanica	1965	104	32	W
Batlava	Batlava	1965	250	40	I,W
Ribničko	Crni Rzav	1972	70	3.5	W
Garaši	Bukulja	1976	22	6.3	W
Gazivode	Ibar	1977	1,060	350	P,F,N,W
Ćelije	Rasina	1978	598	60	W,F
Bovan	Moravica	1978	522	58	W,F
Lisina	Božička reka	1978	182	10.4	P,W
Sjenica	Uvac	1979	920	212	P,W
Radonjić	Prue	1980	34	110	W,N
Pridvorica	Pridvorička r.	1982	21	0.8	W
Vrutci	Đetinja	1984	160	54	W
Gruža	Gruža	1984	317	65	W
Brestovac	Pusta reka	1985	112	9.25	W
Grlište	Grliška reka	1988	191	12.5	W
Zavoj	Visočica	1989	584	170	P,W
Barje	Veternica	1991	233	41	W,F
Prvonek	Banjska	2006	86	20	W
Selova	Toplica	in constr.	349	70.5	W
Stubo-Rovni	Jablanica	in constr.	104	51.5	W

Purpose: *W* water supply, *I* industry supply, *P* power generation, *N* stabilize water flow/irrigation, *F* flood prevention

monofunctional (water supply of the population), 5 bifunctional (3 for water supply of population and flood protection and one for water supply and irrigation, the other one for water supply and energy needs) and one – Gazivode is an example of multifunctional reservoir (water supply, energy, irrigation, flood protection, flow control modes, a biological minimum providing, aquaculture farming and tourism).

Tables 12.5 and 12.6 are showing that some accumulations, which are being used for water supply, have been built out of ranked spring zones of surface water. Thus in a relatively short period of time significant territory *de facto* and *de jure* has got the function of spring zone of surface water. Serbia has already deeply got into the developing phase of these type water management systems. Besides accumulations and spring zones of surface water, there have also been over 4,700 km of main pipelines built for the purpose of spatial re-distribution of water. From 18 regional systems for water supply which were planned, 11 are relying, more or less, on spring zones of surface water and accumulation basins.

Some springs, due to asynchronous development dynamics of some systems' parts and expressed water deficit, are being functionally activated without completion. That is the case in following systems:

246 M. Milinčić and T. Đorđević

- Barje, where parts of the main pipeline for transport of processed water are missing.
- Stubo – Rovni on mt. Jablanica, where exists finished water factory, but there is no functional accumulation.
- Nerezine water factory, which takes water from river Vlasina.
- Rzav from 1993 (factory capacity is $1.2\,m^3/s$), which is providing water for Arilje, Požega, Lučani, Čačak and Gornji Milanovac, and takes fresh water from river Veliki Rzav etc.

These circumstances are partially causing the fact that, according to source of spring zone, in surface water usage structure, in the public companies for water supply in the central Serbia, taking water from river flows is dominant. At national level, watercourses are participating with 10% and accumulations with 12%, and in central Serbia, watercourses are participating with 13% and accumulation with 10%.

Spring zones of surface water on the territory of Serbia are providing over $150\,m^3/h$ of high quality water (annually $1.32 \times 10^6\,m^3$), which is necessary for water management needs. Complexity of potential problems of insufficient water resources and social and ecological consequences are already conditioning radical changes in relations with water management infrastructure. There are some deviations during plans realization in other economic areas which are depending on water resources (energetic – thermal power plants and hydro power plants, agriculture, tourism). Also, over-purpose of some hydro-systems which are already being built or are in the last phase of construction, is often the case, as moving priority of their usage into the course of water supply.

12.8 Conclusion

At the local and regional level, SEE has for a long time faced the problem of water resources' deficit and tradition of transformation (construction of channels, surface water accumulation and spatial and temporal redistribution) of natural waters' regime. Previous efforts with an aim of developing water management systems in the function of managing and securing water resources, although with considerable success, failed to solve the problems of poor quality, improper spatial and temporal distribution and lack of water resources. Usually there is absence of water, particularly in the areas where it is needed the most, and its deficits are the most expressed in periods of greatest demands, in all spheres of consumption (agriculture, domestic use, industry, tourism). In fact, long-lasting development periods of SEE countries did not take into account the determinants of the water resources, which have resulted in emerging of, varied and numerous, chronic and intermittent, ecological risks, which are different in intensity regarding different countries. Temporal and spatial dimensions of ecological risks in whose base are water resources, due to the effects of climate change, consumption growth, declining quality and less absolute and relative availability, are showing a tendency of transition from a linear into logarithmic trend of growth.

Recent experiences in the area of SEE show that improvement of environmental risk on water deficit consequences' was often expensive, inefficient and time-limited, and the overall environmental and socio-economic effects were often harmful. Therefore, the forehand development of water resources management systems in the function of prevention of environmental risks is essentially positive contrast to renewal of consequences that can appeared due to their absence. Such activities, regardless of the potentially significant material, spatial, organizational and other investments, are the imperative of rational and organized societies. In relation to the extremely high percentage of some SEE countries in international river basins, and their high participation to transborder water and mutual dependency, interaction and the risks of water resources safety are becoming wider.

Spring zones of surface water resources and reservoirs are playing an important role in the sphere of total area's water and other resources' sustainable management, and therefore in prevention of environmental risks. Surface water resources or reserves of clean water, are contributing to the preservation of the water resources' quality, and reservoirs, which are usually polyfunctional, are transforming downstream water regimes and their effects from dominantly devastating to exploitable. In a large scale, many other undesirable spatial and natural appearances, whose basis are effects of water resources management (floods, droughts, landslides etc.) so as existing ecosystem's protection are being eliminated. Also, a significant segment of the overall impacts is an actual increase of water resources' volume.

However, surface water resources and reservoirs in the function of water supply are causing numerous negative environmental and socio-economic consequences in the region. Significant spatial and environmental requirements of these systems cause noncomplementarity between management and numerous other activities in the area of surface water resources. Ideal monofunctional usage of space in the case of SEE countries, which is desirable in the function of spring zones of surface water, is problematic and realistically is hardly achievable. This is because the basic characteristics of the SEE region and predominately territorially small countries.

This chapter has proven, especially in the case of Serbia, that the accumulation of surface water resources is relevant and an important factor of various positive impacts on the management of quality and quantity of water resources and prevention of environmental risks. The deficit of water resources on the level of contemporary demographic and economic development, and certainly future problems' multiplication, indicates a potential importance of surface water resources management in the context of prevention of environmental risks and sustainable development.

References

Avakyan AB, Yakovleva VB (1997) New data on world reservoirs. Hydrotechnical Construction 31(3):143–148

Cutter SL, Renwick HL, Renwick WH (1999) Exploitation, conservation, preservation: a geographic perspective on natural resource use. Wiley, New York

Davidson J, Wibberley G (1977) Planning and the rural environment. Pergamon Press, Oxford

Dukić D, Gavrilović Lj (1989) Водни ресурси СР Србије – њихово искоришћавање и заштита (Eng. Water resources in Republic of Serbia – their exploitation and valorisation). *Гласник српског географског друштва*, 69, 7–18

FAO (2002) Press Release 02/33: World water day 2002. FAO: Agriculture needs to produce more food with less water. Resource document. FAO. www.fao.org/wfd/wfd2002/. Accessed 3 Apr 2010

FAO AQUASTAT (2002) Food and Agriculture Organization *of the United Nations:* Review of water resources statistics by country. Resource document. FAO AQUASTAT. www.fao.org/nr/water/aquastat/water_res/index.stm. Accessed 10 Apr 2010

Đorđević B (1990) Vodoprivredni sistemi (Eng. Water management systems). Naučna knjiga, Belgrade

Falkenmark M (1990) Population growth and water supplies: an emerging crisis. People 17(1):18–20

Falkenmark M (1994) Population, environment and development: a water perspective. In: Population, environment and development: Proceedings of the United Nations Expert Group Meeting on Population, Environment and Development. United Nations, New York, pp 99–116

Gavrilović Lj, Đorelijevski S (1996) Површинските води као изборишта за снабдување со вода за пиење во Сербија (Eng. Surface Water as Spring Zones for Water Supply of Population in Serbia). *Качество и пречистване на питејните води в Блгарија* 1:88–97

Gleick P (1993a) Water and conflict: fresh water resources and international security. Int Secur 18(1):79–112

Gleick P (1993b) Water in crisis: a guide to the world's fresh water resources. Oxford University Press, New York

Gorski N (1965) Voda čudo prirode (Eng. Water – miracle of nature). Vuk Karadžić, Belgrade

Hauchler I, Messner D, Nauscheler F (1998) Global Trends, Fakten, Analiysen, Prognosen. Ficher taschenbuch Verlag, Frankfurt

International Journal on Hydropower and Dams (1997) Hydropower & Dams World Atlas and Industry Guide. Aqua-Media International Ltd., United Kingdom

Komatina M (1990) Hidrogeološka istraživanja – primenjena hidrogeologija III (Eng. Hydro geological surveys – applied hydrogeology III). Geozavod, Belgrade

Markov J (1986) *Социальная экология* (Eng. Social ecology). USSR Academy of Sciences, Novosibirsk

Milinčić M (2001) *Србија – геополитика животне средине* (Eng. Serbia – environmental geopolitics). Српско географско друштво, Belgrade

Milinčić M (2004) Економско – географска и еколошка поларизација простора као фактор нових функционалних односа међу просторним целинама (Eng. The economic – geographical and environmental spatial polarization as a factor of new functional relations). *Гласник српског географског друштва* 84:157–164

Milinčić M (2005) Дефицит вода и географија жеђи (Eng. Water deficit and "geography of thirst"). *Земља и људи* 55:89–100

Milinčić M (2009a) *Изворишта површинских вода Србије – еколошка ограничења и ревитализација насеља* (Eng. Spring zones of surface waters in Serbia – ecological limits and revitalization of settlements). Univerzitet u Beogradu, Geografski fakultet, Belgrade

Milinčić M (2009b) Ekološka bezbednost postmodernog društva u kontekstu evolucije i koevolucije ekosfere (Eng. Ecological safety of the post-modern society in the context of evolution and coevolution of the ecosphere). Ekološka bezbednost u postmodernom ambijentu 1:105–120

Milinčić M, Jovanović B (2008) Resurs sveže vode kao determinanta bezbednosti i kvaliteta životne sredine (Eng. Fresh water resources as determinants of safety and quality of environment). Bezbednost u postmodernom ambijentu 2:300–326

Milinčić M, Šabić D (2009) Geoekološke determinante ekonomske bezbednosti (Eng. Geo-ecological determinants of economic security). Bezbednost u postmodernom ambijentu 6:193–206

Milinčić M, Videnović D (2009) Управување со свежите водни ресурси и одрживиот развој (Eng. Fresh water resources managment and maintained development). Proceedings of the International Symposium Geography and Sustainable Development 1:123–134

Milinčić M, Lješević M, Pecelj J (2007a) Проблем свјеже воде као фактор друштвеног деловања и организације простора (Eng. The problem of fresh water as a factor for social action and spatial organization). *Радови Филозофског факултета* 9(2):263–274

Milinčić M, Ratkaj I, Pecelj M (2007b) Основна обележја геопростора и стање животне средине – оквир одрживог развоја Србије (Eng. The main attributes of geographic space and the environmental condition – the frame of Serbia's sustainable development). *Гласник српског географског друштва* 3:345–352

Milinčić M, Carević I, Jovanović Lj (2009) Regional cooperation and its importance in prevention of emergencies in water resources domain. Civil Emergencies 1:319–329

Osnove dugoročnog snabdevanja vodom stanovništva i industrije na teritoriji SR Srbije van teritorija autonomnih pokrajina (Eng. The basics of long-term water supply of population and industry on the territory of Serbia without the territories of autonomous provinces) (1977) Resource document. http://www.paragraf.rs. Accessed 28 Mar 2010

Ponting C (2007) A new green history of the world – the environment and the collapse of great civilisations. Toronto: Penguin. Prostorni plan RS (Eng. Spatial Plan of the Republic of Serbia) (1996) Resource document. http://www.iaus.ac.rs. Accessed 15 Apr 2010

Shiklomanov I (2000) Appraisal and assessment of world water resources. Water Int 25(1):11–32

Stevanović Z (1991) Hidrogeologija karsta Karpato-Balkanida istočne Srbije i mogućnosti vodosnabdevanja (Eng. Karst Hydrogeology of Eastern Serbia's Carpatho Balkanides and the Possibility of Water Supply). Rudarsko-geološki fakultet, Belgrade

Strategija dugoročnog razvoja i unapređenja vodoprivrede RS (Eng. The long-term strategy of development and improvement of water management in the Republic of Serbia) (1997) Resource document. http://www.paragraf.rs. Accessed 27 Apr 2010

The World Bank (2003) The International Bank for Reconstruction and Development: Water Resources Management in South Eastern Europe. Resource document. The World Bank. www.riverbasin.org. Accessed 25 Mar 2010

The Third Ministerial Conference "Environment for Europe" (1995) Resource document. UNECE. http://www.unece.org/env/. Accessed 25 Mar 2010

UN WWAP (2006) World Water Assessment Programme. The 2nd UN World Water Development Report: Water, a Shared Responsibility. Resource document. UNESCO. www.unesco.org/water/wwap. Accessed 24 Apr 2010

Vasileva D (1977) Osnovni principi prognoze podzemnih voda (Eng. Basic principles and forecasts for the underground waters). Časopis Hrvatskog saveza građevinskih inženjera 12:471–473

Vodoprivredna osnova Republike Srbije (Eng. Water management basics of Serbia) (2001) Institut za vodoprivredu "Jaroslav Černi". Resource document. http://www.jcerni.rs. Accessed 11 Mar 2010

Vojnović R (1995) *Воде Србије, Планови развоја и неке реализације у водопривреди* (Eng. Waters in Serbia – plans for development of water management). Građevinska knjiga, Belgrade

Voropaev GV, Vendrov SL (1979) Reservoirs in the world. Nauka, Moscow

White CL, Griffin PF, McKnight TL (1964) World economic geography. Wadsworth Publishing Co., Inc., Belmont

Wittfogel K (1956) The hydraulic civilizations. Chicago University Press, Chicago

World Water Council (2007) The Right to Water in Kosovo. News Update, 22. February 2007. Resource document. http://www.worldwatercouncil.org/fileadmin/wwc/News/Weekly_News_Update/WWC_news_update_22.pdf. Accessed 11 Mar 2010

Zakon o iskorišćavanju i zaštiti izvorišta vodosnabdevanja (1977) (Eng. Act on the Exploitation and Protection of Spring Zones of Water Sources). Službeni glasnik Republike Srbije, br. 27/1977, 29/1988

Zeman C, Rich M, Rose J (2006) World water resources: trends, challenges, and solutions. Rev Environ Sci Biotechnol 5(4):333–346

Chapter 13
Environmental Risks to Air, Water and Soil Due to the Coal Mining Process

Ivica Ristović

Abstract The coal exploitation and transportation, its combustion, and deposit of coal combustion products have adverse effects on soil, water and air. In addition to environmental degradation and pollution with uncontrolled utilization and overuse of the non-renewable energy sources we deprive coming generations' rights and possibilities to use them for their needs. Modern society has admitted its mistakes with respect to the environment. As a result we are accepting our responsibilities for environmental conditions in general. Nowadays, numerous coal mining practices that enable environment protection are available.

Significant numbers of technical papers have assessed coal combustion in the energy generation practices adverse influence on the environment by investigating individual sectors of the mining-energy industry, i.e., exploitation, transportation and utilization. In this study the intersectoral approach is applied to emphasize adverse effects on the environment during the all phases of the coal utilization, i.e., excavation, transportation, and coal combusting in thermal power plants within the South Eastern Europe.

13.1 Introduction

Expected energy demand and consumption growth due to development and availability of new technologies very likely would increases coal utilization globally. Coal reserves at the global level indicate that it will be one of the main energy resources in the future, thus it plays an essential part in electricity

---

I. Ristović (✉)
Faculty of Mining and Geology, University of Belgrade, Belgrade, Serbia
e-mail: ivica@rgf.bg.ac.rs

G. Meško et al. (eds.), *Understanding and Managing Threats to the Environment in South Eastern Europe*, NATO Science for Peace and Security Series C: Environmental Security, DOI 10.1007/978-94-007-0611-8_13, © Springer Science+Business Media B.V. 2011

generation and general development worldwide (World Coal Institute 2008; EURACOAL 2008, 2009). The energy mining practices, i.e., exploitation, transportation, combusting, and slug and ash depositing, etc. have adverse effects on the environment. Developed countries implement advanced technologies that lower environmental risks associated with coal mining industry practices. However, technologies used in coal and energy production within developing countries are below the level that provides satisfactory environmental protection.

The objective of this chapter is to emphasize all adverse effects on the environment due to coal exploitation, transportation and processing in energy generation, and to underline the necessity of the application and implementation of available environmental protection practices to decrease the adverse effects of the mining industry on natural resources, namely air, soil and water, within the South Eastern Europe (hereafter SEE) region.

Since its beginning the mining industry has been incompatible with environment. In fact, from the commencements of the mining activities and mineral resources exploitation it was "stigmatized" as the worst environmental polluter. In the books from the sixteenth century G.B. Agricola ("De Re Metallica" 1556) wrote: "Opponents of the mining processes have argued that, mining allegedly devastates land, increased logging of the woods and groves..." However, this famous and well known 'father of mineralogy', defends mining activities: "The damage which results from the mining industry is not more influential than the damage generated by other human activities, habitats and wars" (Agricola 1912).

Despite the fact that the above sentences were written in sixteenth century, they reflect the general opinion that the mining industry is a dominant environmental polluter. Since the mining industry (especially big opencast mines) practices jeopardize the environment by the overuse of non-renewable resources, depletion of the environment and pollution, denial of the adverse environmental effects of the mining industry is impossible.

However, the modern way of life, level of the development, and availability of the technical attainments and new technologies are closely associated with mining practices and technology. It is easy to conceptualize, without thorough evaluation, the number of people that would be affected by a shortage of energy supply if opencast mines and thermal power plants are closed (Vujic et al. 2009).

It is indubitable that achievement of a proper balance among economic development, environmental protection and social responsibility is very important. This required balance is reachable by the implementation and utilization of the advanced coal excavation and conveyance practices, and implementation of clean coal technologies that are based on the recent scientific body of knowledge and accomplishments.

There are many ways to solve the problem of land degradation, and pollution of soil, water and air (Strzyszcz 1996; Huttl 1998; Krönert et al. 2001; Ling et al. 2003; Ibarra and Heras 2005).

13.2 Coal: The Main Energy Resource at the Regional and Global Level

Coal is one of the main energy resources used for the generation of the electricity within the SEE region and worldwide. Total Global Hard Coal Production in 2009 was 5,990 Mt and Total Global Brown Coal/Lignite Production in 2009 was 913 Mt. According the World Coal Institute "coal provides 26% of global primary energy needs and generates 41% of the world's electricity" (World Coal Institute 2008).

IEA World Energy Outlook 2007 (WEO 2007) Reference Scenario study indicates that coal will be the main energy resource in the future at the global level (Coal Industry Advisory Board 2008). Hence, it is crucial to implement mining excavation and processing practices that are based on new, clean coal technologies which are in line with sustainable energy sector development.

Based on the reports of the Global and European agencies that evaluate energy mineral resources trends and consumption, it is noticeable that there are a few countries with significant lignite and hard coal deposits and production within the SEE region. Table 13.1 provides resources, reserves and production of coal in SEE region countries (EURACOAL 2008, 2009).

Data presented in the Table 13.1 indicate that Serbia has the largest lignite reserves, i.e., 15,926 Mt, followed by Bulgaria, Bosnia and Herzegovina and Romania, while the largest reserves of hard coal belong to Bosnia and Herzegovina, namely, 1,068 Mt followed by Romania and Bulgaria. Additionally, the greatest lignite production within the SEE region has Serbia, i.e., 36.4 Mt, followed by Romania, Bulgaria, FYR Macedonia, Bosnia and Herzegovina and Albania. Finally, the greatest hard coal production is from Bosnia and Herzegovina i.e., 3 Mt, followed by Bulgaria, Romania, Serbia and Albania (EURACOAL 2008, 2009).

Vast amounts of the coal production in the SEE region is combusted in thermal power plants, for the electricity generation. The number of the installed capacities for the electricity generation that utilize coal varies for countries within the SEE region. Romania has 15,000 GW installed capacities that use coal, Bulgaria has 11,935 GW installed capacities for electric energy and 4,845 GW utilize coal out of the total.

Table 13.1 Resources, reserves and production of coal within the countries in SEE region (EURACOAL 2008, 2009)

Country	Resources (Mt)		Reserves (Mt)		Production (Mt)		Power plant capacity (GW)	
	Hard coal	Lignite	Hard coal	Lignite	Hard coal	Lignite	Coal	Total
Albania	–	–	–	–	0.1	0.1	–	–
B & H	1,360	2,634	1,068	1,698	3	2	1,947	4,331
Bulgaria	706	3,710	64	1,928	2.7	26.1	4,845	11,395
FYRM						7		
Serbia	–	–	–	15,926[a]	1	36.4	5,171	8,355
Romania	936.8	3,537.9	801	1,364	2.7	32.6	15,000	15,000

[a]BGR

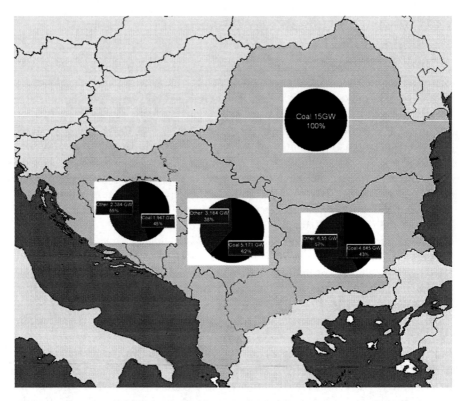

Fig. 13.1 Comparison of the total installed capacities and energy capacities that utilizes coal as a fuel (EURACOAL 2009)

Serbia has a total of 8,355 GW installed capacities for electric energy generation out of which 5,171 GW utilizes coal. Bosnia and Herzegovina have 4,331 GW installed capacities, out of which 1,947 GW consumes coal (EURACOAL 2008, 2009). Figure 13.1 demonstrates a comparison of the total installed capacities for the electric energy generation and thermal power plants that utilize coal as a fuel in SEE region.

In Serbia, open cast lignite mining are developed in two mining basins, namely, Kolubara and Kostolac, while in Kovin basin underwater mining is present. Since 1999 Kosovo and Metohija mining basin do not operate as an integral part of EPIS (Electric Power Industry of Serbia) and at the moment EPS does not manage capacities in Kosovo and Metohija (EPIS 2008). Underground mining utilization is present in eight mines. Table 13.2 displays coal and overburden production in Serbia in 2008 (EPIS 2008; ERM 2008).

EPIS (Electric Power Industry of Serbia) thermal power sector that use lignite consists of eight thermal power plants with 25 blocks that have in total 5.171 MW installed capacities out of which two thermal power plants with seven blocks and 1.235 MW total installed capacities are in Kosovo and Metohija (Table 13.3) (EPIS 2008).

Table 13.2 Coal and overburden production in Serbia in 2008

Open pit	Realized coal production (t)	Realized overburden production (m³cm)
Kolubara	30.583,976	77.167,612
Kostolac	7.367,518	30.079,629
Kosovo	7.000,000[a]	–
Underground coal	600,000	–

[a]Electric Power Industry of Serbia and Draft Final Sesa Report

Table 13.3 Thermal power plants in Serbia (Electric Power Industry of Serbia)

Thermal power plant	Power (MW)
Nikola Tesla A	1502
Nikola Tesla B	1160
Kolubara	245
Morava	108
Kostolac A	281
Kostolac B	640
Kosovo A[a]	617
Kosovo B[a]	618

[a]As of June 1999, EPS does not operate its facilities on the territory of Kosovo and Metohija

Figure 13.2 exhibits open pit lignite mine and thermal power plants in Serbia (EPIS 2008).

13.3 Air, Water and Soil Environmental Risks as a Result of Mining-Energy Process

Widespread opinion exists that the environment is at risk due to natural resources overuse, depletion and pollution. Accordingly, it can be deduced that mining activities, e.g., excavation, preparation, conveyance and primary extraction contributes to all above mentioned aspects of environmental degradation. Furthermore, the mining industry practices adverse effects on the environment might remain for a long period of time within the area of the activities after the mine closure. Thus, natural resources, namely, air, water and land are exposed to the pollution during the all segments of the coal mining practices, i.e., excavation, preparation, transportation and primary extraction (Duyverman 1981; Singh 2006).

Open-cast lignite mining activities generate pollution due to the change of the land use practices and land depletion (Spasic et al. 2009). Additionally, to the terrain changes open-cast coal mining exploration cause changes in the ground and surface water regime. These changes are more extreme for former than for latter and local air pollution.

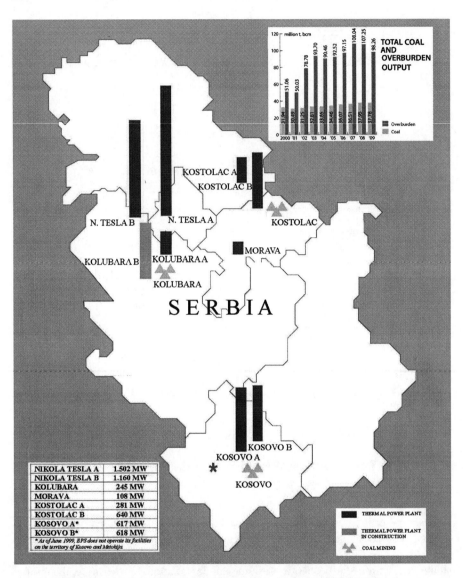

Fig. 13.2 Open pit lignite mine and thermal power plants in Serbia (EPIS 2008)

Application of the new practices and technologies in the mining industry accompanied with accelerating energy demand as a consequence, has increased the quantity of excavated coal, and as a final result expanded surface coal conveyance length and capacities. The conveyance length from the mine to the primary processing facilities, transshipment, disposal, end users, etc., might vary from a few hundred meters up to tens of kilometres and generates effects on the environment that are mainly adverse (Grujić et al. 2010). There are many sources of these negative effects, e.g., conveyance route construction, transportation means operation, coal and environment uncontrolled contacts, etc. In generally, during the coal conveying air, land and water are more vulnerable to the pollution due to the previously indicated potential sources of the pollution.

Coal combustion is a process that generates air pollution by the CCP (Coal combustion products) and polluting of land and water by stockpiles percolating water and filtrates from the ash and slag landfills. Most of the coal constituents remain in CCP, slag and ash, consequently the atmosphere is overloaded by gases and particles. The coal processing is the process that converts primary coal energy into the second energy and it generates environmental pollution in general, but the most severe consequences are on the air quality. Mining waste deposits that result from the combustion have adverse effects on the air, water and land resources (Simonovič et al. 2003).

13.3.1 Air, Water and Soil Environmental Risks Due to Coal Excavation

Lignite opencast coal mining induces tremendous degradation of the environment, e.g., increase in both surface and ground water pollution, air pollution, disturbance of ecosystems, etc. Coal excavation can result in a number of adverse effects on the environment:

- Devastation of the fertile agricultural or forestry land
- Change of natural landscape characteristics, destruction of the old and creation of the new geological forms
- Natural habitats destruction
- Change of surface and groundwater regimes and currents, and their pollution by the opencast waste waters, oil derivatives, etc.
- Air pollution by coal dust from the open cast, and by the gas emission from mechanization work
- Land pollution by the opencast waste waters, oil derivatives, etc.
- Noise and vibration extend, etc.

Air pollution within the area of the open cast mine is created by the suspended particles (mineral particles). All open cast mining technological processes phases are the sources of the floating dust. Typical sources of the air pollution by the suspended particles are: excavators, auxiliary mechanizations, belt conveyors, transportation roads, active mining surface. This emission is limited to a relatively small area.

Open cast coal mining excavation induces changes in hydrogelogical regime within the area of the mining activities. Protection of the open pits coal mine from the atmospheric and ground water are accomplished by the drainage from the mining filed. These protection practices cause a decrease in groundwater levels and have adverse effects on surface water and vegetation.

Some of the environmental risks to air, water and soil resulting from coal excavation are presented in Fig. 13.3.

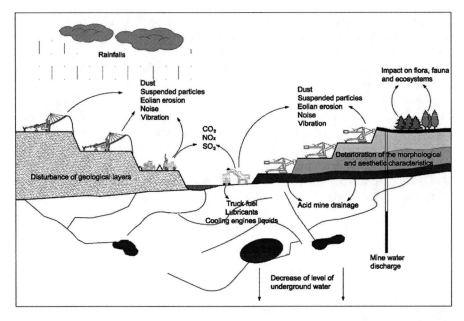

Fig. 13.3 Some of environmental risks to air, water and soil resulting from the coal excavation

Lignite open cast mining is specific since it spreads over sizable areas. Since the lignite is coal with relatively low quality, to accomplish efficient production a large amount of the lignite and significant excavation of the overburden are needed (Danicic et al. 2009). Mining activities occupy large areas both within the open cast and adjacent regions by the construction of the overburden dump. Figure 13.4 exhibits land degradation due to lignite open cast mining in Kolubara and look of the coal and slag dump levels with formed overburden dump in Kolubara, respectively.

The state manages environmental security by the multidisciplinary approach that includes various aspects e.g., organizational, scientific, technical, ecological, financial, educational, informational, etc. The coal mining companies' responsibility is huge with respect to the existing policies and regulations for the environmental protection.

13.3.2 Environmental Risks to Air, Water and Soil Due to Coal Conveyance

Coal conveyances in coal mining process that generates electrical energy can be divided into interior and exterior practices. The former presents transport at the open cast, while the latter refers to coal transport from the open cast to the preparation and coal combustion facility. Additionally, there is a coal combust products conveyance, i.e., slag and ash, from thermal power plant (TPP) to ash and slag dumps.

Fig. 13.4 Land degradation due to lignite open cast mining Kolubara (Google Earth 2010)

Each of these before mentioned conveyance practices have adverse effects on the environment in general.

Within most of the SEE region diverse transportation means for coal conveyance, either interior or exterior, are employed. With respect to coal transportation means, we can separate them in a few categories, namely truck, rail, and conveyors belts transportation (Ristović 2006). Combustion products are conveyed by wet or dry transportation in different ways, i.e., tank truck, conveyor belts, and most frequently by hydraulic conveyance pipe.

Transportation costs contribute to the total coal production costs by 50–60%. As a result, thermal power plants are built close to the open cast mines. However, if other circumstances are unfavourable (e.g. lack of the water necessary for the cooling), the TPP are placed in the nearby locations where all required conditions are met. Transportation practice selection process is influenced by conveyance route length, and many other aspects. Examples of two large open cast coal mines in Serbia, specifically Kolubara and Kostolac, demonstrate feasibility of various coal conveyance practice combinations.

The TPP Kolubara with installed capacity of 245 MW is constructed close to the open cast coal mine Kolubara with total yearly coal production of approximately 30 Mt that exceeds the capacity of the TPP. Given that, TPP Nikola Tesla A (NTA) and Nikola Tesla B (NTB), with installed capacity of 1.502 MW and 1.160 MW, respectively were constructed in Obrenovac, approximately 30 km from the open cast mine to enable entire coal quantity utilization. Location specificity, and open cast mine and TPP distance result in coal conveyance combined practices. At the open cast and from there to the transfer point, coal is conveyed by the conveyor belts. While at the transfer point the coal is loaded into the wagon and conveyed by the rail to TPP NTA and NTB. In the open cast mine Kostolac, that has yearly production of approximately 9 Mt, the conveyor belts are used for conveyance from open cast to the crusher plants, as well from there to the TPP Kostolac. Both types of transportation have significant adverse impacts on the environment.

Coal conveying in the open cast mine has local influence on the environment within the micro area of the open cast and adjacent area. However, the combined conveyance practices in open cast Kolubara have adverse environmental effects within the micro area, nearby localities, and along the railway route from Lazarevac to Obrenovac which passes through the areas that consists mostly of the agricultural lands on which the coal conveyance has important adverse effects. Figure 13.5 exhibits transportation route from open cast Kolubara to TPP Nikola Tesla, Obrenovac.

Adverse effects on environment, i.e., soil, water and air notably vary as a result of different conveyance practices.

Soil degradation is evident as a result of transportation routes construction, particularly construction of the rail trucks and apparent structures, i.e., railway stations, transfer and loading points. In addition to soil degradation and pollution, conveyance route construction has adverse effects on flora and fauna due to the disconnection and division of the natural habitats (Ristović and Djukanović 2008).

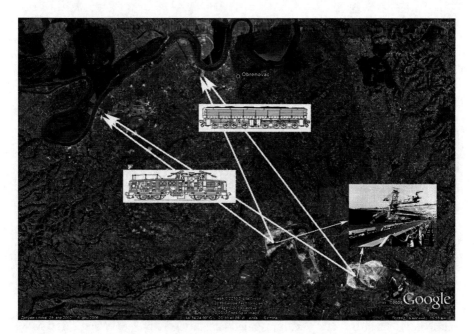

Fig. 13.5 Transportation route from open cast Kolubara to TPP NT A & B (Google Earth 2010)

13 Environmental Risks to Air, Water and Soil Due to the Coal Mining Process

Adverse effects of the coal open cast mining on the air are manifested during all transportation phases, from transportation routes construction to utilization of the conveyance routes and transportation means. The most important adverse effects on the air occur during the classical belt conveyers and truck transportation, while coal transportation by rail does not have significant negative influence on the air since coal conveyance is utilized by specially constructed wagons that have minimum influence on the environment in general (Ristović and Djukanović 2008).

Surface and ground waters are endangered by all coal conveyance types and practices. During the transportation routes construction, streams allocation might be needed, which has adverse effects on the water bodies, and micro and macro areas. During the conveyance phase, the surface waters are the most significantly jeopardized by the raw material aeolian erosion of raw material (Ristović and Djukanović 2008).

Besides previously listed adverse effects on air, soil and water, utilization of the conveyance means generates additional adverse effects: noise and vibrations, increase in temperature, pollution by released gases due to the internal combustion engine worked.

There is a possibility of the reduction and remove of the environmental pollution and adverse effects of the open cast coal mining conveyance practices by application of the new coal conveyance practices. In addition to application of ecologically acceptable conveyance practices (pipe conveyor, capsule pneumatics conveyance, coal briquette hydraulic conveyance) particular modification of the existing transportation practices and equipment would reduce open cast coal mining adverse effects on the environment (Ristović 2006; Ristović and Djukanović 2008; Fedorko et al. 2007).

13.3.3 Environmental Risks to Air, Water and Soil Due to the Coal Processing

Burning coal for the production of electrical energy creates huge pollution and environmental risks (Špeh and Plut 2001). The combustion process generates many polluting substances, including CO_2, SiO_2, Al_2O_3, CaO, MgO, TiO_2, SO_3, P_2O_5, Fe_2O_3, Na_2O, K_2O, Cl, dust, soot, smoke and other suspended particles. The main problem with burning coal is carbon dioxide (CO_2).

As a result of coal combustion process a part of the heat is released to the environment and increase in the air temperature extends environmental risks. Furthermore, operation of the transformation stations, electro-power lines and other electrical equipments as a result has a creation of the electrical filed and heating of the surrounding air.

Coal Combustion Products (CCPs), fly ash, bottom ash and slag are being produced in thermal power plants. According to the European Coal Combustion Products Association e.V., the annual production in EU 27 is estimated to amount to about 100 million tons (ECOBA 2007). Ash and slug dumps are potential

Table 13.4 The areas occupied by the ash dumps in Serbia

Thermal power plant	Ash dump (ha)
Nikola Tesla A	400
Nikola Tesla B	600
Kostolac A & B	200
Kosovo A[*]	240
Kosovo B[*]	73

[*]ERM (2008) Draft Final Sesa Report

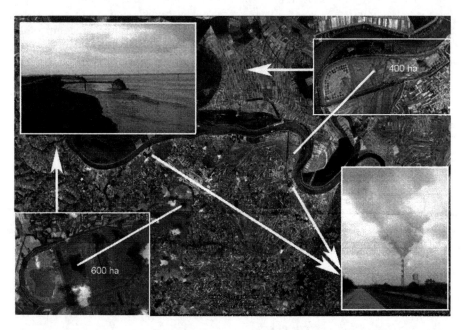

Fig. 13.6 Devastated land by ash dumps for thermal power plant NT A & B (Google Earth)

environmental risks for land, water and air. Disposal of the ash in dumps is a huge problem with respect to the land contamination given the dump area and quantity of the disposed material, no matter which procedure is applied for the disposal. A part of deposited ash is spread by aeolian erosion over the surrounding land and is the source of heavy metals that plants take from the soil. Thus the heavy metals are carried by flora to the fauna. The areas occupied by the ash dumps in Serbia are listed in Table 13.4. Figure 13.6 displays devastated land by ash dumps for thermal power plant Nikola Tesla A & B.

In surrounding area dumps adverse effects on the land resources can be classified in a few different categories. The main categories are: Expropriation and devastation of the fertile agricultural land; Endangering of the surrounding arable land and terrain by the ground waters; Surrounding terrain pollution by the ash particles.

Ash and slug are commonly disposed in dumps that are close to lakes and rivers. Thus, this practice presents environmental risks to surface and ground water quality.

A portion of water used for the ash and slag hydraulic conveyance drains to the dump and through the sand and gravel layers infiltrate into the ground water. This infiltrated water has heavy metals and other harmful substances in it.

Depending on the way of transportation and ash disposal particles are dispersed from the dry sections of the dumps occasionally (Grujić and Ristović 2007). In that case, significant air pollution occurs with harmful and adverse effects on the population and thermal power plant employee's health, and environment in general.

13.4 Conclusions

This chapter presents an overview of a negative environmental impact of coal mining in Serbia and SEE. Despite the fact that the major part of this study reflects conditions in Serbia within the SEE region circumstances in the coal mining industry are similar. Given that, and due to the fact that pollution of the air, soil and water are not local in their nature, we can presume that environmental problems linked to the coal mining industry are regional, or even more global.

To put these findings in context, there is a necessity for implementation of the coal mining practices that are commonly applied in developed countries to reduce environmental risks within SEE region, and other developing countries and regions worldwide.

Comprehensive assessment and identification of all the coal mining practice's adverse environmental impacts and application of the best management practices from the developed countries and clean coal technology would significantly decrease adverse environmental impacts resulting from coal mining practices.

References

Agricola GB (1912) De re Metallica. Translated from the first Latin edition of 1556. Salisbury House, London

Coal Industry Advisory Board (2008) Clean coal technologies. Accelerating Commercial and Policy Drivers for Deployment. http://iea.org/work/2008/asean_training_coal/Clean_Coal_CIAB_2008_WEB.PDF. Accessed 27 May 2010

Danicic D, Mitrovic S, Pavlovic V, Kovacev S (2009) Sustainable development of lignite production on open cast mines in Serbia. Min Sci Technol 19(5):679–683. doi:10.1016/S1674-5264(09)60126-5 DOI:dx.doi.org

Duyverman CJ (1981) Environmental impact through the utilization of coal. In: Proceedings of the international congress advanced energy conversion a challenge to industry. Resource and Conservation, vol 7, pp 145–158

ECOBA – European Coal Combustion Products Association (2007) Statistics/ECOBA Stat 2007. http://www.ecoba.com/evjm,media/statistics/ECOBA_Stat_2007_EU15.pdf. Accessed 3 June 2010

EPIS–Electric Power Industry of Serbia (2008) Annual Report 2008. http://www.eps.rs/. Accessed 13 May 2010

ERM (2008) Draft Final Sesa Report. ERM Italia S.p.A. http://www.erm.com. Accessed 3 June 2010

EURACOAL – European Association for Coal and Lignite (2008) Coal industry across Europe 2008. EURACOAL, Brussels, Belgium. Resource document. EURACOAL. http://www.euracoal.be. Accessed 15 May 2010

EURACOAL – European Association for Coal and Lignite (2009) An energy strategy for Europe: importance and best use of indigenous coal. Resource document. EURACOAL. http://www.euracoal.org. Accessed 15 May 2010

Fedorko G, Kubin K, Ivančo V, Husakova N (2007) A pipe conveyor belt modelling. In: Proceedings of 1st international symposium energy mining 07, Vrnjacka Spa, Serbia, 21–24 Nov 2007, pp 330–335

Grujić M, Ristović I (2007) Some aspects of application of reversible belt conveyors in coal transportation to steam power plant and return transportation of the ashes. In: Proceedings of 2nd Balkan mining congress BALKANMINE 2007, AISS, Belgrade, pp 173–177

Grujić M, Ristović I, Grujić M (2010) Model for the selection of the optimal location of a thermal power unit according to the external coal conveyance criterion. Acta Montan Slovaca 15(1):31–33. http://www.actamont.tuke.sk/pdf/2010/s1/6Grujić.pdf. Accessed 15 May 2010

Huttl RF (1998) Ecology of post-mining landscapes in the Lusatian Lignite Mine District, Germany. Land restoration using an ecologically informed and participative approach. In: Fox HR, Moore HM, Elliott S (eds) Land reclamation: achieving sustainable benefits. Balkema, The Netherlands, pp 187–192

Ibarra JMN, Heras MM (2005) Opencast mining reclamation. In: Mansourian S, Vallauri D, Dudley N (eds) Forest restoration in landscapes. Springer, New York, pp 370–378

Krönert R, Steinhardt U, Volk M (2001) Landscape balance and landscape assessment. Springer, New York/Berlin/Heidelberg

Ling C, Handley J, Rodwell J (2003) Multifunctionality and scale in post-industrial land regeneration. In: Fox HR, Moore HM, Elliott S (eds) Land reclamation: achieving sustainable benefits. Balkema, The Netherlands, pp 27–34

Ristović I (2006) Environmental protection in conveyance of metallic mineral raw materials. In: Grujić M, Boroska J (eds) Environment-friendly external ore conveyance. FMG, Belgrade, pp 71–87

Ristović I, Djukanović D (2008) Impact of external coal conveyance on urban and natural environments. J Transp Logistics 15(8):19–29

Simonovič B, Belić D, Vukmirović Z, Grujić M, Petovar K, Knežević D et al (2003) Study: solving environmental problems caused by the operations of the thermal power plant "Nikola Tesla A & B". Holding Institute of General and Physical Chemistry, Belgrade

Singh G (2006) Environmental and social challenges facing the coal industry. In: Proceedings of international coal congress & EXPO, New Delhi, pp 301–310. http://www.ismenvis.nic.in/My_Webs/Digital_Library/GSingh/ENVIRONMENTAL%20AND%20SOCIAL%20CHALLENGES.pdf. Accessed 13 May 2010

Spasic N, Jokic V, Maricic T (2009) Managing spatial development in zones undergoing major structural changes. Spatium 21:53–65. http://scindeks-clanci.nb.rs/data/pdf/1450-569X/2009/1450-569X0921053S.pdf. Accessed 3 June 2010

Špeh N, Plut D (2001) Sustainable landscape management in Slovenia: environmental improvements for the Velenje Coal Mining Community 1991–2000. GeoJournal 55(2–4):569–578. doi:10.1023/A:1021749229959

Strzyszcz Z (1996) Recultivation and landscaping in areas after brown-coal mining in the Middle-East European Countries. Water Air Soil Pollut 91(1–2):145–157. doi:10.1007/BF00280930

Vujic S, Cvejic J, Miljanovic I, Drazic D (2009) Планирање, рекултивација и Површинском УПрављање Пределима. (Eng. Planning recultivation and open cast landscape management). University of Belgrade, Belgrade

World Coal Institute (2008) Coal facts. 2008 Edition. World Coal Institute, London. Resource document. WCI. http://www.worldcoal.org. Accessed 27 May 2010

WEO – World Energy Outlook (2007) International Energy Agency (IEA). Resource document. WEO. http://www.worldenergyoutlook.org. Accessed 19 May 2010

Chapter 14
Solid Municipal Wastes in Ukraine: A Case Study of Environmental Threats and Management Problems of the Chernivtsi Dump Area

Igor Winkler and Grygoriy Zharykov

Abstract A problem of the solid municipal waste management becomes more acute as numerous municipal dumps are growing and capturing new lands. This problem gains more importance in all of Europe and in the Danube region because free land areas for environmentally safe collection of the waste materials are very insufficient. Landfills of many cities in Ukraine are often overfilled and poorly protected, which makes this issue very troublesome for Ukraine.

Results of the investigation of environmental effects caused by municipal dump of the city of Chernivtsi, Ukraine are reported. It is shown that an influence of the landfill on the nearby atmosphere and soils is moderate but serious worsening of the ground water quality is registered in some areas. Contaminated groundwater can influence water quality in the nearby creeks and result in the long-lasting chemical pollution of the riverbed silt. The silt can accumulate significant amounts of zinc, chromium and copper from wash-offs of the nearby municipal dump area. No accumulation of lead and nickel were found. Concentration of zinc in the silt however can significantly exceed the threshold limit value.

Various methods of the waste materials management have been analyzed and a complex sorting/utilization of the solid municipal waste with partial waste composting is proposed as the most optimal way to mitigate the problem. This solution would ensure better isolation of the toxic waste decomposition products within the landfill. On the other hand, it would not require such serious investment like construction of a waste incineration plant.

I. Winkler (✉)
Department of Physical and Environmental Chemistry,
Chernivtsi National University, Chernivtsi, Ukraine
e-mail: igorw@ukrpost.ua

G. Zharykov
State Department of Ecology in the Region of Chernivtsi, Chernivtsi, Ukraine

G. Meško et al. (eds.), *Understanding and Managing Threats to the Environment in South Eastern Europe*, NATO Science for Peace and Security Series C: Environmental Security, DOI 10.1007/978-94-007-0611-8_14, © Springer Science+Business Media B.V. 2011

14.1 Introduction

An effective solid municipal waste (SMW) management is closely connected with many environment protection issues. This problem becomes increasingly important and cannot be confined within any national boundaries. Many European cities suffer the persistent environmental contamination caused by collection of SMW. Huge dump areas are growing near the cities, seriously worsening the quality of environment. As the average level of worldwide industrial development is growing, the amount of the waste materials is also increasing, which results in new ecological problems. Urbanization leads to the population being concentrated in big cities with higher consumption of various goods and packing materials. The result of these goods can be found at the city dumps. More than 100,000 ha of agricultural lands should be allocated every year for the growing landfills. Waste collection and utilization becomes a vital part in the everyday life of many cities.

A specialized 'waste collectors' quarter exists in modern Cairo (Egypt, population 8 million, 17 million with suburbs). More than 40,000 Egyptian Copts (the Zabaleen) are engaged in this activity, which to a certain extent secures ecological safety of the city. However, environmental conditions in the 'waste quarter' are very unfavourable. There is not any fresh water inside the quarter and a specific objectionable smell can be felt even at some distance from the quarter (Fahmi 2005).

Waste collection and utilization problems have provoked a serious and continuous crisis in Italy, especially in the region of Naples. High population density, low capacity of the available dumps, hot and dry Mediterranean climate and significant criminalization in the field of waste collection and utilization cause serious problems with waste management in the region. These problems reached culmination in the year 2007 when waste collection and disposal had been stopped and the city sunk in its own garbage. The Italian government tried to organize waste transportation to Sardinia but faced raising protest of the local communities and had to resign this plan. Another option was discovered and German waste incineration plants were utilized for waste materials from the region. However, excessive radioactivity had been located in some parts of the waste soon after beginning the incineration and the German incineration plants refused to continue the contract. Waste utilization and disposal issues are still very important for the Naples region and entire Southern Italy (Bonafede et al. 2009; Iovino 2009).

A problem of environmental pollution with various waste materials has grown into a global threat. A unique Great Pacific Garbage Patch was discovered in the northern part of the Pacific Ocean, in the area between 135°–155° West and 35°–42° North, in the late twentieth century. Slow ocean fluxes bring anthropogenic waste, mostly various plastics, to this area and form a huge gyre, which has already collected several million tons of the waste. Similar but smaller structures have also been found in other parts of the Ocean. These structures contaminate the water and air and endanger the life of many living species. Fishes and birds can swallow plastic pieces, which can cause diseases and death (Moore et al. 2005).

There are several approaches to resolve this problem: composting, incineration, simple burial and partial recycling of SMW (Wilson 1976; Tchobanoglous 2009). The latter option is the most effective because it ensures low environment pollution and saves valuable primary resources.

Waste recycling is one of the most effective ways to reduce consumption of the natural resources. Today's technologies can ensure cost-effective recycling of about 90% of municipal and industrial waste materials (Tchobanoglous 2009). A concept of "Design for Recycling" is being actively implemented in many European countries (Ferrao and Amaral 2006). This concept requires manufactures of various goods, especially automobiles and large-size merchandise, to take back "retired" products instead of sending them to the landfill areas. However, implementation of this concept is absolutely insufficient, especially in Ukraine and some of the "new" members of the EU. For example, currently 75–85% of electronics are being sent to the general landfills as unsorted waste instead of recycling (Kang and Schoenung 2005). Partial recycling of the disposed electronics is available and can recover part of non-ferrous metals (tin and lead from the solder, nickel, cadmium and silver from the batteries, gold from the contacts coating, etc.) used in the electronic details. However, partial recycling as described is less efficient than a process which involves preliminary separation of electronic waste. It is interesting to note that the number of obsolete computers in the U.S. in 2003–2005 was greater than the number of the new ones (Kang and Schoenung 2005).

Annual disposal of various kinds of waste worldwide is estimated approximately 1 billion tons. 85% of these materials are produced in the USA, Europe and Oceania. Daily production of the waste materials per person in the USA is about 1.6 kg; in Europe – 1.5 kg; in Oceania – 0.8 kg and only about 0.4 kg in Asia (Williams 2005).

About 12 million tons of SMW are being formed in Ukraine every year. This amount brings serious epidemiological threats and causes numerous negative effects, especially on the ground waters. Expanding waste disposal areas stimulates the growth population of noxious animals and other organisms (rats, mice, pigeons, gulls, etc.). Waste materials can also spontaneously ignite causing emission of dangerous air pollution agents.

Approximately 120,000 t of SMW are being collected annually within the region of Chernivtsi. The recycling ratio is very low and does not exceed 10–15%. Some of the smaller settlements do not have their own landfills and utilize either neighbour dumps or even dispose of waste materials in a disordered manner.

Effective utilization and recycling of SMW remains the most acute problem today. This process requires an accurate inventory of the waste. There are significant difficulties encountered even at this stage because of the wide variety of waste materials which are being produced by thousands of municipal sources.

Therefore, this problem must be resolved since further urbanization and population growth will cause further worsening of the above-mentioned threats. One of the directives of the European Commission (European Parliament and the Council Directive 2006) requires the installation of biogas (landfill gas) collection equipment at any landfill area which consists of at least 10,000 t. Most municipal waste dumps in Ukraine have collected much more waste material but are not equipped with such systems. The potential gas-bearing of 1 t of waste materials containing about 70% of organics is about 200 m³ of methane gas per year. The gas-bearing remains active for about 50 years and releases methane, ammonia, carbon monoxide and other hydrocarbons (Nekrasov and Gobzin 1995). The concept of the landfill gas production and utilization is not new (Farquhar and Rovers 1973) but it is especially important in the light of recent waste crisis and raises attention towards the reduction of the greenhouse gas emissions. Landfill areas in Canada produce about a quarter of all the greenhouse gases emitted in the country. Collection and utilization of the landfill gas ensures the return of investment through the energy production and also clears part of the national quota for the greenhouse gases emission. It should be emphasized that this kind of energy production is quite stable and long lasting. Landfill gas production reaches a stable level in about 1–2 years and can remain practically unchanged for about 20 years. Then, the gas output gradually decreases until the complete exhaustion in 20–40 years (Reinhart 1993; Bogner and Spokas 1993). It is estimated that the gas-bearing period of many commercial gas fields is significantly shorter.

The first landfill gas collection station was installed in the USA in 1937. Currently, about 800 stations are engaged in the landfill gas collection in Europe and about 400 in the USA. Utilization of the landfill gas saved 170 million barrels of oil in the USA in 2006 (U.S. EPA 2009).

Implementation of the landfill gas collection systems requires the construction of a dump with a waterproof bed. Because of this requirement, the dump does not cause any serious groundwater contamination. All products of the waste decomposition stay within the dump area and gradually decompose in the anaerobic conditions forming methane, carbon dioxide, ammonia, small amounts of hydrogen sulphide, nitrogen and some other gases. General composition of the landfill gas is quite similar to the biogas and many well-developed technologies of the biogas production, collection and utilization can easily be applied to the landfill gas utilization.

This chapter describes current conditions of the municipal landfill of Chernivtsi (Ukraine) and discusses possible solutions, which should ensure the reduction of the negative environmental effects caused by the landfill. Similar approaches can be useful for solving the problem of SMW management in other areas. General composition of SMW is similar for many modern cities and our results (especially results of analysis of the ground and surface water pollutants pattern and concentrations) can be applied to other similar cases.

14.2   Current Conditions of the Chernivtsi Dump Area

The Chernivtsi region (Fig. 14.1) is the smallest region in Ukraine and located in the Western part of the country. Its area is 8.1 km² (1.3% of territory in Ukraine). Most of the region (except its northern part) belongs to the Danube river basin. The river Prut flows from North-West to South-East through the city of Chernivtsi.

14 Solid Municipal Wastes in Ukraine: A Case Study of Environmental Threats 269

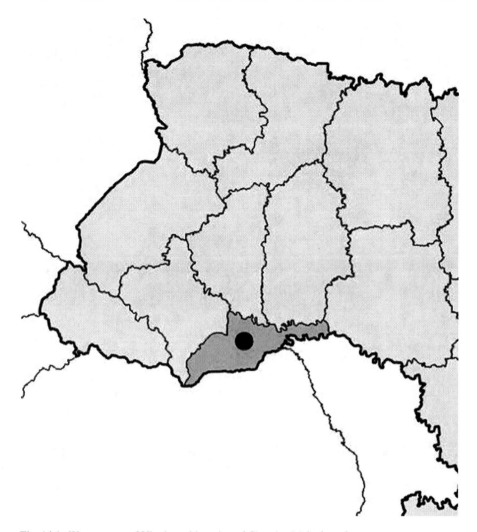

Fig. 14.1 Western part of Ukraine with region of Chernivtsi (*shadowed*)

Water pollutants can travel long distances with the river water and may finally reach Danube.

There are only ten cities in the region and Chernivtsi is the largest (population approximately 260,000). Atmospheric precipitations are quite abundant all over the region and ground water level is usually close to the surface. Because of the high ground water level, strict requirements are placed for water proofing in the waste collection areas and installations.

In Chernivtsi, the local landfill is located on the left bank of the river Prut, only 1 km away from the city's northern limit (see Fig. 14.2). The area is rather steep and the diagonal altitude difference across the dump is about 25 m. Forest-covered

Fig. 14.2 City of Chernivtsi and suburbs. Local dump area is in the *upper right corner* (inside the *white circle*)

hills with numerous creeks confine the outer edges of the area. The upper horizon of the ground waters has been found at a depth of 6.5–7.5 m. At the time of planning and construction of the municipal dump, this depth was considered as sufficient to protect the ground water from excessive contamination with waste material degradation products.

However, serious contamination of the ground water has been determined in several years after opening the landfill. This environmental threat still remains important. Waste decomposition also results in contamination of the atmospheric air. This contamination includes dust, which can be wind-blown from the surface, waste combustion products and toxic gases, which are being emitted after aerobic and anaerobic decomposition of organics.

An average composition of the waste materials from the Chernivtsi landfill are the following:

- Dust, soil, other inert materials (3–15%)
- Organics (20–50%)
- Paper, carton, wood chips and other wood processing wastes (20–50%)
- Metal wastes (1–5%)
- Plastics (1–10%)
- Glass, ceramics (3–10%)

The above composition indicates that such organic enriched waste materials will emit significant amounts of gaseous products from decomposition and also form dangerous wastewaters (so called "filtrate"), which contain very high amount of organic and inorganic pollutants (products of the food waste decompositions, metals, ammonium salts, nitrites, etc.). Filtrate is formed as a result of washing-out the soluble waste components with atmospheric precipitation and an additional extrusion of the liquid occurring at self-compression of the waste. It is a highly toxic liquid and any direct contact between the filtrate and water objects should be avoided. However, filtrate can flow into local creeks and soak into the soil because of the poor condition of the waterproof installation. This process worsens water quality both in the surface and underground sources.

The effect of infiltration of the filtrate on the near-by groundwater quality can be determined through analysis of the groundwater samples taken along the dump perimeter.

14.3 Analysis of Influence of the Dump on the Environment

A series of water samples have been taken from the landfill drainage collector and two test points around the dump in order to determine how the water quality is influenced in the context of possible contamination with organic pollutants. This type of contamination can be characterized by two parameters – chemical and biological oxygen demand (COD and BOD_5 respectively). Both parameters characterize an overall level of the organic pollutant content without a detailed estimation of concentration of each separate type of pollutant. Water quality standards usually regulate overall content of the organic substances and determination of COD and BOD_5 can provide useful information related to this parameter (Laurie 1973).

The filtrate, surface water samples (taken from the creeks) and groundwater samples (taken from the wells or special control ground holes) have been analyzed in order to determine a character of the landfill influence on the environment. Averaged results of the analysis are shown in Table 14.1. Test point 1 was located just near the edge of the landfill, points 2 and 3 were located at two neighbouring streets (point 3 at the upper-level street and 2 – at the lower-level). Samples from area 1 were taken from the small creek flowing near the dump, control ground holes and the drainage collector (filtrate). Samples from area 3 were taken from the small

Table 14.1 Water quality parameters for the area around Chernivtsi municipal dump (Sawyer et al. 2003)

Test point	Sample	pH	COD	BOD$_5$
1 (Dump area)	Filtrate	6.77	**44,800**	**5,800**
	Surface water	8.09	**70**	3.8
	Groundwater	7.60	**100**	4.4
2 (Street, down)	Groundwater	7.40	**15**	2.0
3 (Street, upper)	Surface water	7.80	**800**	**14**
	Groundwater	7.30	5	0.8
Threshold limit value for the fish-production water objects		6–8	5	5–8

creeks flowing along the street and the ground wells. Samples from area 2 were taken only from the ground wells. Both areas 2 and 3 were located approximately 1 km away from the dump within the residential area.

Analysis of COD values show that existing waterproof installations can prevent (to a certain extent) direct contact between the filtrate and neighboring surface water objects. Surface water samples taken near the dump area (test point 1) do not show exceeding limits in the BOD$_5$ value, which would be unavoidable in the case of massive discharge of the filtrate into local creeks.

The data in Table 14.1 indicate that the filtrate is being soaked into the soil around the landfill and then gradually spreads, causing contamination of the soils and groundwater. Bio-contamination level (see BOD$_5$ values in Table 14.1) of the groundwater meets sanitary requirements for both test points, while chemical pollution for the lower-level down test point 2 is three times higher than the threshold level. This parameter remains sub-threshold for the upper-level test point 3. Such contamination of the groundwater can be caused only by the filtrate soaking into the soil and distribution within the soils. This assumption can be supported by a worsening quality of groundwater taken from the dump area, where influence of the filtrate soaking should be stronger.

Analysis of the data from Table 14.1 indicates that the filtrate formed at the Chernivtsi landfill is a highly toxic liquid containing high levels of organic materials. COD of the filtrate is about 9,000 times higher than the threshold value. It is obvious to expect that this substance will travel downstream on the surface and will also soak into the soil and contaminate the groundwater. The latter effect is noted as the COD of the groundwater samples from test point 2 is three times higher than the limit. However, the BOD$_5$ in point 2 is normal. It should be emphasized that BOD$_5$ was within the limits even for the surface and groundwater samples taken near the landfill at test point 1. This value is sub-threshold for the groundwater and quite far from the limit for the surface water. Groundwater quality is worse than the quality of the surface water, which supports the influence of the filtrate soaking into the soil and contaminating the groundwater. The abovementioned results support that the surface waterproof installation around the dump area needs additional protection of the neighbourhood area from the filtrate and the dump wastewater wash-offs.

14 Solid Municipal Wastes in Ukraine: A Case Study of Environmental Threats 273

It is of interest to note that the surface waters were highly polluted, even for the upper creek (test point 3), which cannot be directly influenced by the filtrate or other dump area wash-offs. Pollution level for test point 3 is even higher than the level determined for the landfill surface water samples. This is an unusual situation and it may be caused by an external pollution source, which influences a creek flowing near test point 3. For example, the creek may be polluted with effluents from the nearby sewage collector but this assumption requires further support.

Therefore, our results support that a local landfill affects neighbouring water objects. The process mainly influences the groundwater in the nearby downstream area. While waterproof installations almost secure protection from the direct discharge of the dump area surface filtrate, the filtrate soaks into the soil and seriously contaminates the groundwater instead. Groundwater contamination results in worsening of the well-water quality in the neighbouring downstream area and fuels up continuous public protests against extended exploitation of the landfill.

Another analysis was conducted to determine the content of heavy metals and inorganic ions in the same samples. Table 14.2 represents results of the analysis. Concentrations of the metals have been determined using the atom-absorption photometry method and concentrations of the anions have been determined using relevant standard UV–VIS photometry methods (Laurie 1973).

As seen from Table 14.2, the filtrate shows excessive content of all pollution agents. However, there is not any excessive content of the inorganic pollutants for test point 2. Groundwater from test point 3 also meets all the water quality requirements, while the surface water sample shows excessive content of some pollution agents. Analysis of the pollution profile for the surface water sample from point 3 indicates that the highest exceeding pollutants were nitrites and ammonium ions. This is direct evidence of contact between the water sample and untreated wastewater, which can run through leakages in the sewage system.

The highest levels of limit concentrations in the filtrate were registered for iron (1,000 times), ammonia (348 times), hydrocarbonates and chlorides (24 times) and nitrites (19 times). This pattern is easily explainable. The highest level of iron ions originate from various metal products collected at the dump. Atmospheric precipitation dissolves some of the carbon dioxide and its pH shifts towards the acid reaction. Additional amounts of the hydrocarbonate ions come from the food and organic waste, which results in further shift of pH towards the acid reaction. This causes an active dissolution and wash-off of the iron, which forms a high concentration of iron(II) and iron(III) compounds in the filtrate. Ions of ammonia mostly originate from the food and other organic nitrogen containing waste. The natural nitrogen cycle process gradually transforms ammonia into nitrites, then into nitrate ions. Therefore, concentration of ammonia in the 'fresh' wastewater is usually higher than the concentration of nitrites and nitrates. As the wastewater becomes older, more ammonia transforms into nitrites and nitrates. This dynamic can easily be seen in the changes of concentration for the dump water samples: filtrate-surface-groundwater. The concentration of ammonia in the filtrate is 174 mg/l; in the surface water – 0.06 mg/l and in the groundwater – 0.01 mg/l. The concentration of nitrates shows an opposite tendency: 7.0 mg/l in the surface water and 34.2 mg/l in

Table 14.2 Content of some inorganic compounds in the test sample

Test point	Component content, mg/l										
	SO_4^{2-}	Cl^-	HCO_3^-	Ca^{2+}	Mg^{2+}	Na^+ and K^+	NO_3^-	NO_2^-	NH_4^+	Fe^{2+} and Fe^{3+}	CO_3^{2-}
1 (Dump)											
F	–	**7,200**	**12,204**	**1,463**	**644**	–	–	**1.50**	174	**200**	–
S	86.4	60.4	208	97.2	28.0	4.84	7.0	0.01	0.06	**0.25**	6.0
G	**101**	54.7	244	128	27.4	2.30	34.2	0.02	0.01	0.15	9.0
2 (Down)											
G	60.0	25.5	360	119	26.8	12.0	31.6	0.01	0.08	0.10	15.0
3 (Upper)											
S	52.5	**1,050**	**738**	126	59.6	**745**	37.3	**10.5**	**14.5**	**6.00**	21.0
G	65.2	27.5	458	82.2	57.2	32.2	10.3	0.04	0.03	0.35	12.0
Threshold limit value for the fish-production water objects	100	300	500	150	120	120	40	0.08	0.5	0.2	200

F means filtrate, *S* surface sample, *G* groundwater sample, excessive parameters are in the bold

14 Solid Municipal Wastes in Ukraine: A Case Study of Environmental Threats

the groundwater. Therefore, most of the ammonia nitrogen transforms into the nitrate nitrogen during soaking of the filtrate into the soil. Concentrations of other substances were also found but below corresponding threshold limit values.

It can be concluded that excessive content of inorganic pollutants in the dump filtrate does not result in exceeding the threshold limits for the groundwater in the neighbourhood area. Only the content of sulphates in the surface landfill water was slightly higher than the threshold limit. On other hand, it should be noted that many parameters for both neighbouring test areas are sub-shoulder and can exceed the limits at further intensification of the landfill exploitation. Even by keeping an existing workload level, results can still exceed limitations since soaking of the microelements is often very slow and the responding contamination may be registered in the environmental objects 2–4 years later, after the period of more intense exploitation of the landfill.

Analysis of the surface water sample from test point 3 indicates an anomalous pattern. This test point cannot be influenced by the dump wastewaters and wash-offs but concentrations of chlorides, sodium and potassium, nitrites, ammonia and iron are much higher than in the landfill surface water sample. The concentration of nitrites is even seven times higher than in the filtrate. This pollution pattern is very characteristic for the municipal wastewater (greywater) and is additional evidence that effluents from the local sewage collector contaminate the surface water in area 3. This influence is even stronger than an obvious negative effect of the local dump area. Therefore, immediate efforts should be directed to fixing the sewage collector and elimination of all greywater effluents. The quality of the surface water is expected to improve, but the improvement of the groundwater quality would require an additional step for collection of the filtrate and prevention of its soaking.

Another environmental threat can be caused by the bottom sediments in local creeks. The sediments accumulate water pollution agents, which results in significantly higher concentration of many pollution agents comparing with corresponding averages for the surface water and soil. For example, the concentration of zinc in the bottom silt was found 1.5–2 times higher than the background content in the soil. Concentrations of chromium and copper in the silt were slightly higher than in the background soil, while there was not any difference in the concentrations of lead, nickel and cadmium. These results were similar for all the test points.

The sub-shoulder concentrations of the pollution agents in the surface water can easily result in the excessive content in the bottom sediments. This fact gives us another argument for the urgent need in reconstruction of the waterproof installation at the Chernivtsi municipal landfill.

14.4 Conclusion

The municipal landfill of Chernivtsi is comparatively small but even the current level of its exploitation causes contamination of the near-by water objects. This pollution is mainly caused by general organic contamination from the filtrate, which soaks into the soils and spreads pollution around the dump area.

The current condition of the waterproof installation secures minimal protection of the surface water but it does not provide sufficient protection of the ground water. Furthermore, the population growth and development of the urban infrastructure would result in the higher workload on the landfill and cause even more severe contamination.

Even though waterproof installation renovation/reconstruction seems unavoidable in the near future, more thorough waste sorting can be recommended as an interim alternative. Thus, some reduction in amount of the waste materials collected in the dump can be achieved and, consequently, emission of the pollution agents, which are being formed at the waste decomposition, will be reduced. Waste composting and the dump biogas collection can be applied in a more long-ranged prospect.

Contamination pattern analysis proves that strong pollution of the surface water in area 3 is caused by the effluents from the municipal sewage collector. This problem should be fixed as soon as possible since this pollution is worse than influence of the dump area wastewaters and wash-offs on the downstream water objects.

References

Bogner J, Spokas K (1993) Landfill CH_4: rates, fates and role in global carbon cycle. Chemosphere 26(1–4):369–386

Bonafede G, Marotta P, Schilleci F (2009) Cities and garbage: an un-sustainable relationship. Resource document. http://www.cityfutures2009.com/PDF/43_G_Bonafed_F_Schilleci_P_Marotta.pdf. Accessed 14 July 2010

European Parliament and the Council Directive 2006/32/EC (2006) Resource document. European Parliament. http://eur-lex.europa.eu/LexUriServ/LexUriServ.do?uri=OJ:L:2006:036:0032:01:EN:HTML. Accessed 1 June 2010

Fahmi WS (2005) The impact of privatization of solid waste management on the Zabaleen garbage collectors of Cairo. Environ Urban 17(2):155–170

Farquhar GJ, Rovers FA (1973) Gas production during refuse decomposition. Water Air Soil Pollut 2(1973):483–495

Ferrao P, Amaral J (2006) Assessing the economics of auto recycling activities in relation to European Union directive on end of life vehicles. Technol Forecast Soc Change 73(3):277–289

Iovino S (2009) Naples 2008, or, the waste land: trash, citizenship, and an ethic of narration. Neohelicon 36(2):335–346

Kang HY, Schoenung JM (2005) Electronic waste recycling: a review of U.S. infrastructure and technology options. Resour Conserv Recycl 45(4):368–400

Laurie Y (1973) Unified methods of analysis of water. Department of Chemistry, Moscow

Moore CJ, Lattin GL, Zellers AF (2005) Density of plastic particles found in Zooplankton Trawls from Coastal Waters of California to the North Pacific Central Gyre. http://www.algalita.org/pdf/Density%20of%20Particles%20spellchkd11-05.pdf. Accessed 11 July 2010

Nekrasov VG, Gobzin IM (1995) Waste or new resources? Science, Moscow

Reinhart D (1993) A review of recent studies on the sources of hazardous compounds emitted from solid waste landfills: A U.S. experience. Waste Manage Res 11:257–268

Sawyer SN, McCarty PL, Parkin GF (2003) Chemistry for environmental engineering and science, 5th edn. McGraw-Hill, New York

Tchobanoglous G (2009) Solid waste management. In: Nemerow N, Agardy F, Sullivan P, Salvato J (eds) Environmental engineering: environmental health and safety for municipal infrastructure, land use and planning, and industry. Wiley, New Jersey, pp 177–307

U.S. EPA (Environmental Protection Agency) (2009) Landfill gas modeling. LFG energy project development handbook. 30 January 2009. Resource document. EPA. http://www.epa.gov/lmop/publications-tools/handbook.html#01. Accessed 1 Aug 2010

Williams P (2005) Waste treatment and disposal. Wiley, Chichester

Wilson DG (1976) A brief history of solid-waste management. Int J Environ Study 9(17): 123–129

Chapter 15
Environmental Risk Factors in Connection with Hospital Laundry Effluent

Sonja Šostar-Turk and Sabina Fijan

Abstract This chapter presents research findings on how to optimize a washing process to reduce the effluent burden of hospitals thus achieving a more sustainable laundering procedure. Water is used in large quantities in laundries and currently they are producing relatively high quantities of wastewater. Effluent control is especially important in hospitals where potentially pathogenic microorganisms are present. A program for hospital bed linen was investigated and the first steps of optimization were performed. The disinfection effect of the laundering procedure was determined using standard bioindicators *Enterococcus faecium* and *Staphylococcus aureus*. The washing quality was also determined by measuring the parameters of the RAL-GZ 992 criteria for hospital textiles and the effluent quality was determined by measuring the parameters in the waste water according to the Slovenian regulations for concentration limits of emission into water or waste water treatment plants for waste water from laundries and dry-cleaners, published in Slovenian Official Gazette in 2007 (Uredba o emisiji snovi pri odvajanju odpadne vode iz naprav za pranje in kemično čiščenje tekstilij 2007). The disinfection effect and washing quality were both attained in the optimized program. However due to higher concentrations of laundering agents the quality of the waste water was reduced after optimization of the program. Therefore the effluent waste water treatment for water reuse remained necessary.

S. Šostar-Turk (✉)
Faculty of Health Science, University of Maribor, Maribor, Slovenia
e-mail: sonja.sostar@uni-mb.si

S. Fijan
Faculty of Health Science, University of Maribor, Maribor, Slovenia
and
Faculty of Mechanical Engineering, University of Maribor, Maribor, Slovenia

G. Meško et al. (eds.), *Understanding and Managing Threats to the Environment in South Eastern Europe*, NATO Science for Peace and Security Series C: Environmental Security, DOI 10.1007/978-94-007-0611-8_15, © Springer Science+Business Media B.V. 2011

15.1 Introduction

Environmental friendly industry is of special importance since industrial pollution is a major concern. Textile laundering processes use significant amounts of water (Šostar Turk et al. 2005) as well as chemicals such as detergents, bleaches and disinfectants which cause pollution of waste water and the need for effluent treatment methods before emission into public drain. It is also known that surfactants have a significant impact on the environment (Karahan 2010). Research has shown that most commercial surfactants available today in low concentration ranges have rapid and extensive biodegradation properties in an aerobic environment (Garcia et al. 2009) and most surfactants are not acutely toxic to organisms at normal environmental concentrations (Ying 2006). Disinfectants in laundering agents can also have toxicological effects, especially when using disinfectants such as sodium hypochlorite, which is an oxidant that can result in the creation of mixtures such as adsorbable organic halogen compounds (AOX) in waste water (Emmanuel et al. 2004). On the other hand, the use of ecological disinfectants such as peroxyacetic acid, which is a strong oxidizing agent and disinfectant effective against a wide spectrum of microorganisms (Kitis 2003), is becoming much more common.

Since the very definition of green crime is influenced through collective processes that affect the behaviour of individuals and groups (Lynch and Stretsky 2003), the European Union and its member states have successively over the last three decades implemented European Union wide and national measures to ensure a sustainable water management process. An important outcome is the Water Framework Directive 2000/60/EC (Council Directive 2000) of the European Commission which defines certain environmental targets to be carried out by all EU members. The directive 2000/60/EC requires the creation of incentives for efficient use of water resources before 2010 by the use of differential water charges and improvements of sustainability of production processes as well as the request that in all member countries the inshore waters should be in a "good status".

The achievement of "good waste water quality" for laundry effluent (Leonardo da Vinci project 2010) is reached by acts such as: reduction of the amount of effluent and fresh water consumption, removal of harmful components in waste water by means of treatment, reduction of effluent temperature, reduction of chemistry and use of environmental harmless detergents and washing additives. Other measures for improvement of sustainability include reduction of energy consumption by lower water consumption and low temperature washing processes, use of gentle wash processes with a consequence of increased textile life time, etc. Each member state ministry has issued guidelines for the implementation of the European directive 2000/60/EC (Council Directive 2000). Slovenian laundry effluent should therefore comply with the "Slovenian regulations for concentration limits of emission into water or waste water treatment plants for waste water from laundries and dry-cleaners" (Uredba o emisiji snovi pri odvajanju odpadne vode iz naprav za pranje in kemično čiščenje tekstilij 2007).

15 Environmental Risk Factors in Connection with Hospital Laundry Effluent

On the other hand, a special concern when washing hospital textiles is the appropriate disinfection effect, since hospital textiles may contain various pathogenic bacteria, fungi or viruses (Fijan et al. 2007a). There are several reports of ineffectively washed hospital textiles being one of the possible sources of nosocomial infections of patients with *Streptococcus spp.* (Brunton 1995), *Enterococcus spp.* (Wilcox and Jones 1995), *Bacillus cereus* (Barrie et al. 1994), *Staphylococcus spp.* (Gonzaga et al. 1964) and coliforms (Kirby et al. 1956). There is also a risk of infection of staff in hospital wards and laundries when treating dirty laundry. There are some documented cases of infections with scabies (Thomas et al. 1987), fungi (Shah et al. 1988), salmonellas (Standaert et al. 1994), gastroenteritis viruses (Gellert et al. 1990), hepatitis A viruses (Borg and Portelli 1999), coxiellas (Oliphant et al. 1949), etc. These reports stress the significance of the disinfection effect of the laundering procedure is most important. Disinfection of textiles is achieved either by using higher temperatures or more disinfection agents, sometimes for a longer period of time. However, all these factors create non-ecological procedures as they can cause higher effluent burdens or require more energy (Fijan et al. 2008). Therefore, optimization of laundering procedures, which is a complex process, is compulsory.

Thermal laundering procedures for 10 min at 90°C or 15 min at 85°C according to the Guidelines of the German Robert-Koch Institute (Liste der RKI 2007) are nowadays being replaced by chemo-thermal laundering procedures using officially approved washing processes with exactly defined dosage levels of industrial washing and disinfecting agents, fixed durations, bath ratios and temperatures such as are noted in various lists of disinfectants and disinfectant processes as tested and approved by the Robert-Koch Institute (Liste der RKI 2007) or the German Association for Applied Hygiene (Desinfektionsmittel-Liste des VAH 2009). Another efficient tool for sustainable use of water in laundering programs in continuous batch tunnel washers is the reuse of the water after the press. As large amounts of laundry effluent often carry high bioburden and contamination, waste water treatment of laundry effluent is usually necessary. Waste water treatment can be conducted either by transferring the waters into municipal wastewater treatment plants or in the case of larger laundries the laundry itself conducts some form of waste water treatment before emission into the sewage system.

Another important factor to consider is the wash quality of hospital textiles, such as hospital bed linen, hospital pyjamas, hospital towels, surgical textiles, as excessive use of laundry detergents and disinfecting agents can cause greying, weakening of fibres, damages to the textiles and thus decrease the lifetime of the textiles which results in higher costs (Fijan et al. 2007b). The wash quality can be evaluated by determining the secondary laundering effects, as assessed by criteria such as the German RAL-GZ 992 criteria for the laundering quality and hygiene of hospital textiles (RAL-GZ 992 2001).

When all the above-mentioned aspects are correctly taken into consideration, the processing cost of washing and the quantity of materials released into the environment are reduced and, at the same time, an appropriate level of hygiene and quality is maintained. In our research the program for hospital bed linen was checked for a disinfection effect with at least a five log step reduction aim, wash quality as well as the waste water parameters. An optimisation of the procedure was conducted and the investigations were repeated to confirm that the results were within the tolerance limits.

15.2 Methods

15.2.1 Continuous Batch Washing Machine

A continuous batch washing machine (type Senking, Jensen Group, Denmark) (Fig. 15.1) with 14 compartments, a maximum capacity of 50 kg of dry linen per compartment, a maximum water load of 100 L per compartment and a cycle time of 4 min was used for the test. Pre-washing was conducted in the first three compartments reusing water from the tank (Fig. 15.1).

The water was in co-current flow with a consumption of 4 L/kg of linen. Bleaching was conducted in compartment four followed by a drain from compartment 4. 2 L/kg of linen of water from tank 1 together with 2 L/kg of linen fresh water was then added in compartment 4 for the main wash. The main wash was in compartments 5–10 with a water consumption of 4 L/kg of linen. Water was sent from compartment 10 to the drain. Steam injections were used for heating compartments 5 and 11. Rinsing was in compartments 11–13 using counter-current water flow with fresh water input in compartment 13 of 5 L/kg of linen. Finally neutralization was conducted in compartment 14 using the same amount of fresh water. Water from compartments 11 and 14 were released to tank 1.

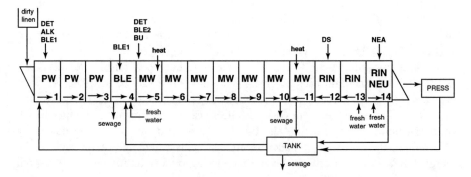

Fig. 15.1 Water consumption and dosing scheme in the continuous batch washer for the hospital bed linen program before optimization. *PW* pre-washing phase, *BLE* bleaching phase, *MW* main washing phase, *RIN* rinsing phase, *NEU* neutralization phase, *DET* detergent, *ALK* alkalis, *BLE1* bleaching agent 1, *BLE2* bleaching agent 2, *BUI* builder, *DIS* disinfecting agent, *NEA* neutralizing agent

15.2.2 Washing Procedure

The washing program for hospital bed linen was studied. The dosages of detergents, alkalis, bleaching, disinfecting and neutralizing agents are noted in Table 15.1 for both the original (BL) and optimized (BLO) program.

The optimized program had a lower temperature of 75°C, while the original program was conducted at 90°C with a thermal disinfection effect. Detergents (DET) were added in compartments 1 and 5 with a somewhat higher dosage in the optimized program. In the main wash zone in compartment 5 a builder was also added in both programs. The original program contained a double dosage of bleaching agent (BLE1) in compartments 1 and 4, bath exchange followed by an additional dosage of another bleaching agent (BLE2) in the main washing compartment 5 for persistent stains. In the optimized program the double dosage of bleaching agent (BLE1) was omitted in C4. A disinfecting agent (DIS) was added in the rinsing zone in compartment 12 in both programs with a somewhat higher concentration in the optimized program in order to prevent recontamination by rinsing water. Both programs ended with neutralization in compartment 14 (NEU).

This process optimization is the first step as the next step is to replace the original bleaching agent (BLE1) that contains sodium hypochlorite and also acts as a disinfectant with a more environmental friendly one. However, currently sodium hypochlorite is sometimes used for bleaching and disinfecting hospital laundry. It should also be possible to reduce the amount of fresh water being used for rinsing and neutralisation.

15.2.3 Determination of the Disinfection Laundering Effect

To determine the disinfection efficiency of the investigated laundering procedure, the following bacteria were used: *Enterococcus faecium*, ATCC 6057 and *Staphylococcus aureus* ATCC 6538. *Both bacteria* are standard bioindicators used in tests for determining bactericidal disinfectant efficiency according to regulations of the RKI (Guidelines of the RKI 1995). The target was to achieve at least a 10^5-fold reduction, as noted in the modified suspension test according to the recommendations of the DGHM-standards (Borneff-Lipp et al. 1998). The method for determining the disinfection efficiency using sheep's blood as a substrate for simulating human excrements has been described previously (Fijan et al. 2005; Fijan et al. 2007a; Gebel et al. 2002). Several pieces of previously sterilized and dried cotton pieces with an area of $1 \, cm^2$ were put into previously autoclaved glass Petri dishes ($\Phi = 15$ cm). Petri dishes with cotton substrates were prepared. Then 200 µL of defibrinated sheep's blood was inoculated onto the cotton pieces in one Petri dish and left to dry overnight in a laminar flow cabinet. The next day 200 µL of a previously prepared concentrated suspension of microorganisms was inoculated onto each cotton piece (this procedure was repeated for both microorganisms).

Table 15.1 Washing recipes for the program for hospital bed linen before and after optimizing

Laundering stage	Dosing	Laundering agents[a]	Laundering agents	Laundering conditions
		BL	BLO	BL and BLO
Pre-wash C1–3	C1	6.5 g DET[b]/kg linen 3.1 g ALK[c]/kg linen 20 mL BLE1[d]/kg linen	6.5 g DET[b]/kg linen 3.1 g ALK[c]/kg linen 20 mL BLE1[d]/kg linen	t=12 min, T=55°C, bath ratio: 4 L/kg linen
Bleaching C4	C4	20 mL BLE1[d]/kg linen	/	
Main wash C5–10	C5	0.4 g DET[b]/kg linen 6.1 mL BLE2[e]/kg linen 1.5 mL BUI[f]/kg linen	**0.6**g DET[b]/kg linen 6.1 mL BLE2[e]/kg linen 1.5 mL BUI[f]/kg linen	T=95°C T=**75°C** t=24 min, bath ratio: 4 L/kg linen
Rinse C11–13	C12	2.3 mL DIS[g]/kg linen	**2.8**mL DIS[g]/kg linen	t=12 min, T=40°C, bath ratio: 5 L/kg linen
Neutralization C14	C14	NEA[h] until pH=6.3		t=4 min, T=40°C, bath ratio: 5 L/kg linen

BL original hospital bed linen program, *STO* optimized hospital bed linen program, *C* chamber

[a] All dosages are values of each active ingredient calculated per kg of linen. The dosage of water is noted in the last column as a bath ratio per kg of linen

[b] DET: detergent containing sodium alkylbenzene sulphonate (2%), fatty alcohol ethoxylate (25%), sodium aminotriacetate (10%), 2-butiloktan-1-ol (5%), sodium carbonate (2%)

[c] ALK: alkalis: sodium hydroxide (35%)

[d] BLE1: bleaching agent containing sodium hypochlorite

[e] BLE2: bleaching agent containing hydrogen peroxide (30%)

[f] BUI: builder containing phosphonates

[g] DIS: disinfecting agent containing hydrogen peroxide (10%), peracetic acid (2.5%), acetic acid (2.5%)

[h] NEA: neutralizing agent containing acetic acid (50%)

The final concentration of bacteria on the cotton pieces after drying overnight was between 10^6 and 10^7 bacteria per cotton piece assessed by serial dilutions and plating on appropriate agars for each microorganism, namely Baird-Parker agar for *Staphylococcus aureus* and kanamycin azide agar for *Enterococcus faecium* after incubation at 37°C for 24 h.

15.2.4 Determination of the Laundering Quality

The laundering quality was determined by using a standard cotton control cloth (DIN 53919-1 1980) and laundering it 50 times in the tested laundering procedure. The following parameters were determined; decrease in tensile strength, chemical wear, ash content and whiteness quality (Ganz whiteness degree, Ganz-Griesser tint deviation and basic whiteness degree). The limit values according to the RAL-GZ 992 criteria are noted in Table 15.2 for hospital textiles (RAL-GZ 992/1) and for textiles from the food processing industry (RAL-GZ 992/3).

15.2.5 Measurements of Waste Water Parameters

The measured waste water parameters were chosen according to the Slovenian Regulation on the substance emission during the removal of waste water from laundries and dry cleaners. The parameters and concentration limitations are noted in Table 15.3.

Table 15.2 Laundering quality parameter limit values according to the RAL-GZ 992

Quality parameter		Standard	Limit value for hospital textiles	Limit value for textiles from the food processing industry
Decrease in tensile strength		DIN 13934-1	<30.0%	<30.0%
Chemical wear		DIN 54270-3	<1.0	<1.0
Ash content		ISO 4312	<1.0%	<1.0%
Whiteness quality	Whiteness	GG-method[a]	>170	>170
	Tint deviation		R1,5–G2,5	R1,5–G2,5
	Y value		>85	>85
No of laundering cycles			50	25

[a]Ganz-Griesser method (Griesser 1994)

Table 15.3 Wastewater parameters according to the Regulation on the substance emission during the removal of wastewater from laundries and dry cleaners (Uredba o emisiji snovi pri odvajanju odpadne vode iz naprav za pranje in kemično čiščenje tekstilij 2007)

	Conc. limit of emission into		
Parameter	Water	Sewage system	Standard
Temperature (°C)	30	40	DIN 38404-C4
pH value	6.5–9	6.5–9.5	SIST ISO 10523
Suspended substances (mg/L)	80	[a]	ISO/DIN 11923
Sedimented substances (mL/L)	0.5	10	DIN 38409-H9
Chlorine-free (mg/L)	0.2[b]	0.5[b]	ISO 7393/2
Nitrogen ammonia (mg/L)	5	[c]	SIST EN 25663
Total phosphorus (mg/L)	2	–	SIST ISO 6878-1
COD (mg O_2/L)	120	–	SIST ISO 6060
BOD_5 (mg O_2/L)	25	–	SIST ISO 5815
AOX (mg/L)	3.0	3.0[d]	SIST ISO 9562
Anionic + non-ionic surfactants (mg/L)	3.0	[a]	Appl. Bulletin No. 233/3 & 230/1 (Metrohm SW)

[a] The limit concentration of suspended substances and surfactants in wastewater is determined by the value at which there is no influence on the sewage system or purifying plant
[b] The limit value is not defined in the case of disinfection of laundry from health care
[c] For wastewater flowing into a purifying plant with capacity less than 2.000 PE, the limit value is 100 mg/L; for wastewater flowing into a purifying plant with capacity of 2.000 PE or more, the limit value is 200 mg/L
[d] The parameter limit value is valid for wastewater originating from washing laundry from the health care sector

15.3 Results and Discussion

15.3.1 The Disinfection Effect of the Program for Hospital Bed Linen Before and After Optimisation

The results for the disinfection effect before and after laundering are summarised in Table 15.4.

The program for hospital bed linen before optimizing had an efficient thermal disinfection effect as none of the chosen bioindicators survived the laundering procedure. The optimized program for hospital bed linen with a maximum temperature of 75°C and an additional amount of disinfecting agent also proved to have a sufficient chemo-thermal disinfection effect. Therefore the implemented improvements of the optimized program were effective for disinfection, which is the most important aspect when laundering hospital textiles, as the primary function of hospital textiles is to serve as protection and not as a vehicle for microbe infection.

15 Environmental Risk Factors in Connection with Hospital Laundry Effluent

Table 15.4 The disinfection effect of the hospital bed linen program before and after optimization

Bacteria	cfu_b	BL cfu_a	RED	BLO cfu_a	RED
Staphylococcus aureus	4.5×10^7	0	>7.65	0	>7.65
Enterococcus faecium	3.3×10^7	0	>7.52	0	>7.52

Where:

cfu_b means the initial number of colony forming units of individual microorganisms per textile swatch

cfu_a means the number of colony forming units of individual microorganisms per textile swatch after laundering

RED means the logarithmic reduction efficiency after laundering (RED = log (cfu_b/cfu_a) or in the case of zero cfu after laundering RED > log cfu_b)

Table 15.5 Results for the laundering quality according to the RAL-GZ 992 quality parameter limit values after 50 laundering cycles

Quality parameter		Standard	Limit value for hospital textiles	BL	BLO
Decrease in tensile strength		DIN 13934-1	<30.0%	**33.3%**	20.6%
Chemical wear		DIN 54270-3	<1.0	**1.3**	0.8
Ash content		ISO 4312	<10%	0.2%	0.3%
Whiteness quality	Whiteness	Ganz-Griesser method	>170	225.4	209.1
	Tint deviation		R1,5–G2,5	N0,4	N0,38
	Y value		>85	95	91

15.3.2 The Laundering Quality of the Program for Hospital Bed Linen Before and After Optimisation

The secondary laundering effects of the standard cotton control cloths of the investigated laundering procedures were determined by measuring the following quality parameters: decrease in tensile strength (F), chemical wear (s), ash residue (A) and whiteness quality, including Ganz degree of whiteness (W_G), lightness (Y) and Ganz-Griesser tint value (TV_{GG}) by standard methods noted in Table 15.5.

From the results it is obvious that the program for bed linen before optimizing did not reach the quality level, according to the RAL-GZ 992 criteria, since the limit values for the decrease in tensile strength and chemical wear were both exceeded due to the double bleaching conducted of a hypochlorite-based bleach in compartments 1 and 4. Decrease in tensile strength of cotton after laundering is a result of a combination of mechanical and chemical factors operating during the laundering procedure and chemical wear of cotton which is caused by the aggressive action of chemical substances (oxidative bleaching agents, such as sodium hypochlorite, peroxides etc.) and which leads to changes in the intrinsic properties of cotton fibers, thus reducing the degree of polymerization of the cellulose constituent.

The results for the decrease in tensile strength and chemical wear after optimizing show that the omission of the double bleaching was an appropriate measure as the whiteness quality was still within the limits. The ash content was also within the limits as softened water was used and further helped by the addition of builders as well as efficient rinsing which prevented redeposition of mineral deposits derived from salts contained in the laundering water or from detergents. The whiteness quality of cotton after laundering is also an important indicator of the laundering quality because inefficient laundering procedures can cause greying or yellowing due to redeposition of coloured pigment soils from soiled white loads on the control cloth, staining from dyes if coloured materials are present in the laundry, deposition of iron salts from the wash bath, deposition of alkaline substances due to inadequate rinsing, development of colour due to fatty soap residues, etc. (Fijan et al. 2007a). Both laundering procedures exhibited adequate whiteness quality.

15.3.3 The Waste Water Parameters of the Program for Hospital Bed Linen Before and After Optimisation

The results of the waste water parameters of the program for hospital bed linen before and after optimization are noted in Table 15.6.

The results show that two parameters of both laundering procedures before and after optimization exceeded the limit values for the emission into the sewage system: pH-value and AOX. The high AOX value is a consequence of the use of sodium hypochlorite as a bleaching agent that also acts as a disinfectant. This disinfectant still has a widespread biomedical use due to its biocide properties

Table 15.6 Results for wastewater parameters of the hospital bed linen program before and after optimizing according t the Regulation on the substance emission during the removal of wastewater from laundries and dry cleaners (Uredba o emisiji snovi pri odvajanju odpadne vode iz naprav za pranje in kemično čiščenje tekstilij 2007)

Parameter	Conc. limit of emission into sewage system	BL	BLO
Temperature (°C)	40	54.8	36.2
pH value	6.5–9.5	10.64	10.97
Suspended substances (mg/L)	–	49	52
Sedimented substances (mL/L)	10	3	4
Chlorine-free (mg/L)	–	49	30
Nitrogen ammonia (mg/L)	–	<0.05	<0.05
Total phosphorus (mg/L)	–	6.5	6.9
COD (mg O_2/L)	–	1,280	1,200
BOD_5 (mg O_2/L)	–	900	990
AOX (mg/L)	3.0	9.7	9.85
Anionic + non-ionic surfactants (mg/L)		180.2	245.7

(Emmanuel et al. 2004), especially for disinfecting hospital effluent. However it is being replaced by efficient and, at the same time, environmentally friendly agents due to the fact that chlorine substances in wastewaters may react with organic matters, giving rise to organic chlorine compounds such as AOX that were also found in our experiment and that are toxic to aquatic organisms and are persistent environmental contaminants (Emmanuel et al. 2004). It was also noted (Emmanuel et al. 2004) that finding an optimum concentration of sodium hypochlorite in disinfecting hospital effluent, i.e. its non-observed effect on algae and *Daphnia magna,* is a research issue that could facilitate the control of AOX toxicity on aquatic organisms.

The temperature of the optimized program for bed linen was within the limit value for emission into sewage system as the chemo-thermal laundering procedure had a 20°C lower washing temperature. Although the following parameters, chemical oxygen demand (COD), biological oxygen demand (BOD_5) and the sum of surfactants were high, there are no limit values in Slovenia for emission into sewage systems. The overall results clearly show that the optimized program produced less burdened effluent; however the results also show that waste water treatment methods are inevitable (Fijan et al. 2008).

15.4 Conclusion

From the research it can be concluded that the optimized program for hospital bed linen demonstrated better overall results and that a transformation from thermal to chemo-thermal laundering procedures can be environmentally friendly and, at the same time, ensure disinfection. The optimized program also showed better laundering quality results as the decrease in tensile strength and chemical wear were within the limit values according to the RAL-GZ 992 criteria while the thermal procedure before optimization did not reach an efficient laundering quality due to the double bleaching.

To further improve laundering procedures and to become even more effective, sustainable and environmentally friendly, the following measure besides optimization of washing agents are suggested: (1) water from tanks (from rinsing, neutralization and after press) are treated before reuse in order to decrease the amount of organic deposits, microfibrils, pills etc. from the dirty laundry while the detergents are also partly recycled thus decreasing the necessary dosages and, in turn, decreasing the costs for chemicals, (2) the energy from the hot effluent after the main wash is pumped via a heat exchanger for heating fresh rinsing water, (3) the water consumption is decreased by increasing the amount of recycled water via water treatment plants, (4) incorporating various sensors to manage the dosage of different washing agents, thus preventing overdosing due to reuse of water with a certain amount of washing agents. Thus potentially excessive damage to the laundered textiles will be presented and unnecessary effluent burden reduced (Fijan et al. 2008).

However, the downside of most laundering procedures in industrial laundries at the moment is the need for waste water treatment. The necessary aim for the future is the creation of such laundering programs, waste water recycling programs, or an individual waste water treatment plants, that demonstrate a more economical and environmentally sound alternative. An alternative which incorporates almost 100% water recycling and reduces sewer discharge to zero.

References

Barrie D, Hoffman PN, Wilson JA, Kramer JM (1994) Contamination of hospital Linen by Bacillus Cereus. Epidemiol Infect 113(2):297–306

Borg MA, Portelli A (1999) Hospital laundry workers – an at-risk group for hepatitis A? Occup Med 49(7):448–450

Borneff-Lipp M, Christiansen B, Eggers HJ, Exner M, Gunerman KO, Heeg P et al (1998) Disinfection commission of the German Society for hygiene and microbiology. Chemotermische Wäschedesinfektion Hyg Med 4:127–129

Brunton WA (1995) Infection and hospital laundry. Lancet 345(8964):1574

Council Directive of 23 October (2000) Establishing a framework for community action in the field of water policy (2000/60/EC). Off J Eur Communities L327

Desinfektionsmittel-Liste des VAH (2009) Mhp-Verlag: Deutschen Gesellschaft für Hygiene und Mikrobiologie. http://www.dghm.org/red/komissionen/desinfekt/index.html?TextID=242. Accessed 5 May 2010

Emmanuel E, Keck G, Blanchard JM, Vermande P, Perrodin Y (2004) Toxicological effects of disinfections using sodium hypochlorite on aquatic organisms and its contribution to AOX formation in hospital wastewater. Environ Int 30(7):891–900

Fijan S, Šostar-Turk S, Cencič A (2005) Implementing hygiene monitoring systems in hospital laundries in order to reduce microbial contamination of hospital textiles. J Hosp Infect 61(1):30–38

Fijan S, Koren S, Cencič A, Šostar-Turk S (2007a) Antimicrobial disinfection effect of a laundering procedure for hospital textiles against various indicator bacteria and fungi using different substrates for simulating human excrements. Diagn Microbiol Infect Dis 57(3):251–257

Fijan S, Šostar-Turk S, Neral B, Pušić T (2007b) The influence of industrial laundering of hospital textiles on the properties of cotton fabrics. Text Res J 77(4):247–255

Fijan S, Fijan R, Šostar-Turk S (2008) Implementing sustainable laundering procedures for textiles in a commercial laundry and thus decreasing wastewater burden. J Cleaner Prod 16(12):1258–1263

Garcia MT, Campos E, Marsal A, Ribosa I (2009) Biodegradability and toxicity of sulphonate-based surfactants in aerobic and anaerobic aquatic environments. Water Res 43(2):295–302

Gebel J, Werner HP, Kirsch-Altena A, Bansemir K (2002) Standard Methoden der DGHM zur Prüfung chemischer Desinfektionsverfahren. Mhp Verlag GmbH, Wiesbaden

Gellert GA, Waterman SH, Ewert D, Oshiro L, Giles MP, Monroe SS et al (1990) An outbreak of acute gastroenteritis caused by a small round structured virus in a geriatric convalescent facility. Infect Control Hosp Epidemiol 11(9):459–464

Gonzaga AJ, Mortimer EA, Wolinsky E (1964) Transmission of Staphylococci by fomities. JAMA 189:711–715

Griesser R (1994) Assessment of whiteness and tint deviation of fluorescent substrates with good instrument correlation, colour research and application. Ciba-Geigy, Basle, Switzerland

Guidelines of the Robert-Koch Institute (1995) Guidelines for hospital hygiene and infection prevention. Robert-Koch Institute, Berlin

Karahan O (2010) Inhibition of linear alkylbenzene sulphonates on the biodegradation mechanisms of activated sludge. Bioresour Technol 101(1):92–97

Kirby WMM, Corporon DO, Tanner DC (1956) Urinary tract infections caused by antibiotic-resistant Coliform bacteria. JAMA 162:1–4

Kitis M (2003) Disinfection of wastewater with peracetic acid: a review. Environ Int 30(1):47–55

Leonardo da Vinci project, number: 146 360 (2010) Training modules on the sustainability of industrial laundering processes. E-learning tool for trainees and employees in laundries. http://www.laundry-sustainability.eu/. Accessed 5 May 2010

Liste der vom Robert Koch-Institut geprüften und anerkannten Desinfektionsmittel und – verfahren (Eng. List of the Robert Koch institute tested and approved disinfectants and procedures) (2007) Robert-Koch Institute, Bundesgesundheitsbl - Gesundheitsforsch – Gesundheitsschutz. Springer Medizin Verlag 50:1335–1356. doi: 10.1007/s00103-007-0341-4

Lynch M, Stretsky PB (2003) The meaning of green: contrasting criminological perspectives. Theor Criminol 7(2):217–238

Oliphant JW, Gordon DA, Meis A, Parker R (1949) Q-fever in laundry workers, presumably transmitted from contaminated clothing. Am J Hyg 47:76–81

RAL-GZ 992 (2001) Sachegemёße Waschepflege, Gütezicherung RAL-GZ 992 (Eng. Proper linen care, quality assurance RAL-GZ 992). RAL, Deutsches Institut für Gütezicherung und Kennzeichnung e.V, Sankt Avgustin

Shah PC, Krajden S, Kane J, Summerbell RC (1988) *Tinea Corporis* caused by *Microsporum canis*: report of a nosocomial outbreak. Eur J Epidemiol 4(1):33–38

Šostar Turk S, Petrinič I, Simonič M (2005) Laundry wastewater treatment using coagulation and membrane filtration. Resour Conserv Recycl 44(2):185–196

Standaert SM, Hutcheson RH, Schaffner WA (1994) Nosocomial transmission of salmonella gastroenteritis to laundry workers in a nursing home. Infect Control Hosp Epidemiol 15(1):22–26

Thomas MD, Giedinghagen DH, Hoff GL (1987) An outbreak of scabies among employees in a hospital-associated commercial laundry. Infect Control 8:427–429

Uredba o emisiji snovi pri odvajanju odpadne vode iz naprav za pranje in kemično čiščenje tekstilij (2007) (Eng. Regulation on the substance emission during the removal of wastewater from laundries and dry cleaners). Uradni list Republike Slovenije, št. 41/2007

Wilcox MH, Jones BL (1995) Enterococci and hospital laundry. Lancet 345:1574–1575

Ying GG (2006) Fate, behaviour and effects of surfactants and their degradation products in the environment. Environ Int 32(3):417–431

Chapter 16
Risk Assessment of Chemicals in Food for Public Health Protection

Elizabeta Mičović, Mario Gorenjak, Gorazd Meško, and Avrelija Cencič

Abstract World climate, environmental and other global changes would have wide-ranging and, mostly, adverse consequences for human health. These changes are a significant and emerging threat to public health, including the growing season, variability of crop yields and water demands. Above mentioned factors have been influenced in areas of food security and food safety and moreover the food industry struggle for profits. These factors control the conditions and affect the use of chemicals in the agriculture and food industry. In this chapter, the risk assessment of chemicals in food as a part of risk analysis process have been addressed. We would like to point out the importance of this process in the area of public health and consumer protection. Through food and nutrition products and most often misleading advertisements, people are exposed to different hazards that could have severe harmful consequences. Among the consumer groups, children are mostly vulnerable because of their adult dependencies, age and body weight. In this chapter, an example of risk assessment of exposure to food additives among preschool children in Slovenia is evaluated.

E. Mičović (✉)
Faculty of Criminal Justice and Security, University of Maribor, Kotnikova 8, 1000 Ljubljana, Slovenia
e-mail: elizabeta.micovic@gmail.com

M. Gorenjak
Faculty of Medicine, University of Maribor, Maribor, Slovenia
and
Faculty of Agriculture and Life sciences, University of Maribor, Maribor, Slovenia

G. Meško
Faculty of Criminal Justice and Security, Institute of Security Strategies, University of Maribor, Kotnikova 8, 1000 Ljubljana, Slovenia
e-mail: gorazd.mesko@fvv.uni-mb.si

A. Cencič
Faculty of Agriculture and Life sciences, University of Maribor, Maribor, Slovenia
and
Faculty of Medicine, University of Maribor, Maribor, Slovenia

---

G. Meško et al. (eds.), *Understanding and Managing Threats to the Environment in South Eastern Europe*, NATO Science for Peace and Security Series C: Environmental Security, DOI 10.1007/978-94-007-0611-8_16, © Springer Science+Business Media B.V. 2011

293

16.1 Introduction

Living in a risk society as a parallel to the rapid changes witnessed in all spheres of life indicates the growth of various threats. Constantly changing circumstances in a society are increasing the number of hazards. As humans convert more land, water, and ecosystem services for their own use, the environmental changes resulting from these activities are combining to magnify several serious public health threats including exposure to infectious disease, food scarcity, water scarcity, air pollution, natural disasters, and population displacement. Taken together, these represent the greatest public health challenge of the twenty-first century.

One of them is climate change, which threatens human health in numerous and profound ways. Large segments of the population will experience more heat waves, altered exposure to infectious disease and more frequent natural disasters. Climatic disruption threatens health for large populations around the globe: sufficient food and nutrition, safe water for drinking and sanitation, fresh air to breathe, and secure homes to live in. Such a society with such a lifestyle would certainly have to adjust the responsiveness of a good organization – coordinated team of experts in various fields, in addition to a strategic consideration (thinking), should agree measures of the ability to realize them operational (Heines et al. 2000).

Although altered exposure to infectious disease receives much of our attention, the most important impacts of global change on human health are likely to result from reduced access to food and safe water. As the human population expands, world agricultural production will roughly double over the next 50 years to keep up with projected demand (Myers and Patz 2009). Food production is already facing significant ecological constraints – including limits to arable land, water scarcity, soil nutrient depletion, and biological limits to increasing crop yields. Climate change will further challenge food production through myriad mechanisms and is expected to reduce yields significantly in many regions of the world – particularly those where food scarcity is already endemic (McMichael et al. 1996; Heines et al. 2000; Gregory et al. 2005; WMO 2007).

One of the results of all these new conditions in food production is also using chemicals with the aim to achieve sufficient quantities of food (food security), which has to be of good quality and safe (food quality and food safety) (FAO/WHO 2002; Unnevehr 2003). Worrying risk factors, such as chemicals that are in the food are invisible, just as their effects are not immediately visible. These risks are set out as acceptable risks, especially given the fact that it has not developed adequate methods of determining the exact exposure to these risk factors and their impact on health (Beck 2001).

It is urgent to reduce ecological disruption while simultaneously strengthening the resilience of populations to withstand the impacts of unavoidable environmental change (Vanderlinden et al. 2005). Therefore, it is important to develop a strong tool to analyse risk, special hazards and effects on health of consumers. In response to some unwanted and unexpected incidents of non-food, safety system is a risk analysis developed to be a tool for early detection of risks, facilitating the prompt effective action. Risk analysis is based on a database of research results to scientific and sociological sciences. Results of risk analysis contribute to trust all those

16 Risk Assessment of Chemicals in Food for Public Health Protection 295

involved in the system and to ensure safe food (Kleter and Marvin 2009). Management risk analysis involves combining activities related to risk assessment, risk management and risk communication, and is used to guide the adoption of administrative decisions at the national and international level, including in the areas of trade (Buchanan et al. 2004).

Risk assessment is a scientific process that usually begins in order to define the reasons for the request, risk assessment and efficiency of implementation. Risk assessors are responsible for scientific evaluation whether a formal risk assessment is necessary or not (Benford 2001; Mack et al. 2006).

The challenge of risk is that these risks represent a potent political charge, and potential political challenge. Identifying the risks, risk assessment and risk management can lead to a reorganization of power and responsibilities in the society. Therefore, the development of the risk of conflict developed between those who seek to risk and those who profit from them. It is important to have sufficient information to enable it to research potential risks carefully investigated and can be regarded as acceptable or unacceptable (Teuschler and Hertzberg 1995; Beck 2001; Newburn 2007).

It is very important to change way we protect the vulnerable populations. Children are one of the most vulnerable populations as they are more heavily exposed to environmental toxins than adults. In this chapter, an example study among preschool children's exposure to food additives in food they consumed and risk assessment of such exposure have been addressed. Food intake of preservatives, polyphosphates, sweeteners and colours was observed with aim to find out if such exposure is an acceptable or unacceptable risk for children included in this study.

16.2 Consumer as Possible Victim of Invisible Threats

"Consumers by definition, include us all. They are the largest economic group, affecting and affected by almost every public and private economic decision. Furthermore, consumers are the only important group whose views are often not heard" (Kennedy 1962). A consumer is a person who buys goods or services for personal needs and not for resale or use in the production of other goods for resale (Zakon o varstvu potrošnikov [ZVPot-UPB2] 2004).

Routine Activity Theory (RAT) is one of the main theories of "environmental criminology". Criminologists Lawrence Cohen and Marcus Felson have worked for a number of years on crime prevention theory. RAT states that for a crime to occur, three elements must be present at the same time and in the same place: a suitable target is available, there is the lack of a suitable guardian to prevent the crime from happening and likely a motivated offender is present (Cohen and Felson 1979). All three elements from the theory in the area of food safety are recognized – the motivated offender as the food industry, the suitable targets as consumers and the lack of a suitable guardian as insufficient official control. It is also important that opportunity for the commission of misleading, abuse or even crime occur. Opportunity is crucial in many deviant activities and can be related to a potential victim (a consumer), routines (practices) or places (interactions).

Consumers also expect a wide range of competitively priced, highly processed and convenient food products of consistently high quality. They expect it to be fresh, good looking, nutritious, wholesome, tasty and it must be primarily and absolutely safe. On the other hand, consumers have no direct means for the verification of their expectancies and have to rely completely on the food legislators and enforcement agencies (Anklam and Battaglia 2001). Consumers could be victims of food poisoning, food adulteration and food frauds, misleading food content (labelling), misleading indications, misleading descriptions, misleading pictures, and food packaging (Jin and Kato 2004; Gibson and Taylor 2005; Tombs 2008; Croall 2009). In today's technological age, a reactive response to the consumer fraud is neither efficient nor effective (Holtfreter et al. 2005).

A guideline for consumer protection provides eight basic consumer rights interpreted from the food safety point of view (Harland 1987; Hogarth and English 2002):

- The Right to Safety – to be protected against threats that are hazardous to health or life.
- The Right to Information – to be protected against fraudulent, deceitful or grossly misleading information, advertising, labelling, or other practices and to be given the facts s/he needs to make an informed choice.
- The Right to Choose – to be assured, wherever possible, access to a variety of products and services at competitive prices: and in those industries where competition is not workable and Government regulation is substituted, an assurance of satisfactory quality and service at fair prices.
- The Right to be heard – to be assured that consumer interests will receive full and sympathetic consideration in the formulation of Government policy, and fair and expeditious treatment in its administrative tribunals.

Consumers are privileged to have human rights. However, they come with certain responsibilities too – to seek, to evaluate and to use available information on products and services in order to make healthier and better decisions. In case of exposure to different chemicals in food, consumers know that food products consist of additives, but they do not know the quantity. Total intake is unknown, so how could consumers be sure that the consumed food is safe? It is very important to consider different sensitivities of populations, especially among infants and children.

16.3 Food Safety

The main challenge in the area of agriculture is to provide sufficient quantities of food, which have to be of good quality and safe. Food production and food consumption are the primary aspects of our lives and are therefore subject to our care. To achieve these goals the agriculture and food industry has to use in their production different chemicals such as pesticides and food additives. Without the use of chemicals the yields of fields will reduce, food production and the earnings will be considerably smaller. With using chemicals, the benefits are visible, but the risks are often invisible. Chemicals can have negative effects on us and make us vulnerable (Beck 2001).

16 Risk Assessment of Chemicals in Food for Public Health Protection

Safe food is food that is free not only from toxins, pesticides, chemical and physical contaminants, but also from microbiological pathogens such as bacteria and viruses that can cause illness (Golob and Jamnik 2004). There are main concerns regarding food safety. Consuming food is a daily routine activity throughout lifetime. It is very important that such food does not cause health risks to consumers. Possible hazards in food are microorganisms, viruses, contaminants (toxins, heavy metals), pesticides and other chemicals. Consumer could be concerned and afraid of such hazards, but in many cases one does not know the exact quantity of exposure and exact effect on his health. Therefore, it is very important to be aware of possibility of invisible threats in food, especially because of the fact that all three elements from RAT in the food safety area have been recognized. The food industry has a strong motive to make profit and many opportunities to manage it. The food operators try to convince consumers that they need their products. The food industry uses all sorts of food additives, ingredients and advertising tactics to achieve better sales (Cheftel 2005). The food industry can also be involved in corporate crimes committed directly against the consumers. Examples include the sale of unfit goods, the provision of unfit services, false/illegal labelling or information, selling unsafe food, poisonings, etc. (Jin and Kato 2004; Gibson and Taylor 2005; Tombs 2008; Croall 2009).

On the market, companies make money by providing goods to consumers, and consumers pay for goods received. Consumers have the legal right to be protected (Xu and Yuan 2009). The long-term health of consumers is also jeopardised by the use of foods and other consumer products of a vast range of chemicals and other substances that, while associated with long- term health risks, do not result in imme-diate harm. While there is a growing public concern about the number of foods and consumer issues, these facts have a lower political and governmental profile than the occupational health and safety or the safety of the environment (Croall 2009).

The main authority concerns, regarding food safety, are to protect interests of public health, interests of food producers (economic view) and the consumer interests and their rights. It is not easy to make the right decision and to achieve all that goals in practice at the same time. Recognizing possible invisible threats could assure better consumer protection. Furthermore, knowing all the risks of invisible threats in the food area help us to make corrective measures on time. These measures could make the system for food safety more effective and give consumers better protection. The big challenge in the area of food safety for consumer health is recognition of possible invisible threats on time. This can help us to set up effective response regarding those threats and risk assessment. Identifying all potential hazards that have to be assessed, eliminating or reducing them to acceptable levels, are the most important activities for achieving consumer protection, specially their basic right to safety.

16.3.1 Risk Analysis

Public health decisions on the plausible risks of chemical exposures can include several possible outcomes. The ultimate goal is to implement a risk management action that will produce the desired reduction of risk. A risk analysis paradigm is a formal representation of a process that distinguishes the scientific bases from the risk management objectives and generally contains a component where the probability of harm is estimated. The overall risk analysis process includes risk assessment, risk management and risk communication, and involves political, social economic and technical considerations. Moreover, there is consensus among scientists that risk assessment should be an independent scientific process, distinct from measures taken to control and manage the risk. Risk assessors are responsible for scientific evaluation whether a formal risk assessment is necessary or not (Benford 2001). As a probability calculation, a risk assessment will include both a statement of the objective under consideration (harm) and the basis for the assertion that the harm may occur (probability). Risk management is the decision-making process involving the consideration of political, social, economic and technical factors with relevant risk assessment information relating to a hazard so as to develop, analyse, select and implement appropriate risk mitigation options. Risk management is comprised of three elements: risk evaluation, emission and exposure control and risk monitoring (WHO 2004).

Risk management strategies may be regulatory, advisory or technological and take into account factors such as the size of the exposed population, resources required and available, costs of implementation and the scientific quality and certainty of the risk assessment. Risk managers are responsible for judgements concerning the acceptability of risk; they have to weigh risk against other factors including costs, benefit and social values, so called risk – benefit approach.

Risk communication should include interactive exchange of information and opinions among risk assessors, risk managers, consumers and all other interested parties, often called stakeholders (Benford 2001).

16.3.1.1 Risk Assessment – Exposure to the Chemicals in the Food

Risk is defined as the chance or probability of an adverse health effect occurring and severity of that effect (Benford 2001). Risk assessment is a scientific process, conducted by scientific experts, who may begin with a statement of purpose intended to define the reasons that the risk assessment is required and support the aims of the subsequent stages of risk management. Chemical risk assessment often does not have a formal statement of purpose (Benford 2001). Generally, formal risk assessments are preceded by preliminary risk assessments. These are usually subjective and informal and could be initiated from inside or outside the risk assessment and scientific communities. A key consideration of these preliminary risk assessments is whether a formal risk assessment is necessary or not (WHO 2004).

However, risk assessment could be defined implicitly in a generic form, as in the terms of reference of an expert committee, such as Joint FAO/WHO or JECFA. In this context, it concerns the definition of acceptable or tolerable levels of intake for a chemical in food that may require review and revision in the light of new information. It is a very important fact that risk assessment may need to be quantified differently for persons with different degrees of susceptibility. Risk assessment for chemical agents requires consideration of the factors mentioned before and these

16 Risk Assessment of Chemicals in Food for Public Health Protection

are generally encompassed within the stages of the overall risk assessment process, defined as hazard identification, hazard characterisation, exposure assessment risk-characterisation (Benford 2001).

Hazard identification is the process of identifying of the type and nature of adverse health effects [human studies-epidemiology, animal – based toxicology studies, in-vitro toxicology studies, structure-activity studies (cell cultures, tissue slices)]. Hazard characterization involves the derivation of a level of exposure at or below which there would be no appreciable risk to health if the chemicals were to be consumed daily throughout life. Exposure assessment is evaluation of concentration or amount of a particular agent that reaches a target population: magnitude, frequency, duration, route, extent. Risk characterization is the stage of risk assessment that integrates information from exposure assessment and risk characterization into advice suitable for use in decision-making (Benford 2001).

The fact is that people could be exposed to different chemicals that pass from environment to food such as contaminants, food additives, pesticides, heavy metals, toxins, detergents etc. (De Muelenaer 2006). They could represent a threat to humans, especially, because they are invisible. The vast majority of toxicity studies and risk evaluations deal with single chemicals. In reality, people are exposed to large numbers of chemicals via multiple routes (Feron and Groten 2002). So far, possible effects of a mixture of all chemicals and interactions between them, so called 'cocktail effect', have not been understood fully. Food additives are typically used in combination within processed foods and therefore collectively may have some adverse effects at the cellular level, even if their individual concentrations are below the recommended acceptable daily intake (ADI) value (Lau et al. 2005).

The basic concept underlying any chemical risk assessment is the dose-response relationship. As described by Paracelsus nearly 500 years ago, "All substances are poisons; there is none which is not a poison. The right dose differentiates a poison and a remedy" (Winter and Francis 1997). This means that any chemical substance is likely to produce some form(s) of harmful effect, if taken in sufficient quantity. Experts refer to a potential harmful effect as a hazard associated with that substance. The definition of hazard is "a biological, chemical or physical agent with the potential to cause an adverse health effect" (Unnevehr 2003; Raspor 2004; Armstrong 2009). While this may be appropriate with respect to pathogenic organisms, chemical substances may be associated with a number of different adverse health effects, not all of which would necessarily be expressed in a specific exposure scenario. Therefore, experts dealing with chemical substances prefer to define the potential health effects as individual hazards, which need to be considered separately during the evaluation. The likelihood or risk of that hazard actually occurring in humans is dependent upon the quantity of chemical encountered or taken into the body, i.e. the exposure. The hazard is an inherent property of a chemical substance, but if there is no exposure, then there is no risk that anyone will suffer because of that hazard.

Risk assessment is the process of determining whether a particular hazard will be expressed at a given exposure level, duration and timing within the life cycle, and if so the magnitude of any risk is estimated (Benford 2001). Among the first

300 E. Mičović et al.

objectives of a risk assessment is the determination of the presence or absence of a cause-effect relationship. If there is sufficient plausibility for the presence of such a relationship, then dose-response modelling (DRM) information is needed.

The procedures used to estimate exposure to chemical contaminants in food are essentially the same as those used for food additives (DiNovi and Kuznesof 2006). Exposure assessment should cover the general population, as well as critical groups that are vulnerable or are expected to have exposures that are significantly different from those of the general population, for example infants, children, pregnant women, or the elderly (WHO 2005).

16.3.1.2 Acceptable Daily Intake and Estimated Daily Intake of Food Additives

The concept of the ADI is internationally accepted today as the basis for estimation of safety of food additives and pesticides, for evaluation of contaminants and by this, for legislation in the area of food and drinking water. The ADI (Tables 16.1–16.3) is

Table 16.1 Acceptance daily intake of preservatives and polyphosphates

Preservatives	E number	ADI (mg/kg bw/day)
Sorbic acid	E 200	0–25 (JECFA[a], 1973)
Bensoic acid	E 210	0–5 (JECFA, 2002)
Nitrate	E 251	0–3.7 (SCF[b]; JECFA 2002)
Nitrite	E 250	0–0.07 (JECFA, 2002)
Sulphur dioxide	E 220	0–0.7 (JECFA, 1973)
Polyphosphates	(E 450-452)	0–70 (JECFA, 1982–2001)

[a, b]Acceptable daily intakes have been derived from toxicological studies by the former EU Scientific Committee on Food (SCF) and by the WHO/FAO Joint Expert Committee on Food Additives and Contaminants

Table 16.2 Acceptable daily intake of colours

Colours	E number	ADI (mg/kg bw/day)
Tartrazine	E 102	0–7.5 (EFSA[a], 2009)
Quinolyne yellow	E 104	0–0.5 (EFSA, 2009)
Sunset yellow FCF	E 110	0–1.0 temporary (EFSA, 2009)
Azorubine	E 122	0–4 (EFSA, 2009)
Amaranth	E 123	0–0.5 (JECFA, 1975)
Ponceau red 4R	E 124	0–0.7 (EFSA, 2009)
Erythrosine	E 127	0–0.1 (JECFA)
Allura red AC	E 129	0–7 (EFSA, 2009)
Patent blue V	E 131	0–1 (JECFA, 1969)
Brilliant Indigotine	E 132	0–5 (JECFA, 1974)
Blue FCF	E 133	0–12.5 (JECFA, 1969)
Brilliant black PN	E 151	0–1 (JECFA, 1981)

[a]European Food Safety Authority

16 Risk Assessment of Chemicals in Food for Public Health Protection 301

Table 16.3 Acceptable daily intake of sweeteners

Sweeteners	E number	ADI (mg/kg bw/day)
Saccharin	E 954	0–5 (JECFA, 1993)
Cyclamate	E 952	0–11 (JECFA, 1982)
Acesulfame K	E 950	0–15 (JECFA, 1990)
Aspartame	E 951	0–40 (JECFA, 1981)

an estimate of the amount of a food additive, expressed on a body weight basis that can be ingested daily over a lifetime without appreciable health risk (WHO 1987).

Although ADI is derived from the safety assessment of each food additive, their combined adverse effects are unclear and have not been widely studied. Food additives are typically used in combination within processed foods and therefore, collectively may have some adverse effects at the cellular level, even if their individual concentrations are below the recommended ADI value (Lau et al. 2005).

Dose response modelling (DRM), used as quantitative risk assessment tool for public health recommendations about chemical exposures, can be described as a six-step process. The first four steps-data selection, model selection, statistical analyses, and parameter estimation-constitute dose–response analysis. The fifth step involves the integration of the results of the dose–response analysis with estimates of human exposure. The final step involves an assessment of the quality of the dose–response analysis and the sensitivity of model predictions to the assumptions used in the analysis.

Extrapolation is a fundamental problem in the quantitative health risk assessment of exposure to chemicals that are toxic to human in experimental systems. Adverse health effects of chemicals are, in the absence of human data, typically evaluated in laboratory animals at significantly higher doses than the levels to which humans may be exposed. Moreover, the data obtained in animals are very often misleading, as the animals used usually do not respond to the toxic compounds in the same way as humans (WHO 2004).

The acceptable daily intake (ADI) is used widely to describe "safe" levels of intake; other terms that are used are the reference dose (RfD) and tolerable intakes that are expressed on either a daily (TDI or tolerable daily intake) or weekly basis. The weekly designation is used to stress the importance of limiting intake over a period of time for such substances (Herman and Younes 1999).

In order to calculate an ADI using the data from toxicity studies, the lowest dose should ideally result in no effects under the conditions of the particular study. That dose may be termed as the No Observed Effect Level (NOEL). Observed effects are referred to because assumptions cannot be made about effects not detectable by the methods used. Some effects observed in toxicity studies may represent adaptive responses with no implications for the health status of the animal and would generally not be used as the basis for establishing an ADI. Effects that are considered to result in harm to the animal are referred to as "adverse", and therefore some expert committees use the expression No Observed Adverse Effect Level (NOAEL) (Benford 2001). When using this approach no-observed-adverse-effect levels (NOAELs) are identified in the critical studies, to which appropriate safety or uncertainty factors are applied. Although the value of safety factors varies depending upon a number of factors, 100 is most often used, which is designed to account for interspecies and interspecies variations (Herman and Younes 1999).

$$ADI = NOAEL/100$$

Estimated daily intake (EDI) is value of chemical exposure, which can be determined by combining food consumption data with data on the concentration of additives in food.

The resulting dietary exposure estimate is afterwards compared with the relevant toxicological or nutritional reference value for the food additive concern, for example acceptable daily intake (ADI) or tolerable daily intake (TDI) (WHO 2005). Usually EDI is expressed as percentage of value of ADI for each food additive.

If EDI is lower or the same as ADI, there is no concern regarding such exposure and we can assume that food additive intake among the observed population represent no risk. If EDI is higher or the same as ADI, than exact risk assessment should be done on a case-by-case base, and risk managers should decide and take effective measurements.

16.4 Example of Risk Assessment Among Preschool Children in Slovenia

The purpose of research was to assess if food additives intake could be a possible threat to the health of preschool children. It is assumed that intake of each food additive separately could not be so high or greater than ADI. However, the total intake of all food additives per day is unknown. It is assumed that higher exposure to food additives is in relation to appearance with possible harm reactions among observed children. In the first place, it is necessary to find out the exact exposure to food intake and consumed food additives: preservatives, polyphosphates, sweeteners and colours as they shown in Figs. 16.1 and 16.2.

16.4.1 Methodology

16.4.1.1 Observed Population

Among randomly selected regions in Slovenia, we selected kindergartens and all children aged 2–6 years. Initially, the study included 250 children, representing all children of one section of selected kindergartens. Due to incomplete data, 60 cases were excluded from analysis. Therefore, the total number of children used in this study were 190, 98 boys and 92 girls. Anthropometric measurements of experimental children groups were conducted as for example data regarding their age and gender, weight and height. In the selected group, boys and girls were divided into two groups regarding their age: 2–3.9 years and 3.9–6 years.

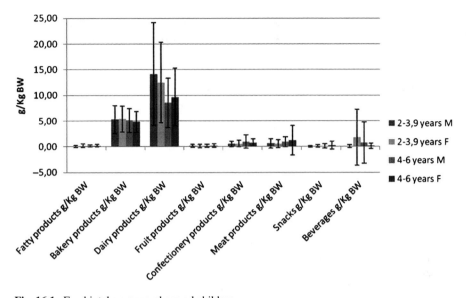

Fig. 16.1 Food intake among observed children

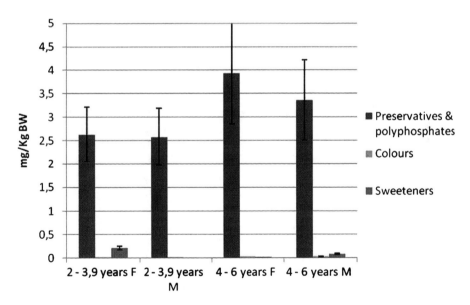

Fig. 16.2 Intake of food additives: preservatives, polyphosphates, colours and sweeteners

16.4.1.2 Food Consumption Data

The dietary intake is generally based on the 3-day-weighed record method in the kindergartens and food diary at homes. In the kindergartens subject to the research, we have weighed quantities of individual foods and dishes, randomly chosen and

Table 16.4 Food categories

Food category	Food product
Non-alcoholic beverages	Coca-cola, ice-tea
Fat spreads	Margarine, butter
Bakery products	Bread, cookies, pastry
Fruit products	Dried fruits, jams, candied fruits
Confectionary products	Chocolate like products, cream cakes, milk cakes
Meat products	Salami, pate, hot-dogs
Snacks	Salty snacks

offered to the observed children. Data were calculated by average day and body weight of each child included into research.

Food intake (g) = food offered (g) – waste food on the plate (g)

From these data, we have obtained the exact amount of food eaten by each child in three days. Specific types of food it has divided regarding to their characteristics in the following categories as shown Table 16.4. It was considered those food categories in a particular food or otherwise adds any additives, which are the subject of our research. These are mainly processed industrially prepared food.

16.4.1.3 Estimated Daily Intake of Food Additives

Food additives set in Tables 16.1–16.3 are included in the research. These are preservatives: sorbic acid (E 200), benzoic acid (E 210), nitrate (E 251), nitrite (E 250), sulphur dioxide (E 220); emulsifiers: polyphosphates (E 450-452); sweeteners: cyclamate (E 952), saccharin (E 954), aspartame (E 951), acesulfame K (E 950) and colours: tartrazine (E 102), quinoline yellow (E 104), sunset yellow FCF (E 110), azorubine (E 122), amaranth (E 123), ponceau red 4R (E 124), erythrosine (E 127), allura red AC (E 129), patent blue V (E 131), brilliant indigotyne (E 132), brilliant blue FCF (E 133), brilliant black PN (E 151).

To estimate the daily intake of food additives we had to combine food consumption data with data on the concentration of additives in food. Calculation of the average daily consumption of specific groups of food per body weight for each child was made. The data from databases obtained from the official control and monitoring on food additives content in different food categories were used.

EDI = food additive concentration x food consumption per body weight

To calculate how much of each additive, the children daily ingested per kg body weight, food additives were added together, equal and shared with the child's body weight expressed in kg and thus received the additive value in mg/kg body weight.

16.4.1.4 Comparison Between ADI and EDI

To assess safety of food additives exposure, we compare acceptable daily intake (ADI) of each food additive with estimated daily intake (EDI). The estimated daily intake (EDI) of each additive is acceptable and safe if its value is lower or the same as ADI. To determine safety assessment among observed children EDI was compared with ADI, and expressed in percentage of ADI.

16.4.2 Results

16.4.2.1 Food Consumption Data

Estimated dietary intake of food is the first important data for making successful assessment of exposure to any agent in the food (Fig. 16.1). The most eaten food categories among observed children are milk products and bakery products. Dietary intake of non-alcoholic beverages, meat products and confectionary products is not so high than consumption of milk and bakery products.

16.4.2.2 Estimated Daily Intake of Food Additives

From results of analytical tests from official control, we can calculate how much exactly consumed food contains each food additive. All additives were divided into group of preservatives and polyphosphates, sweeteners and colours. The conclusion is that observed children eat more preservatives and polyphosphates than sweeteners and colours. As Fig. 16.2 shows, the higher intake of food additives expressed in mg/kg body weight, are among older children, 4–6 years old girls.

16.4.2.3 Results of Comparison Between EDI and ADI of Each Group of Food Additives

Results of safety assessment regarding relation between EDI and ADI of food additive show that intake of each group of additive is acceptable and safe. Figure 16.3 presents estimated daily intake of sweeteners – EDI expressed as percentage of ADI. The average exposure to sweeteners does not achieve 1% of ADI. Average exposure to colours reaches very low values from 0 to 1,1% ADI. It is interesting that average exposure to preservatives and polyphosphates (EDI) achieves almost 30% of ADI. It is obvious that the higher exposure among food additives belongs to preservatives and polyphosphates. It could be concluded that basic food products that are eaten very often by children (few times a day) contain preservatives and polyphosphates. Such food products are bakery products, meat products and fruit products.

Regarding exposure to different sweeteners as Fig. 16.3 shows, the highest intake is intake of cyclamate, intake of aspartame is the lowest. Neither one average exposure of sweetener exceeds 1% ADI. Exposure to sweeteners is very low and only cyclamate is consumed by almost all groups of children.

Figure 16.4 presents EDI–ADI relationship of preservatives and polyphosphates. It is evident that average exposure to nitrites and sulphite dioxide is relatively high,

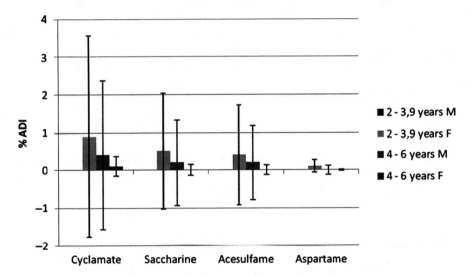

Fig. 16.3 EDI–ADI relationship of sweeteners

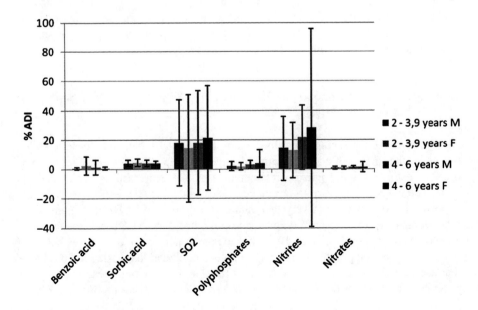

Fig. 16.4 EDI–ADI relationship of preservatives and polyphosphates

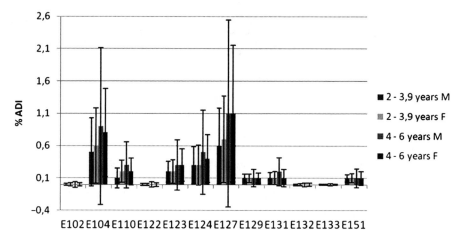

Fig. 16.5 EDI–ADI relationship of colours

the highest among older children. Intake of benzoic acid, sorbic acid, nitrates and polyphosphates is not so high. In compare to sweeteners and colours, the highest level of exposure belongs to preservatives and polyphosphates.

Figure 16.5 presents intakes of food colours expressed as EDI–ADI relationship. Exposure to colours among observed children are low. It is apparent that intake among older children is higher especially for erythrosine (E 127) and quinoline yellow (E 104). Such average exposure (around 1% of ADI) is very low and does not represent risk for observed children.

16.5 Conclusion

The human health risk assessment of food constituents is an internationally agreed and well-established process, being an integral part of the risk analysis process, which also includes risk management and risk communication. These three elements are separate tasks, performed by different player, and are parts of an interactive process.

The risk assessment of chemicals in food is a purely scientific process that requires expertise in toxicology, nutrition and exposure assessment. The risk management includes an identification of the food safety problem, consideration of its magnitude, seriousness, and consequently the way of handling it. In this process, the risk manager may include cost-benefit considerations before deciding how to manage the case (ban the compound, introduce limitations, provide specific dietary advice or accept the status quo). Finally, the risk analysis has to include a clear and interactive risk communication with consumers, industry and other stakeholders.

In this contribution an example of risk assessment is presented, special exposure assessment of food additives among preschool children. In research 190 children

308 E. Mičović et al.

were included, 92 girls and 98 boys, aged 2–6 years. It is apparent that such exposure does not present any harm or threat to observed children, regarding comparison between estimated daily intake (EDI) and acceptable daily intake (ADI) of each food additive.

Other question is the total intake of all food additives daily consumed by observed children: preservatives, polyphosphates, sweeteners and colours together. It is the fact that acceptable daily intakes for sum of food additives have not been set yet. As we can see from results, exposure to preservatives and polyphosphates is the highest among all observed food additives.

We strongly believe that future study should lead to exposure assessments of mixture of chemicals and interactive influence among them. We must not forget, that children are vulnerable and the most sensitive population and could become victims of invisible threats.

References

Anklam E, Battaglia R (2001) Food analysis and consumer protection. Trends Food Sci Technol 12(5–6):197–202

Armstrong DJ (2009) Food chemistry and U.S. food regulations. J Agric Food Chem 57(18):8180–8186

Beck U (2001) Critical theory of world risk society: a cosmopolitan vision. Constellations 16(1):3

Benford D (2001) Principles of risk assessment of food and drinking water related to human health. International Life Sciences Institute, ILSI, Washington, DC

Buchanan RL, Dennis S, Miliotis M (2004) Initiating and managing risk assessments within a risk analysis framework: FDA/CFSAN'S practical approach. J Food Prot 67(9):2058–2062

Cheftel JC (2005) Food and nutrition labelling in the European Union. Food Chem 93(3):531–550

Cohen LE, Felson M (1979) Social change and crime rate change: a routine activity approach. Am Sociol Rev 44(August):588–608

Croall H (2009) White-collar crime, consumer and victimization. Crime Law Soc Change 51(1):127–146

De Muelenaer B (2006) Chemical hazards. In: Luning PA, Devlieghere F, Verhé R (eds) Safety in the agri-food chain. Academic, Waggeningen, pp 145–209

DiNovi MJ, Kuznesof PM (2006) Estimating exposure to direct food additives and chemical contaminants in the diet. Office of Premarket Approval, Center for food safety & Applied Nutrition, U.S. Food & Drug Administration, Washington, DC

FAO/WHO (2002) Pan European conference on food safety and quality: final report. FAO/WHO, Rome

Feron VJ, Groten JP (2002) Toxicological evaluation of chemical mixtures. Food Chem Toxicol 40(6):852–839

Gibson JE, Taylor DA (2005) Can claims, misleading information, and manufacturing issues regarding dietary supplements be improved in the United States. J Pharmacol Exp Ther 314(3):939–944

Golob T, Jamnik M (2004) Vloga senzori ne analize pri zagotavljanju varnosti živil. (Eng. The role of sensory analysis in ensuring food safety). In: Gašperlin L, Žlender B (eds) Varnost živil/22. Biten evi živilski dnevi. Biotehniška fakulteta, Oddelek za živilstvo, Ljubljana, pp 101–115

Gregory PJ, Ingram JSI, Brklacich M (2005) Climate change and food security. Philos Trans R Soc Biol Sci 360(1463):2139–2148

Harland D (1987) The United Nations guidelines for consumer protection. J Consum Policy 10(3):254–266

Heines A, McMichael AJ, Epstein RR (2000) Environmental and health: 2. Global climate change and health. Can Med Assoc J 163(6):729–734

Herman JL, Younes M (1999) Background to the ADI/TDI/PTWI. Regul Toxicol Pharmacol 30(2):109–113

Hogarth JM, English MP (2002) Consumer complaints and redress: an important mechanism for protecting and empowering consumers. Int J Consum Stud 26(3):217–226

Holtfreter K, Van Slyke S, Blomberg TG (2005) Social change in consumer fraud: from victim-offender interactions to global networks. Crime Law Soc Change 44(3):251–275

Jin GZ, Kato A (2004) Consumer frauds and the uninformed: evidence from an online field experiment. Department of Economics, University of Maryland, College Park

Kennedy JF (1962) Special message on protecting the consumer interest statement read by President John F. Kennedy Thursday, 15 March 1962. Resource document. http://www.consumersinternational.org/shared_asp_files/uploadedfiles/4F4F223B-73E3-4F19-85C6-E705AD922376_kennedy.pdf. Accessed 20 Feb 2008

Kleter GA, Marvin HJP (2009) Indicators of emerging hazards and risks to food safety. Food Chem Toxicol 47(5):1022–1039

Lau K, McLean WG, Williams DP, Howard CV (2005) Synergistic interactions between commonly used food additives in a developmental neurotoxicity test. Toxicol Sci 90(1):178–187

Mack A, Schmitz T, Schulze Althoff G, Devlieghere F, Petersen B (2006) Steps in the risk management process. In: Luning PA, Devlieghere F, Verhé R (eds) Safety in the agri-food chain. Academic, Wageningen, pp 355–397

McMichael AJ, Haines A, Slooff R, Kovats S (1996) Climate change and human health. WHO, Geneva

Myers SS, Patz JA (2009) Emerging threats to human health from global environmental change. Annu Rev Environ Resour 34:223–252

Newburn T (2007) Criminology. Willan Publishing, Cullompton, Devon

Raspor P (2004) Sedanji pogled na varnost živil (Eng. Present view about the food safety). In: Gašperlin L, Žlender B (eds) Varnost živil/22. Biten evi živilski dnevi. Biotehniška fakulteta, Oddelek za živilstvo, Ljubljana, pp 1–14

Teuschler LK, Hertzberg RC (1995) Current and future risk assessment guidelines, policy, and methods development for chemical mixtures. Toxicology 105(2–3):137–144

Tombs S (2008) Corporations and health and safety. In: Minkes J, Minkes L (eds) Corporate and white-collar crime. Sage, Los Angeles/London/New Delhi/Singapore, pp 18–38

Unnevehr LJ (2003) Food safety in food security and food trade. In: Unnevehr LJ (ed) 2020 Focus 10: food safety in food security and food trade. International Food Policy Research Institute, Washington, DC, pp 1–2

Vanderlinden L, Cooper K, Sahni V, Campbell M (2005) Environmental threats to children. Toronto Public Health, Environmental Protection Office, Toronto

WHO (1987) Principles for the safety assessment of food additives and contaminants in food. Environmental Health Criteria No. 70. WHO, Geneva

WHO (2004) Principles for modelling dose-response for the risk assessment of chemicals. International programme on chemical safety. WHO, Geneva

WHO (2005) Dietary exposure assessment of chemicals in Food. Report of a Joint FAO/WHO Consultation Annapolis. WHO, Maryland

Winter CK, Francis FJ (1997) Assessing, managing and communicating chemical food risks. Food Technol 51(5):85–92

WMO (2007) Intergovernmental panel on climate change. Climate change 2007: impacts, adaptation and vulnerability, summary for policy makers. IPCC secretariat, c/o WMO, Geneva

Xu Z, Yuan Y (2009) Principle-based dispute resolution for consumer protection. Knowl-Based Syst 22(1):18–27

Zakon o varstvu potrošnikov [ZVPot-UPB2] (2004) (Eng. Consumer Protection Act). Uradni list Republike Slovenije, št. 98/2004, 46/2006, 126/2007, 86/2009

Chapter 17
Possibilities of Risk Quantification in the System of Save-for-Health Food Production

Midhat Jašić and Dejana Dimitrijević

Abstract All foods are made of nutritive and non-nutritive compounds. Nutritive compounds contribute to maintenance of homeostasis and metabolism of the human body. On the other hand, the non-nutritive components are often hazardous and used in excessive used doses can cause diseases. Estimation of non-nutritive harmful substances intake is a very complex process based on risk assessment.

The elements of risk are assessed and can be graded according to degree of hazards, size of exposure, and the consequences that may arise. Using method of risks quantification it is possible to evaluate each elements of risk and in this case, risk assessment is more reliable. The methods of assessing and quantifying risk in the food chain can be used in the implementation of the food safety management system.

The modern approach to risk analysis requires monitoring of hazards in all of the food chain, from the farm to the table. There are limitations in the hazards when monitoring the size of the food chain. It is one of reason why risk assessment and validation processes are performed. Risk assessment in food, either chemical or microbiological is made at the level of the company or at the state level implementation of the food safety management system.

The aim of this chapter is to present a contemporary approach in theoretical and practical sense of risk assessments in the system of food production. This chapter analyses the most common contaminants in food as well as reviewing methods and approaches to the process of assessment and quantification of the risks in the save-for-health food production.

M. Jašić (✉)
Faculty of Pharmacy, University of Tuzla, Tuzla, Bosnia and Herzegovina
e-mail: jasic_midhat@yahoo.com

D. Dimitrijević
Department of Environmental Protection, Faculty of Security Studies, University of Belgrade, Belgrade, Serbia

G. Meško et al. (eds.), *Understanding and Managing Threats to the Environment in South Eastern Europe*, NATO Science for Peace and Security Series C: Environmental Security, DOI 10.1007/978-94-007-0611-8_17, © Springer Science+Business Media B.V. 2011

17.1 Introduction

A food safety management system is required at each step of food production chain to ensure safety of food, show compliance with legislation, and meet customer requirements. Establishing a food safety management system is becoming more and more of an obligatory requirement by the authorities in developed and developing countries. The establishment of this system is based on legislation. The increased demand for safer food has resulted in the development and introduction of new food safety standards and regulations in order to reach a higher level of food safety. Implementation of a food safety management (FSM) system starts with the requirements of good manufacturing practice (GMP) in forms of perquisite (PRP) and the perquisite operative program. It is pro-active risk identification in the form of a system with the selection of critical factors related to the occurrence of hazards. The next step is hazards analysis and establishment of critical control points (CCP) in the chain of production. However during this CCP process the personnel responsible for building the system are often missing key skills and knowledge regarding food toxicology, risk assessment, and risk analysis. Required knowledge is difficult to integrate in one person, causing teamwork difficulties during the establishment of the system. The risk assessment structure approach is not simple and is faced with a lack of data, capable infrastructure and is in need of trained personnel and resources (Sherif et al. 2009). The most common is a lack of competence in the area of risk assessment.

The food safety implies a satisfactory level of safety that food will not cause adverse impacts on human health (WHO 2002). The food safety can be achieved if food is processed, packed, distributed, prepared and consumed in accordance with safety purpose. Food safety policy is based on a comprehensive, integrated approach of risk analysis through the food chain production with mandatory requirements of transparency and traceability (Jašić et al. 2005). Risk analysis has three main components: risk assessment, risk management and risk communication. The risk analysis is the scientific method as used in implementation of food safety management system. It consists of risk assessment, communications related to the risk and risk management. The risk assessment represents hazard identification, the size and circumstance under which it becomes hazardous. This is the systematic, scientific process based on food toxicology, microbiology and statistics. Communication about risks is the process of selection and implementation of specific measures with the goal to reduce the risks. Risk analysis is the term that has evolved over the past decade to indicate the methodologies to approach to food-related risks in an objective manner rather than on the basis of feelings and beliefs (Klapwijk et al. 2000).

The safe framework for improved risk analysis of foods shift the focus of food safety governance from reducing risks to human health to optimizing health outcomes (König et al. 2010). The application of risk analysis has the potential to allow an overall assessment of risks and benefits in food hygiene programmes as well as to improve the scientific elaboration of standards and guidelines for food safety. Additionally, inspection and monitoring resources can be allocated in such

17 Possibilities of Risk Quantification

a manner that is proportional to the greatest ability of both the inspection and the monitoring body to ensure food safety.

Risk assessment provides the scientific foundation upon which the risk analysis process is built. Risk assessment is the process of evaluation including the identification of the attendant uncertainties. It is the likelihood and severity of an adverse effect occurring to man or the environment following exposure under defined conditions to a risk source (EC 2000). Risk assessment is comprised of hazard identification, hazard characterization, exposure assessment and risk characterization steps (Lammerding 1997).

17.2 Food Contaminants, Residues and Hazards

Before implementation of food safety management systems, it is necessary to know the possible contaminants which can get into food. There are a number of chemical substances in food with potentially harmful effects on metabolic processes in the human body. Some of them come from sources created by human technology, but some are naturally present in food. Chemical, biological and physical contaminants are the hazards that may harm health of the food consumers. A hazard is a biological, chemical or physical agent in the food, or a condition of food with the potential to cause an adverse health effect (CAC 2004). Food safety hazards include bacteria, viruses, parasites, hazardous chemicals, and foreign materials that can cause an adverse health effect to a consumer. Hazards carry certain risks that indicate the likelihood and severity of possible hazards as well as potential adverse effects on human health.

Risk is a function of the probability of an adverse health effect and the severity of that effect, consequential to a hazard(s) in food (CAC 2004). Risk is estimated because it is difficult to measure all of its elements such as severity, exposure and consequences of its effect. The risks to food safety include allergens, but also the content of substances which can harm the health of specific population groups such as people with celiac disease and lactose intolerance.

The most common hazards are the different types of contaminants that can get into the food, from the uncontrolled environment, residues from treatment plants or animals in various stages of agricultural production or food processing. These substances are not intentionally added to food but appear as a result of treatment during primary agricultural food production, processing, packaging, transportation, or are sometimes caused by natural processes (Jašić and Begić 2008).

In some segments of production, food can be contaminated with biological, chemical and physical agents. The causes of contamination can be quite diverse. Thus, the chemical hazards in food in terms of the cause of their occurrence during the production processing can be: natural hazards present in food as its legal ingredient (Essers et al. 1998), hazards in the form of residues from pesticide (Boobis 2008), hazards in the form of residues from animal treatment (WHO 2004a), hazards in the form of toxic substances that are formed by thermal processing of food (Tritscher 2004), food additives in amounts exceeding the ADI value (WHO 2004b), residues of hygiene and sanitation, migrating groups from the packaging material and others.

Toxic substances naturally present in foods can be from plant (examples amygdaline in seeds and solanine in potatoes) or animal origin (neurotoxins in fish from the warm sea). In certain foods they are present as natural toxins in plants, one example being mushrooms (natural toxins that can accomplish as pesticides, toxins animal origin, etc.). In addition, toxic substances which are produced by algaes can be transferred from the fish to human beings (Jašić and Begić 2008).

17.2.1 Environmental Hazards

The causes of contamination of food can be air and soil pollution and waste water. Those sources of contamination due to the global impact are very difficult to control. The main air pollutants are toxic gases from industrial plants, houses and gases from motor vehicles. Typical air pollutants include nitrogen oxides, carbon monoxide, sulphur dioxide, hydrocarbons and particulate matter. Toxic particles from the air are absorbed into grain, fruits, and vegetables and continue to spread through the chain of processing and are distributed to the consumers. Furthermore, current industrial and household waste often ends up in the water, where the toxic chemicals are absorbed into the eco system and enter into the food chain.

The contamination of soil is mostly affected by the use of pesticides and fertilizer (Molloy et al. 2005). Some of these components can be retained by soil for many years. Toxicant which causes contamination of food originating from industrial waste and natural environment are: chlorinated hydrocarbons, polycyclic aromatic hydrocarbons, dioxins and heavy metals, radioactive and other elements.

Dioxins are usually located in the air through which they enter the soil, water and plants. They can be found in the various cycles of the food chain. Dioxins enter the human food chain through the consumption of contaminated plants and animals. Once ingested it is difficult for the organism to metabolize them. Dioxins may then even be present in breast milk. The human body then gradually accumulates the contaminants in adipose tissue, gradually excretes them from the body.

Another set of environmental contaminant present in the food can be polychlorinated biphenyls (PCBs). They show a similar toxicity as dioxins causing a decline in immunity, hair loss, disturbances in the nervous system. They may be present in small concentrations of meat, fish and dairy products. They are then absorbed into the human body and accumulated in adipose tissue. Polychlorinated biphenyls are widely used as plastics and rubber additives, as transmitters of heat in industrial oils, paints and varnishes.

One often mentioned contaminant found in food is polycyclic aromatic hydrocarbon (PAH) which is a group of organic compounds containing two or more joined aromatic rings. PAH are formed during the incomplete combustion or pyrolysis of organic substances in industrial processes but may also be found in the households. They have expressed carcinogenic and genotoxic potential.

17.2.2 Hazards of Microbial Origin

All phases of production in the food chain carry certain biological (macrobiological and microbiological) hazards. Human nutrition has always been connected with the danger of contamination which can lead to the diseases by microorganisms and their products such as bacteria, yeasts, fungi, protozoa and viruses belonging to different systematic categories of organisms.

Large numbers of pathogenic bacteria and viruses can be found in food. Their enterotoxins are often resistant to treatment conditions at elevated temperatures. Food control and monitoring for the presence of bacteria and viruses in most countries is regulated by law. The presence of mycotoxins in food is more difficult to control, even though monitoring in this area is well regulated by laws. However, the processes of validation are being done with a smaller degree of frequency. A single intake of foods with a high concentration of mycotoxins may lead to acute poisoning, while consuming food with low concentrations of mycotoxins for longer periods leads to cancer and other diseases due to cumulative effects.

17.2.3 Hazards That Can Arise During Production and Processing of Food

In recent decades, chemical food safety issues that have been the center of media attention include the presence of natural toxins, processing-produced toxins (e.g., acryl amide, heterocyclic aromatic amines, and furan), food allergens, heavy metals (e.g., lead, arsenic, mercury, cadmium), industrial chemicals (e.g., benzene, perchlorate), contaminants from packaging materials, and unconventional contaminants (melamine) in food and feed (Jackson 2009). Due to the global nature of the food supply and advances in analytical capabilities, chemical contaminants will continue to be an area of concern for regulatory agencies, the food industry, and consumers in the future (Lauren 2009).

Pesticides are applied primarily to agricultural production in order to protect plants from destruction by pests. According to the origin they can be inorganic materials, and substances from plants, bacteria and fungi. Pesticides can be organic or synthetic materials such as chlorinated and phosphorus compounds, triasines, phenoxy-derivatives of carbonic acid, synthetic pyrethroids, etc. Most often they contain toxic elements like mercury, arsenic, phosphorus, and others. Some of them are carcinogenic after prolonged consumption. The origin of pesticide residue in food may be tied to the treatment of the environment. Indirectly, pesticides can come through with streams of treated crops in the river, but also from food that is fed to domestic animals, etc. If the use of pesticides in food production remains uncontrolled, consequences could be disastrous for human health. Therefore, attempts to use of pesticides should be controlled by providing integrated production and monitoring of their proper implementation. Producers of agricultural products are required to comply with the required waiting period in order to reduce the content of pesticides in food to a minimum (MRL). Commission Regulation (EC) No 1881/2006 of 19 December 2006 sets maximum permitted levels for certain contaminants in foodstuffs (EC 2006). Pesticides in the food chain can be transferred to animals that man uses to obtain food. Therefore, monitoring the presence of pesticides becomes mandatory for plants and animals used for food.

For health safety concerning food of animal origin it is particularly important to use veterinary drugs to accomplish the treatment and control of animal health. The most often residues of treated animals are: antibiotics, anti-parasitic medications, hormonal preparations, antiseptics, disinfectants and sedatives. Clearance of medication is the period of its excretion from the body of animals to allowed level (MRL). It is most important parameter to control the appearance of drug residues in food from animal origin. In terms of health safety, food of an animal origin, veterinary drug residues in edible tissues of animals can cause toxic effects if ingested over a long period of time.

Also, heat treatment of some kinds of food at high temperatures brings a number of risks because of the possible creation of certain toxic substances. In addition, different chemical reaction, during the food processing treatments such as the Maillard reaction, lipid oxidation, alkali and acid treatment, fermentation, curing, and ionizing radiation, can cause an array of different toxins.

During food processing different means of hygiene and sanitation are used which can be harmful to health. Soaps and detergents contain surfactant with different forms of manifestation of toxic effects, and residues above MRL can be harmful for human health. As for sanitizers, there are commonly used products based on chlorine, per acetic acid, hydrogen peroxide, alcohol, different acids and the base.

Metals that are referred as heavy metals can be found in food as part of environmental contaminant. The toxic metals include lead, mercury, cadmium, arsenic, thallium uranium and others. Some heavy metals are essential for living organisms (biogenic) such as zinc, iron, molybdenum, manganese, cobalt and selenium. Major sources of pollution with heavy metals are traffic agents, metal industry, mining, metal smelting plant, organic and mineral fertilizers and urban waste.

17.3 The Process of Assessment and Quantification of the Risks in the Save-for-Health Food Production

At the all levels of food production, such as primary agricultural production, processing, packing and distribution it is necessary to establish various forms of a food safety management system. Contemporary food safety management is based on risk assessment. The risk is assessed (CAC 1999) by a combination of exposure assessment and a dose–response relationship. The exposure is characterized as a distribution and variability of ingested doses. The dose–response relationship is combined with other factors such as food consumption frequency.

The risk assessment can use several models. The food chain model, or Process Risk Model (PRM), describing the changes in prevalence and concentration over a series of processing steps representing the consecutive stages of the food chain (Cassin et al. 1998; Nauta et al. 2009). If possible, the description should be based on the available data and include the dynamics of the processing step modeled. Factors like the number of food chain participants, processing steps, and raw materials, number of suppliers of raw materials, logistics and destination are relative to the food chain complexity. The more complex the system, the higher the possibility of errors resulting in food safety risks.

The aim of risk assessment is to estimate the level of illness that may be expected in target population from food product or group of food products. There are several types of risk assessment. Most of them belong to three categories: qualitative risk assessment, quantitative risk assessment and semi-quantitative risk assessment (Sumner et al. 2004). All categories provide useful information and choice of assessment will depend on the speed and complexity from assessment required.

Quantitative risk assessments (QRAs) are a risk assessment that provides numerical expressions of risk and indication of the attendant uncertainties (CAC 1999) can be divided into two categories: deterministic and stochastic (Lammerding and Fazil 2000). Quantitative risk assessments are done for specific purposes and provide numerical risk estimates to answer questions that were posed by the risk managers who originally commissioned the assessment.

Qualitative risk ranking can be made with assessment based on factors which are linked with exposure assessment and with hazard characterization. The qualitative risk assessment estimated risk according to subjective terms such as high, low or medium. Every system of food safety management contains a simple plan of qualitative risk assessments.

In semi-quantitative risk assessment obtain a numerical risk estimate based on a mixture of qualitative and quantitative data. These types of assessment need much of the data that will be used in a full quantitative risk assessment. There is a great deal of work involved, but not as much as for a full quantitative risk assessment (Sumner et al. 2004). Quantitative risk assessment could be used in food industry context to prioritize the safety management measures needed according to real importance of the main hazards identified for a particular food processing, which simultaneously would better protect the consumer's health and be cost effective and efficient for the food industry (Doménech et al. 2007). Moreover, the risk assessment of genotoxic and carcinogenic substances has been an area of much debate over the years (Barlow and Schlatter 2010).

Present-day concepts about food safety are the outcome of earlier and recent historical negotiations, confrontations and decisions. There is no universal norm with regard to safe food production and safe food risk management (Houghton et al. 2008) and who is responsible for food risk types and risk characteristics (Leikas et al. 2009). As a consequence, consumer reactions may differ widely according to time and place (Scholliers 2008).

17.3.1 Chemical Risk Assessment

Regarding the necessity for farming measures in primary agricultural production and processing, chemical agents have to be present in the entire food chain causing the presence of unavoidable harmful substances in food. Risk assessment of food chemicals is based on general toxicological principles: hazard identification and characterization, and exposure assessment. The harmful food chemicals include compounds like pesticides, and other agrochemicals, additives and veterinary drugs, residues of cleaning, product of interaction packaging material and food. Harmful food chemical may also be chemical compounds that occur naturally in certain food like glycosides, alkaloids etc.

A number of methods are used for chemical toxicity evaluation. These are all inter-related, but some are mathematically more complex than others. A lot of criteria can be used for chemical toxicity evaluation like hazard index, the reference point index, the relative potency factor (Moretto 2008).

Most toxicological data for risk assessment are derived from animal studies, supplemented with in vitro studies to aid with mechanical interpretation of results. It is essential for such studies to be conducted according to internationally accepted test protocols and methodologies that are appropriately validated. Validation in this context has multiple dimensions; one dimension is that the in-vivo studies should identify both the major toxic effects of the test compound and the intake level that does not result in adverse effects, that is actually the no observed-adverse-effect level (NOAEL). The acceptable daily intake (ADI) may then be derived from the NOAEL by using safety factors that take into account both the uncertainty of extrapolating results from animals to man, and individual variation among humans. The ADI is an estimation of the amount of a compound that can be ingested daily over a lifetime without appreciable health risk. Typically, no quantitative assessment of risk is made for the setting of the ADI, the risk is considered to be so small as to be negligible from the public health perspective.

For exposure scenarios it is relevant to know cumulative risk assessment, acute and chronic exposure in the context of maximum residue level (MRL) in the exposures relations from the actual use patterns, respectively. Each can be addressed either deterministically or probabilistically. Cumulative risk assessments have been performed in a number of countries, on organophosphate insecticides alone or together with carbamates, triazines, chloroacetanilides, carbamates alone, and all pesticides.

Exposure to pesticides can occur via a number of pathways like food, drinking water, residential, occupational, oral, inhalation, and dermal routes. Assessment by each route is generally undertaken independently. The totality of exposure can determine risk, and included so-called aggregate exposure in the risk assessment of pesticides, through multiple routes, and multiple pathways. In considering the joint actions it is important to consider possible exposure levels. In practice relevant information on exposure by a number of these pathways is not available.

The risk assessment of pesticide residues in food is currently performed on a compound-by-compound basis. If potential exposure of consumers is below the relevant health-based guidance value (ARfD and ADI, for acute and chronic exposure, respectively), the use of that pesticide in crop protection is considered acceptable (Boobis et al. 2008).

There is currently no internationally agreed upon methodology to assess risks from combined exposures to pesticide residues. However, a number of international activities are ongoing in this or related areas. These include the development of a framework for cumulative risk assessment by the International Programme on Chemical Safety (Boobis et al. 2008; WHO 2008).

17.3.2   *Microbiological Risk Assessment*

A number of examples of microbiological risk assessment have been published over the intervening years by governments, international organizations, academia, research organizations and companies (Lammerding 2007). Attention is now focused on how microbiological risk assessment can best be utilized in food safety management to understand the magnitude of prevailing risk regarding a pathogen/ product combination, to design risk intervention scenarios, when appropriate, and to help identify and select risk management options for implementation (Lammerding and Fazil 2000).

As part of the microbiological risk management process, risk assessment is a science based tool that serves to answer questions posed by risk managers. Quantitative microbiological risk assessment QMRA is important tool to support food safety control. Its use has been promoted by international bodies because it offers a structured, unified approach to complex problems, as well as a scientific basis for risk management decisions (CAC 1999).

Different modeling approaches can be applied in a QMRA food chain model. A model of a processing step within the food chain can be primarily based on a statistical description of data, with linear or log-linear relationship between input and output. If data on the concentration of a pathogen before and after a given processing step are available, the change in mean and variance can be indicative for the effect of the processing step on the concentration (Nauta et al. 2009). Alternatively, the model can be primarily based on the mechanistic of the processing step and microbiological responses to the process conditions. The model describes the expected change in microbial prevalence and concentration, based on the type of food processing and the characteristics of the microorganism. The model can be a (secondary) predictive model for growth or inactivation of a microorganism in a food, or a descriptive model of cross-contamination. Ideally, a model combines the benefits of both modeling approaches, i.e. including the process mechanistic and fitting it to the available data. This may be complex or even unachievable due to either lack of appropriate data or limited insights into the mechanistic and dynamics of the process described (Nauta et al. 2009). For public health authorities risk assessment may serve as a means to quantify the risks attributable to certain food products. In that case, the purpose of quantitative risk assessment is not so much the production of safe food, but an evaluation of the health status of the population. Quantification of risks is therefore more important when QMRA is applied for public health purposes. By applying QMRA and using integrated public health measures, risks of a different nature can be compared (Havelaar et al. 2010). The industry and public health workers may use QMRA, HACCP and predictive modeling techniques. Important concepts in risk assessment are "risk", "probability" and "probability distribution". In QMRA, such probability distributions play an important role.

Risk assessment is use due to epidemiology can not always provide sufficient sensitivity to measure risks directly using human health data. There are many possible permutations of various sources of food contamination and not practical to consider all via epidemiology. Risk assessment is used to predict relative risks for future scenarios and/or evaluate efficacy of alternative management actions (Soller et al. 2004).

17.3.3 Risk Assessment Procedure

Risk is the probability of an adverse health effect and the severity of that effect, consequential to a hazard(s) in food. Risk assessment, due to the diversity of possible adverse factors on the human organism, has been developed in the specific estimates, depending on whether they are microbial or chemical agents. The different assessments carried out according to whether it is a virus, bacteria, parasites, fungi, mycotoxins, etc. The process of assessment is applied to chemical assessment in the case of pesticides, additives, and other hazards. Special assessments are conducted for GMO obtained foods (Kuiper and Kleter 2003), novel foods (Knudsen et al. 2008), as well carcinogens in food (Barlow and Schlatter 2010).

The question is about level of possibility to quantify risks, in general, because of different levels of an event probability. In point of evaluation the risk would be equal to the product of the hazard rate size in terms of severity and the likelihood that the hazard causes consequences for the health of consumers.

$$R = P \times C$$

Where:

R = risk,
P = probability (likelihood) of events with consequences,
C = consequences or severity of hazards.

In addition, use the other form, which includes other important parameters for the size of the risk:

$$R = P \times C \times E$$

17 Possibilities of Risk Quantification

Where:

R = risk,
P = probability (likelihood) of events,
C = consequences,
E = exposure.

The amount and level of hazard present, dangerous substances, and exposure to these substances during the intake of food determine the possible consequences. These may be due to acute intake of toxic components in a shorter time period, or caused by a gradual accumulation of toxicants in the body that can occur over a longer period of time in small quantities.

Nowadays various tools for risk assessment are developed. Some of them grade the various elements of risk, and some use fuzzy logic (Valerie et al. 2006). The user answers questions related to hazard severity, probability of exposure and effects of processing steps by selecting qualitative statements that appear in the user interface or by entering specific values for the hazard of interest. The qualitative statements are converted to numeric values and used to calculate four measures of risk. Comparative risk defined (Ross and Sumner 2002) as a measure of risk that is adjusted by the proportion of the population consuming is independent of population size:

Comparative risk = probability of illness over all servings × exposures/person day × hazard severity.

The assessment tools use fuzzy values for system parameters and interval arithmetic to characterize hazards to compute risk. A fuzzy model can be useful for food safety risk assessment in total food supply chains. The tool is useful for early stage microbial risk assessment in food systems, in particular, for ranking risks based on total illness and severity of illness (Valerie et al. 2006).

Any health risk assessment contains four analytical steps: hazard identification, hazard characterization, exposure characterization and risk characterization.

17.3.3.1 Hazard Identification

The Hazard identification presents the identification of biological, chemical, and physical agents capable of causing adverse health effects and which may be present in a particular food or group of foods (CAC 1999). Hazard identification is the process of determining whether exposure to an agent can lead to adverse health outcomes. It is based on analyses of a variety of data that may range from observations in humans and animal data to an analysis of mechanisms of action and structure–activity relationships (Sherif et al. 2009). It is the qualitative indication that a substance/agent may adversely affect human health. This is the first stage in risk assessment and screening process to make certain that the hazard really does exist in particular product or process. Three types of agents are priority of the identifications: physical, chemical and microbiological, capable of causing adverse health effects and that may be present in a particular food or group of foods. Hazard identification involves: identifying the hazard, the nature of the hazard, known or potential health effects associated with the hazard, and the individuals at risk from the hazard.

For hazard identification is possible to use methods of animal based toxicology as well methods of in vitro toxicology.

17.3.3.2 Hazard Characterization

Hazard characterization is the qualitative and quantitative evaluation of the nature of the adverse effects, and may include dose/response. The qualitative or quantitative evaluation of the nature of the adverse health effects associated with biological, chemical and physical agents that may be present in food. Hazard characterization is considering what happens when there is large-scale food poisoning.

There are two parts to hazard characterization: a description of the effects of the hazard and the dose–response relationship (Sumner et al. 2004). The severity of hazard can be the qualitative or quantitative evaluation of the adverse health effects like estimation of the dose and severity of adverse health effect.

Dose–response is a measure of how much disease agent is required to cause illness. Dose–response assessment determinate relationship between the magnitude of exposure (dose) to a chemical, biological or physical agent and the severity and frequency of associated adverse health effects (CAC 2004). For any particular individual, dose–response links the amount of the hazard you ingest (dose) with the chance of your becoming infected and the scale of the illness.

17.3.3.3 Exposure Assessment

For any component in human diet, exposure to toxin or micro-organism in that component depends on factors like the level of the agent in food, the amount of food intake and the frequency of consume that component (Sumner et al. 2004). Safe levels of exposure are usually determined by extrapolation of mammalian toxicological data to the human situation. Exposure assessment is the qualitative and/or quantitative evaluation of the likely input of biological, chemical, and physical agents via food as well as exposures from other sources if relevant. Exposure assessment is evaluation of the likely input of the hazard via food as well as other sources. The exposure scenarios are of relevance in a cumulative risk assessment. These assessments in practice can have acute and chronic exposure and must to be based on monitoring data from supervised trials. It is possible to address each scenario either deterministically or probabilistically, in which exposure assessment can be increasingly refined (Boobis et al. 2008).

Microbial exposure assessment is the estimation of how likely it is that an individual or a population will be exposed to a microbial hazard and what numbers of the microorganism are likely to be ingested.

The early identification of emerging hazards is important because it reduces the effects of hazard to human health (Gijs and Marvin 2009) as well selection of critical factors for identifying emerging food safety risks in dynamic food production chains (Van Asselt et al. 2010).

17.3.3.4 Risk Characterization

Risk characterization integrates hazard identification, hazard characterization and exposure assessment. Risk characterization is the process of estimating the probable incidence of an adverse health effect to humans under various conditions of exposure including a description of the uncertainties involved (Sherif et al. 2009).

The output of risk characterization includes recommendations on dealing with variability and uncertainty on how the uncertainty can best characterized to aid the risk manager's decision making. Risk characterization is the process of determining the qualitative and/or quantitative estimation, including attendant uncertainties, the probability of occurrence and severity of health effects in a given population (CAC 2003). This definition recognizes that the ultimate goal of a risk assessment process is to estimate the probability of risk occurrence and this may be based on qualitative and/or quantitative information. It also stipulates that uncertainty is recognized and included in any estimates of risk (Valerie et al. 2006). Quantitative analytical tools are used to synthesize data and information, to represent complex relationships to describe the probability and severity of adverse health effects and to inform decision-making processes related to risk reduction and management (Valerie et al. 2006).

Risk estimation is the output of risk characterization (CAC 2003). The qualitative risk assessments estimate risk as high, low, medium. The ranking is simple and it is mostly traditional for the implementation of HACCP systems. This procedure is carried out as recommended by the Codex Alimentarius or ISO food safety management related standards.

For semi quantitative assessment of risk is necessary to use or develop a database. Three data sets required for risk assessment: type of food, possible type of hazard and possible hazard location (primary agricultural production, processing, packaging, distribution, and serving). The semi quantitative risk assessment provide where predicted the number of people expected to become ill from the particular food product.

The semi-quantitative risk may ranking like number in a specific range like predicted illnesses per annum in the population, or the probability of becoming ill from eating a serving of the product. The result of risk characterization is estimation of how many people will get disease, and how serious it disease.

17.4 Conclusion

This work presents the process of assessment and quantification of the risks in the segment of the risks management in production of food. Risk assessment capabilities are limited due to the complexity of the system of food production because of the complexity of the hazards that can occur during food production. Risk assessments can be on different level of complexity from qualitative, through semi-quantitative to quantitative.

Risk assessments in the food chain are very often expensive and take long time to complete. However, there is almost no food that contains no harmful substances.

At the state level as well at company's level it is necessary to continuously carry out food risk assessment and determine the maximum permissible concentration, which may be found in food. The methodology of risk assessment is constantly improving, becoming more specialized, and a large number of methodologies vary from case to case.

For this purpose it is necessary to continuously develop systems that prevent the possible entry of harmful substances in food at the national level, on the basis of risk assessment. The risk assessment can provide significant elements of uncertainty, as well basis for effectiveness considerations relevant to decision-making related to food safety management system.

References

Barlow S, Schlatter J (2010) Risk assessment of carcinogens in food. Toxicol Appl Pharmacol 243(2):180–190

Boobis R, Ossendorp BC, Banasiak U, Hamey PY, Sebestyen I, Moretto A (2008) Cumulative risk assessment of pesticide residues in food. Toxicol Lett 180(2):137–150

CAC (1999) Principles and guidelines for the conduct of microbial risk assessment. Secretariat of the Joint FAO/WHO Food Standards Programme Rome, Rome

CAC (2003) Food hygiene basic texts, 3rd edn. Joint FAO/WHO Food Standards Programme Rome, Rome

CAC (2004) Procedural manual, 14th edn. Joint FAO/WHO Food Standards Programme Rome, Rome

Cassin MH, Lammerding AM, Todd ECD, Ross W, McColl RS (1998) Quantitative risk assessment for Escherichia Coli O157:H7 in ground beef hamburgers. Int J Food Microbiol 41(1):21–44

Doménech E, Escriche I, Martorell S (2007) Quantification of risks to consumers' health and to company's incomes due to failures in food safety. Food Control 18(12):1419–1427

EC (2000) EC White Paper on Food Safety. Resource document. EC. http://ec.europa.eu/food/food/intro/white_paper_en.htm. Accessed 10 July 2010

EC (2006) Maximum levels for certain contaminants in food. Resource document. EC. http://ec.europa.eu/food/food/chemicalsafety/contaminants/legisl_en.htm. Accessed 10 July 2010

Essers AJA, Alink GM, Speijers GJA, Alexander J (1998) Food plant toxicants and safety risk assessment and regulation of inherent toxicants in plant foods. Environ Toxicol Pharmacol 5(3):155–172

Gijs AK, Marvin HJP (2009) Indicators of emerging hazards and risks to food safety. Food Chem Toxicol 47(5):1022–1039

Havelaar H, Brul S, Jong DA, Zwieteringe MH, Kuilec BH (2010) Future challenges to microbial food safety. Int J Food Microbiol 139(1):S79–S94

Houghton JR, Rowe G, Frewer LJ, Van Kleef E (2008) The quality of food risk management in Europe: perspectives and priorities. Food Policy 33(1):13–26

Jackson LS (2009) Chemical food safety issues in the United States: past, present, and future. J Agric Food Chem 57(18):8161–8170

Jašić M, Begić L (2008) Biohemija hrane (Eng. Biochemistry of food). Printcom, Tuzla

Jašić M, Đonlagić N, Šubarić D, Keran H (2005) Contemporary principles of politics and legislation in food production. In: Proceeding of consortium of EU TEMPUS JEP project Nr. 16140-2001, Banja Luka, 2005

Klapwijk PM, Jouve JL, Stringer MF (2000) Microbiological risk assessment in Europe: the next decade. Int J Food Microbiol 58(3):223–230

17 Possibilities of Risk Quantification

Knudsen I, Sȓborg I, Eriksen F, Pilegaard K, Pedersen J (2008) Risk management and risk assessment of novel plant foods: concepts and principles. Food Chem Toxicol 46(5):1681–1705

König A, Kuiper KHA, Marvin HJP, Boon PE, Buck L, Cnudde F et al (2010) The SAFE FOODS framework for improved risk analysis of foods. Food Control. doi:10.1016/j.foodcont. 2010.02.012

Kuiper HA, Kleter GA (2003) The scientific basis for risk assessment and regulation of genetically modified foods. Trends Food Sci Technol 14(5):277–293

Lammerding AM (1997) An overview of microbial food safety risk assessment. J Food Prot 60(11):1420–1425

Lammerding A (2007) Using microbiological risk assessment (MRA) in food safety management. Summary report of a ILSI Europe risk analysis in microbiology task force and the international association workshop held in October 2005 in Prague. ILSI Europe Risk Analysis in Microbiology Task Force and The International Association, Brussles

Lammerding AM, Fazil A (2000) Hazard identification and exposure assessment for microbial food safety risk assessment. Int J Food Microbiol 58(3):147–157

Lauren SJ (2009) Chemical food safety issues in the United States: past, present, and future. J Agr Food Chem 57(18):8161–8817

Leikas S, Lindeman M, Roininen KL (2009) Who is responsible for food risks? The influence of risk type and risk characteristics. Appetite 53(1):123–126

Molloy R, McLaughlin M, Warne M, Hamon R, Kookana R, Saison C (2005) Background and scope for establishing a list of prohibited substances and guideline limits for levels of contaminants in fertilizers. CSIRO Land and Water Centre for Environmental Contaminants Research Canberra, Canberra

Moretto A (2008) Exposure to multiple chemicals: when and how to assess the risk from pesticide residues in food. Trends Food Sci Technol 19(1):56–63

Nauta M, Hill A, Rosenquist H, Brynestad S, Fetsch A, Van der Logt P, Fazil A, Christensen B, Katsma E, Borck B, Havelaar AA (2009) A comparison of risk assessments on Campylobacter in broiler meat. Int J Food Microbiol 129(2):107–123

Ross T, Sumner J (2002) A simple, spread-sheet based, food safety risk assessment tool. Int J Food Microbiol 77(1–2):39–53

Scholliers P (2008) Defining food risks and food anxieties throughout history. Appetite 51(1):3–6

Sherif OS, Emad ES, Mosaad AAW (2009) Mycotoxins and child health: the need for health risk assessment. Int J Hyg Environ Health 212(4):347–368

Soller JS, Olivieri AW, Eisenberg JNS, Jeffrey A, Sakajii R, Danielson R (2004) Evaluation of microbal risk assessment techniques and applications. IWA Publishing Colchester, Ockland

Sumner J, Ross T, Ababouch L (2004) Application of risk assessment in the fish industry. FAO, Rome

Tritscher AM (2004) Human health risk assessment of processing-related compounds in food. Toxicol Lett 149(1–3):177–186

Valerie J, Davidsona JR, Fazil A (2006) Fuzzy risk assessment tool for microbial hazards in food systems. Fuzzy Sets Syst 157(11):1201–1210

Van Asselt ED, Meuwissen MPM, Van Asseldonk MAPM, Teeuw J, Van der Fels-Klerx HJ (2010) Selection of critical factors for identifying emerging food safety risks in dynamic food production chains. Food Control 21(6):919–926

WHO (2002) WHO global strategy for food safety. Geneva, Switzerland. Resource document. WHO. http://www.who.int/foodsafety/publications/general/en/strategy_en.pdf. Accessed 10 June 2010

WHO (2004a) Evaluation of certain veterinary drug residues in food. Sixty-second report of the Joint FAO/WHO Expert Committee on Food Additives. WHO, Geneva

WHO (2004b) Evaluation of certain food additives and contaminants. Sixty-first report of the Joint FAO/WHO Expert Committee on Food Additives. WHO, Geneva

WHO (2008) WHO food safety. Resource document. WHO. http://www.who.int/foodsafety/en/. Accessed 10 June 2010

Chapter 18
Solutions to Threats and Risks for the National Security of Slovenia

Teodora Ivanuša, Matjaž Mulej, and Iztok Podbregar

Abstract Either peace with no crisis, or war – this is the current dilemma, again. Over the recent decades, this dilemma was solved in both developed and undeveloped countries. The main threat and risk today is the current general social crisis, largely due to the lack of requisite holism (RH) in values/culture/ethic/norms (VCEN) and behaviour, hence based on one-sided and short-term behavior of the influential people and their organizations, including enterprises and governments. Problems can be solved by innovation of VCEN of one-sided behaviour toward RH under the label of social responsibility (SR) and of its legal framework, leading to realization of shared interests of all humankind, such as peace, justice, and survival. The current threats and risks to security of any country, including Slovenia, can best be avoided in this way.

18.1 Introduction

In the recent centuries, especially in the past 10 years, humankind has been, and is moving from, the *hard-physical-cum-routine-loving-work* via *knowledge* to *creative/innovative* society. The latter requires a new economy with new values/culture/ethic/norms (VCEN): equipment replaces muscles and routine, but not thinking and creativity. For the current civilization of humankind to survive, humans' self-interest is to be realized – rather than the abuse of nature, (including humans) in everyone's and global social responsibility (SR) and therefore in requisitely holistic (RH) behavior

---

T. Ivanuša (✉)
Faculty of Criminal Justice and Security, University of Maribor, Maribor, Slovenia
and
Faculty of Logistics, University of Maribor, Maribor, Slovenia
e-mail: teodora.ivanusa@fvv.uni-mb.si

M. Mulej
Faculty of Economics and Business, University of Maribor, Maribor, Slovenia

I. Podbregar
Faculty of Criminal Justice and Security, University of Maribor, Maribor, Slovenia

G. Meško et al. (eds.), *Understanding and Managing Threats to the Environment in South Eastern Europe*, NATO Science for Peace and Security Series C: Environmental Security, DOI 10.1007/978-94-007-0611-8_18, © Springer Science+Business Media B.V. 2011

(made of monitoring, perception, thinking, emotional and spiritual life, decision making, communication, and action). SR can, and must, reach far beyond charity toward the end of abuse of power/influence of the influential persons/organizations in their relations with their co-workers, other business and personal partners, broader society, and natural environment as the unavoidable and terribly endangered precondition of human survival, at least in terms of the current civilization (Božičnik et al. 2008; EU 2000a, b; Harris 2008; Hrast et al. 2006, 2007; Hrast and Mulej 2008, 2009, 2010; IRDO 2006). SR supports innovation also by upgrading the criteria of business excellence and by supporting RH behavior; thus it is also a form of innovation of human VCEN and knowledge, resulting in a RH of behavior. In a most optimistic scenario, SR can also provide a way toward peace on Earth (Gorenak and Mulej 2010). SR can lead to covering all these urgent humankind's needs by making co-workers and other people happier, because it provides to them a greater sense of being considered equal and creative rather than abused and/or misused by power-holders (Šarotar-Žižek and Maučec 2009). Thus, the chapter's thesis reads: innovation of VCEN, not the technological innovation alone, can bring humankind out from the crisis that has surfaced only in 2008, having its roots in the prevailing VCEN of the decades after the second World War, known as neo-liberal economics (Ećimović et al. 2009; Goerner et al. 2008; Logan 2009; Mulej 2009; Mulej et al. 2009a, b, c; Nystrom 2009; Pavkov and Kulić, 2009; Shaz 2009). The life-cycle of the neoliberal ideology that has banned RH and SR has lasted one two-generation cycle (about 70 years) of VCEN, as predicted (Mulej et al. 2000); it must be replaced. Its extreme application of the old Roman-law definition – that ownership gives the owner the right of use and abuse – must be replaced with the right of use including SR instead of abuse. A new world-wide legal power is needed to support this replacement and related innovations (Agnew 2009; Avadhuta 2009; Banerjee 2009; Chatopadhyay 2009; Martin 2006; Menon 2009; Misra 2009; Sinha 2009; Singh 2009; Shankar 2009; Štibler 2008). Security issues are unavoidably a part of the same process.

18.2 Some Data About the Current Threats to Humankind's Security

The recent two centuries have brought both unseen socio-economic development and destruction of humankind's future (Božičnik et al. 2008; Brown 2008; Ećimović 2008, Ećimović et al. 2009; Korten 2009; Stern 2006; Targowski 2009; Taylor 2008; Wilby 2009).

Before industrialization the rate of economic growth used to be 3(three)% per millennium, after 1820 5500% (fifty-five times) in less than two centuries. Since then there are 6 times more humans on the Earth, every person using on average +5 times more energy, having 17 times more wealth, and 1,000 times more mobility. Humankind can no longer afford to emit *every hour* 4 million tons of CO_2, by burning fossil fuels, cut 1,500 ha of wood, and add 1.7 millions tons of nitrogen by mineral dunging into the soil, like today.

Humans are all on the same (sinking) boat, but on different decks. The poor ones cannot change the current trend, while the rich ones are not willing to change it. Climatologists warn: one must reduce emissions of toxins in the air, water, and soil by 80%. It can be attained with the given technologies, but only with a critical innovation of the current consumption patterns and big structural changes in production and use of energy. Alone, a renewal of natural preconditions for the current civilization to survive, after decades of competition by destruction of nature, would cost more than both world wars combined, in a best case scenario and if the action is undertaken immediately. Postponing the action may increase the cost up to +20% of the world-wide GDP. These processes cause global threats.

Centuries of the human practice of one-sided rather than holistic behaviour are also visible in transnational sources of threats and risks for the national security. These threats include terrorism, illicit activities in the field of conventional arms, WoMD and nuclear technology, organized crime, illegal immigration, cyber-threats and misuse of information technologies and systems, the foreign intelligence services' activity, military threats. National sources of threats and risks for the national security include threats to public safety, natural and other disasters, scarcity of natural resources and the habitat degradation, health-epidemiological threat, uncertainty factors. None of these threats are independent from human behaviour and its background – VCEN. Unfortunately, VCEN includes very little or even no SR, so far.

18.3 Social Responsibility: A New VCEN to Be Realized to Fight the Current Threats and Risks

SR can and should be combined with ethics of interdependence, because all human beings are complementary to each other as specialized professionals and as humans, and therefore interdependent. There is also the fact that one lives increasingly on creativity, including innovation. Thus, SR may help humans innovate society to include social efficiency, social justice and similar VCEN. Such VCEN, have for millennia, e.g., made the core of all social teachings called religions, philosophy of moral and ethical behavior, etc. (Avadhuta 2009; Hrast et al. 2006, 2007; Hrast and Mulej 2008, 2009; Martin and Murphy 2009). But power-holders have tended to abuse their influence and avoid their SR. Technological innovation has been used as a remedy for all troubles so far, although technology supports rather than creates future and development, and can be used either with SR or abused/misused with detrimental consequences (Collins 2001; Collins and Porras 1994; Galtung 2009). This includes innovation in security affairs.

The choice depends, now, on the most influential people and how they define their self-interest as a background of the new economy and humankind's future. Innovation of VCEN is unavoidable for the current civilization to survive; they may be well supported by a next step in human integration (Harris 2008; Martin 2006; Potočan and Mulej 2007). The process towards it may be leading from small groups via tribes and city-states prevailing before the industrial/entrepreneurial times, via nation-states prevailing in the twentieth century, and international federations coming into being over the recent decades, such as EU, NAFTA, LAFTA, OAU, etc., and agreement-based international organizations such as GATT, World Bank, UNO and its specialized agencies, etc. The General and Dialectical Systems Theories should be applied to make it happen with RH (Mulej et al. 2000). This chapter's thesis reads: VCEN of SR, including a world-wide federation, might be a potential innovation to help humans switch from a too narrow and dangerous/detrimental behaviour to RH and thus enable survival of humankind's current civilization or, at least, its way out of the current crisis, including security issues.

18.4 Requisite Holism as an Innovation of VCEN and Behaviour: A Basic Precondition for Ideas to Become Innovations

According to many international statistics, less than one percent of ideas become innovations (Nussbaum et al. 2005). This fact must not cause discouragement: (business) success with no innovation is even harder to attain. Data applies to technological innovations only. The socio-economic rather than financial, only, 2008 economic crisis results, from this viewpoint, from fictitious innovations; they do not match the definition that innovation is a novelty beneficial to its users, except for a few persons and organizations and for a short period of time, e.g. the new banking instruments of recent years. Now, the crisis-related problems are finally letting superficial economic behaviour surface. The one-sided opinions are another part of the same case: in reality, good air, arable land, and drinking water are not abundant and free commodities – there is hence a crucial need for environmental care, meaning care for human preconditions of survival. Another part of the same case is the neo-liberal abuse of Adam Smith's economic theory by one-sided interpretation of it; etc. (Božičnik et al. 2008; Brown 2008; Korten 2009; Toth 2008). In practice, there are so many factors of success in the invention-innovation-diffusion process that one-sidedness cannot lead to success, while the total holism (being the optimal theoretical way) cannot be attained, but the RH can: see Fig. 18.1.

Obviously, in each and every case a decision has to be made on: *which level of holism is good enough* to solve the given dilemma. One should avoid exaggerations and select the middle way – the RH (Fig. 18.1).

Fictitious holism/realism (inside a single viewpoint)	Requisite holism/realism (a dialectical system of all essential viewpoints)	Total = real holism/realism (a system of all viewpoints)

Fig. 18.1 The selected level of holism and realism of consideration of the selected topic between the fictitious, requisite, and total holism and realism

Take a look at experience around you and discover (again): *Success* has always resulted from the requisite holism (RH), i.e. *absence* of *oversights* with crucial impact. And failure has always resulted from crucial oversights, be it in business, scientific experiments, education, medical care, environmental care, invention-innovation-diffusion processes, etc., or wars, all the way to World Wars of the twentieth century and the current close to 30 wars, or the world-wide economic crises, including the 2008 crisis. RH of behaviour is aimed at avoiding crucial oversights. Systems thinking should better be called holistic thinking and be the world-view and methodology of holism, or better and more realistic: RH. Thus, the name of the crucial innovation to be attained now is the new economy based on RH of behaviour that can be attained, hopefully, with transformation of social responsibility (SR) from a much-talked about word into reality (On Google – SR offered beyond 350 million suggestions in September 2010). RH and/via SR might lead to a new economy and survival. Security issues also require SR.

18.5 The New Economy and Social Responsibility

A discussion about the "new economy" (Ing 2008) brought several insights that can be summarized as follows: the "new economy" faces:

- Property revolution (because ownership of knowledge and creativity differs from ownership of tangible properties)
- Information revolution (due to information/communication technology, education of many, many research centres, etc.)
- Serious new problems (due to piling up, rather than covering, the cost of care for natural preconditions of humankind's survival beyond the cost of both world wars combined or even much more)
- The need for much more transparency and participatory democracy in all organizations from families via enterprises, countries, to international associations (for RH in behaviour), and SR (for less of the detrimental abuse/misuse of Adam Smith's concepts of self-interest and invisible hand, of the laws of external economics, market, and trust)

Based on rules of economics and economy, according to official data, 15% of humankind – the so called West and Japan and Pacific Rim Tigers – enjoy results of the end of monopolies of the 1870s much more than the other 85% (Nixon 2004). They are much richer because they innovate much more, but they are not holistic enough to avoid the danger of a blind alley. Thus, the pre-market monopolies are renewed under new political and economic names, although they have produced the highest amount and impact of technological innovation ever. The resulting benefit belongs to a very small percentage of humans. The resulting current crisis seems to require innovation of the concept of innovation to include RH. (See also several contributions in ISSS 2008; Mulej et al. 2008). SR may support RH better than the practice of human relations, which so far have been based on exaggerated selfishness and greed, narrow-minded and short-term behaviour (Štibler 2008).

The big depression of 1930, which was quite similar to the current crisis, was not simply resolved with Keynes's economic measures, but continued as the second World War in order for humankind to resolve the problems left over after the first World War. Similar problems are here again. And so are nuclear weapons able to destroy the Planet Earth several times. People forgot that organizations, including enterprises and states, are their tools rather than authorities above people, and they are only more or less tools of those in the positions of higher authorities.

In other words, the lack of SR that has destroyed the slave-owning and feudal societies and created room for democracy and free-market economy is back, called financial, neo liberal or feudal capitalism. The legal names are different, but everything else remains similar. This is why SR is needed and discussed so much today. Security is threatened/at risk due to lack of SR and RH.

However, the content of SR is understood differently. The simplest version of SR is charity, but it might be only a mask for real one-sidedness rather than RH of behaviour of influential persons and their organizations. European Union (EuroCommerce 2001) officially mentions four contents of SR (of enterprises): the point is in a free acceptance of the end of abuse of (1) employees, (2) other business partners, (3) broader society, and (4) natural preconditions of humankind's survival. In literature on business excellence one requires more upgrading of its measures with SR (For an overview, see Gorenak and Mulej 2010). In further literature one sees the connection between systemic thinking and SR (Cordoba and Campbell 2008). A fourth group of references links SR with world peace (Crowther and Caliyurt 2004).

If consideration is limited to the EU's definition and enterprises, SR is found to have uncovered and avoidable costs only fictitiously and in a short term. Costs of honest behaviour replace the cost for mistrust, double-checking of creditworthiness, dissatisfaction, strikes, loss and regaining of high-quality co-workers and other business partners, their routine-loving rather than creative/innovative behaviour, misery and poor health (which are cured rather than prevented), remediation of consequences of natural disasters, terror, and wars, etc. Thus, SR changes the practice of ownership as defined by the (still accepted) Roman law saying that ownership gives to the owner the right of use and abuse; abuse must be replaced with SR/RH for humankind – and its organizations, for that matter – to survive as the current civilization facing the problems of extreme division and affluence.

Development of SR is therefore aimed to be an innovation of human behaviour towards ethic of interdependence.

18.6 Ethic of Interdependence, New Economy, Affluence, and RH by Social Responsibility

In preparation, passing, and realizing of decisions one succeeds if one has attained RH. This does not depend on knowledge alone, but an equal importance belongs to VCEN, because they direct the application of knowledge. The RH of specialists who need each other is expressed in their ethic of interdependence (Mulej and

Kajzer 1998). It expresses the specialists' feeling that they make up for each other's differences to make the RH; therefore success is attainable. Due to these differences, clear boundaries and isomorphisms (which many in systems theory emphasize – see François 2004) are not enough: viewing the world with the concept "through the eyes of the others extends vision" is needed (Churchman 1991, quoted by Antonio Lopez Garcia 2008); it leads to the dialectical systems approach (Mulej 1974, 1979) and the resulting RH of behaviour and requisite wholeness of outcomes. RH is in line with EU's definition of systemic thinking (EU 2000a: 6).

EU is trying to become a sustainable and knowledge/innovation-based society; the concept includes SR. As quoted above, in its document EuroCommerce (2001) defines SR as the integration of the care for society and environment in the daily business of enterprises and their relations with stake-holders, on a voluntary basis. Its messages include the crucial statement that in a longer run the economic growth, social cohesion, and environmental protection complete and support each other. EU stresses, too, that SR-behaviour reaches beyond matching the legal obligations; hence it reflects organizations' additional efforts to meet expectations of numerous/all stake-holders. EU also passed several other documents that support the development of SR (EU 2000b, 2006a, 2006b). These documents only partially cover the real contemporary needs: (1) the creativity-based society is replacing the knowledge-based, which replaced the routine-based one (Chesbrough 2003); (2) the concept of sustainable future needs to replace the concept of sustainable development (Ećimović 2008; Goerner et al. 2008; Hrast and Mulej 2008; ISSS 2008), for humankind to survive. The long-term and broader view is able to contribute more to the daily business success, too (Branson 2009; Meško-Štok 2008; Quinn 2006): it makes employees and other stakeholders more interested, motivated, creative, and loyal to their organization by providing them with means for well-being (Prosenak and Mulej 2008; Prosenak et al. 2008).

EU defined the period until 2010 "A European Roadmap", stressing the sustainable and competitive enterprise, which considers both the short-term and long-term creation of value (Knez-Riedl 2007b). The corporate SR can fortify the competitive position of single enterprises as well as local and regional communities, countries and EU (Knez-Riedl 2007a). Authors prefer no limitation of SR to companies: companies follow influential humans' decisions. For SR to become more than a word, a strategy of promotion of SR as a potential innovation might be needed (Knez-Riedl and Hrast 2006; Hrast and Mulej 2008).

18.7 Making SR an Attainable (Potential) Innovation – Some Practical Economic Preconditions

As components of making such equilibrium attainable, one can use three essential recent findings in economic literature, in a new synergy:

– Florida (2005) found in a comparative analysis of US regions that the best development had been attained in regions with the highest 3 T: *tolerance* for differences between habits of people attracts *talents,* and thus it makes sense to invest in *technology* there Malačič et al. (2006) found an equal situation in Slovenia. The creative class is growing beyond 35% and becoming essential, the (routine-loving) working class is diminishing, and the service class only works on preconditions for the creative class to create for all. (In addition, this percentage does not include people who must be creative to survive with their poor incomes.)

- Porter (1990, 2006) pointed out that the basis of competitiveness evolves in four phases: from natural resources via investment to innovation and then to affluence, which people have always wished to have. But affluence has a crucial side-effect: affluent people have no motive any longer to work in order to have, which results in a growing need of many citizens for solidarity, etc. In affluence sources are not scarce, but real needs. Marketing and advertisement try to persuade people to have wants and to buy these wants: greed would be needs (see also James 2007; Baumol et al. (2007) do not even mention or quote Porter, but they remind of this danger with a single quote on p. 288.)
- The innovation of the traditional incentives for Total Quality as a way to innovation that are often taken in too bureaucratic a way to really work as incentives for contemporary excellent quality as an incentive for innovation and RH to flourish (Pivka and Mulej 2004; Škafar 2006) and practice systemic thinking (SZK 2007). The reality is that the governments are covering the public sector, which makes them big buyers in a modern buyers' market, giving them the bargaining power. Governments can then demand that only the suppliers attaining the highest innovation, quality, and SR based on RH may supply the public sector with everything, from toilet paper to scientific novelties (Mulej 2007b)

One should add creativity-based ambition and life in both working and free time and a shorter work week for people to have more time for their families and creativity in all its forms, from gardening and house-keeping to Nobel-prize winning results and artistic achievements.

The problem lies in the transition of mentality towards cherishing creativity very much – in humans' thinking and worldview, as well as other values/emotions (Brown 2008; Ećimović et al. 2002; Harris 2008; Korten 2009; Martin 2006; Mulej 1979, 2007a, b). One-sidedness results in a lack of contemporary excellence, which requires more RH of behaviour for the humankind's future to exist. Baumol et al. (2007) fail to see this. Thus, it is the practice of interdependence that makes people aware of their need for each other due to their differences in specialization, and their mutual complementary relations on the same basis; this leads them to use the briefed synergy.

Porter (1990) and Brglez (1999) speak of four phases of development of bases of competitiveness – natural resources, investment, innovation and affluence. Mulej and Prosenak (2007) extend this idea to development and add ideas about the related culture and phase 5. Obviously, the affluence phase is not only the highest development phase so far, it is also the phase of growing problems of employment, supporting everybody, growing lack of ambition and related drug etc. abuse, etc.

Conclusion: one must attain and keep capacity of RH in order to enter the innovation phase quickly and remain in it as long as possible, and/or renew its culture; the latter may make room for a fifth phase, which is needed: the fourth phase can hardly be avoided. Porter and Kramer (2006) do not mention phase 5.

One-sidedness, thus, has proved to be a crucial danger. In other words, informal systems thinking in the form of RH/SR behaviour is the back-ground of the innovative society and solving the 2008 crisis. But innovation causes differences, because not all people are equally capable of RH and creation, including innovation. Though difference in wages is one topic, differences based on innovation need not be limited; they always take a percentage of the newly provided benefit (Mulej et al. 2008). Government should foster creativity and innovation – as a big buyer covering the entire public sector – by allowing only the most innovative organizations, including RH and SR into its criteria, to supply the public sector (Mulej 2007b). Experience shows that it makes sense: it makes people more innovative, successful and rich all the way to affluence, while remaining creative and ambitious (Dyck et al. 1998).

However, the affluence phase might be a dead alley if people lose ambition for creation (so far throughout history they have). People, therefore, need either a prolonged innovation phase based on RH invention-innovation-diffusion rather than one-sided processes, or a new phase, a fifth one, of creative happiness based on ethics of interdependence and interdisciplinary creative co-operation with SR replacing the phase of affluence (Prosenak and Mulej 2008).

To make this innovation of culture and economy happen, a part of the population must become the core of the creative class. Lester (2005) found authors detecting that about 15–20% of people are willing to take a risk and cooperate, about the same percentage want to be (abusing) free-riders, and the majority just waits to see what the opinion makers undertake. However, this majority includes many humans with creative potential. Leaders providing a role-model of interdisciplinary creative co-operation can activate this potential rather than the commanding managers who do not. This would make humans happy and society prosperous. But it requires RH/ SR of behaviour.

The USA's General Creech showed his experience: his success as an air-force pilot and commanding officer made him climb the ladder to a 3-star general on his basis of cooperating and role-modelling leadership rather than commanding without listening (Creech 1994).

This might lead to RH in society and economy by SR (Hrast et al. 2006, 2007; Hrast 2007; Knez-Riedl et al. 2001; Knez-Riedl 2003a, b, c, 2006; Knez-Riedl et al. 2006). Such attributes of behaviour create new ambition, reaching beyond complacency of the affluent ones. No short-term efficiency, including e.g. abuse of external economics, or of the law of supply and demand, is enough. Then, a new economy can succeed. Who can start the process? Many influential persons made history by making their individual values a culture, shared by a group of their followers who then diffused this culture to make it a socially acceptable ethic, resulting in the social norms, influential over individual values of others who faced a dilemma: follow the novelty and be accepted or refuse it and be an outlaw (see Fig. 18.2.)

Individual values (interdependent with knowledge)	↔	Culture = values shared by many, habits making them a rounded-off social group
↕	╫	↕
Norms = prescribed values on right and wrong in a social group	↔	Ethics = prevailing values about right and wrong in a social group

Fig. 18.2 Interdependence of values, culture, ethics, and norms

18.8 Legal Preconditions to Be Innovated in Order to Support RH/SR and Resulting Survival of Humankind

Contributions by Harris (2008), Martin (2006), several authors in Martin and Murphy (2009), Letnar-Černič (2009), etc. clearly demonstrate that survival of humankind cannot be taken care of as long as the international law has its legal basis on agreement without legal enforcement, thus denying itself as law. Similar problems show up with all other existing international organizations, including the United Nations. Countries/states obviously tend to prefer their (businesses') narrower and short-term interests to their broader and more long-term ones, thus ruining their basis of existence all the way to threatening the survival of their people.

Governments might innovate their minds, if they used the following historical experience.

– Humankind started living in organized groups when experiencing ethics of interdependence with one another (based on combinations of love, other positive emotions and practical needs for each other). Over time they developed states, but these were mostly city-states (except a few imperial ones) for millennia. Only in the nineteenth and twentieth centuries did the notion of nation-states develop, but they were rarely, if ever, really based on a nation. Rather, they were based on a prevailing nation subordinating others. Colonies were freed once experience showed that colonial resources could be traded, and colonialists therefore had no need to own areas to attain their benefits. In the twentieth century, several international federations were formed in addition to the ones created earlier, including nation-based ones such as the USA. United Nations is not a very effective organization, but it does solve several crucial problems, if its member states find their shared interest, such as keeping nuclear weapons out of use for more than six decades.
– The next step could (and should) be for states of the world to decide that survival of humankind in the current civilization has enough importance to depend on a legal body with authority beyond case-by-case agreement and beyond forgetting about the signed agreements when the short-term pressures of biased interests try to prevail, even when leading to the end of humankind. This might be a World/Earth Federation with a World Government, with its authorities being limited by the principle of coordinated decentralization and based on attainment of RH via SR concerning the topics that cannot be taken care of well by

18 Solutions to Threats and Risks for the National Security of Slovenia   337

individuals, families, enterprises and other local organizations, regional, national and similar bodies and agencies of individual governments.

In terms of a theoretical background, no single traditional science is either unnecessary to or sufficient for RH/SR and requisite wholeness to be attained and kept so humankind does not return to feudal capitalism. All of those in synergies based on RH/SR can be helpful, while systems science can provide their synergy as soon the selected systems-science version is not aimed at description from a single viewpoint. The Dialectical Systems Theory and some others are more appropriate for this purpose than others, which serve other purposes (Mulej et al. forthcoming; Mulej et al. 2009d). Interdisciplinary cooperation is usually a precondition for RH, and SR might support it, including in topics of innovation (Mulej et al. 2006, 2007). There are methods supportive of this effort (Mulej and Mulej 2006).

Obviously, technological innovation is far from enough for humankind in the current civilization to survive rather than perish in the next war and/or destroyed natural preconditions of human survival. Let it be repeated: the crisis of the 1930s resulted in the second World War with many ten of millions of deaths, including development of nuclear weapons and their further sophistication and diffusion to many nation states. These and other weapons would not be needed with more ethics of interdependence, RH and SR, if the (C)SR was not questioned by the political decision that it is not legally obligatory, but beyond legal obligations of enterprises.

18.9 Technology Matters, But as a Human Tool

Collins (2001) and Collins and Porras (1994) concluded from their large field research on the reasons for the long-term top companies to be the best, and on their way to becoming and remaining so, among other crucial attributes, that technology matters, but as a human tool, not as an independent cause of economic success. Rather, one should better speak about the entrepreneurial revolution than about the industrial revolution. The most entrepreneurial individuals have created all new technology and have helped it become an innovation rather than invention/toy of weird persons. Have these people attained RH in their monitoring, perception, thinking, emotional and spiritual life, decision making, communication, and action, i.e. behaviour, for their influence to be beneficial, only, rather than detrimental, too? Not really (Bourg 2007; Božičnik 2007; Ećimović 2008; Goerner et al. 2008; Hilton 2008; Mulej et al. 2008; Stern 2006, 2007): the dangerous consequences of their lack of holism result from one-sidedness causing a too narrow view and the resulting assessment of what is essential in the current conditions.

All specialized knowledge is both beneficial and unavoidably narrow, but no knowledge is either self-sufficient or sufficient (Metcalf 2008; Mulej 1974; Mulej 1979, 2007a, b; Potočan and Mulej 2007). With a lack of RH, it helps less than in interdisciplinary creative co-operation. Perhaps SR can show a new way out of the current world-wide economic crisis. It may matter, because the use of knowledge, including technology, depends a lot on users' values. This is true of VCEN, knowledge, and practice of security provision, too.

18.10 Security in the Case of Non-Transparent Threats

A special viewpoint of SR and RH inside the security forces surfaces when non-transparent threats appear and require attention of many subjects. Problems emerge in the early phases when facing a complex national security issue, where cooperation of more subjects in the national security system is needed, or it is also necessary to include other entities (e.g. health system) in the tasks and problem-solving process. Namely, despite very clear tasks which have to be undertaken by individual subjects due to their statutory competences, the national security problems are mostly positioned in such a manner that it is quite impossible to identify the bearer or the one person who is competent and responsible for solving the problem. Such problems are usually solved by co-ordinations and searching for better synergies between individual subjects of the national security system. This is an extremely challenging task when facing complex issues, where defenders deal with unrecognized threats, fully unqualified and even unequipped individual subjects. In these activities, defenders usually lose an enormous amount of time that is important precisely because of the complex threats.

When shaping the legal normative regulation of the national security system and organizing the subjects of the national security system, considering their qualification, training and equipment, it would be considerate to follow the principles of the utmost synergy between these subjects by following the organizational integration of Systems Theory. (Let us repeat: No Systems Theories versions apply that are aimed at precise description of a selected problem from a single viewpoint, but a version that is aimed at holism of approach and wholeness of outcomes, e.g. Mulej's Dialectical Systems Theory.) When facing any national security threat, countries should reach a state, in which their "newly organized security forces" would be competent, responsible, and trained for the immediate response to every potential non-transparent threat, which causes harm. It is necessary to reach a state in which only the unavoidable number of forces is optimally included in the problem-solving process, in order to optimize the number of problem-solvers throughout the entire period of the crisis. It is necessary to avoid circumstances, in which in the initial situation defenders experience a lack of engagement due to a long decision-making process, followed by an engagement avalanche of too many forces, and followed by a shortage of resources. A balanced engagement throughout the entire period is required.

When non-transparent threats induce crises, it is necessary to determine the right ratio between generalists and specialists in a particular security problem. Generalists need to be trained well enough to identify, define, and submit the request for proper engagement of individual specialists during the security problem solving process. In certain situations, defenders can solve the ratio between generalists and specialists with smaller multi-disciplinary teams, which are simply trained for the initial solving of the individual complex problem and are able to define the inclusion of new additional subjects in the further crisis resolution process.

In such situations, manuals, which are more directional in nature, are much more useful than crises' rescue plans. Plans with their precision require a constant upgrading, while manuals only broadly offer a general guidance for the professional teams of problem-solvers in a given crisis. Proper informing and notifying, both within teams as well as to all the others involved in the crisis, is of utmost importance. Therefore, it would be proper to include such knowledge and skills in smaller multi-disciplinary teams. Where it is possible, television picture broadcasting should be included in such situations; it transmits real time image, sound, and data to higher levels, thus reducing the possibility of an error at the micro-locations where individual teams sometimes lack a comprehensive picture and overview.

It is also very important that a successful completion of the intervention is followed by a timely re-development, where competences and responsibilities of individual subjects generally change, and the local level usually becomes more important. It is also necessary to conduct a qualitative analysis of the intervention, which is intended not only to complement the crises' manuals, but also to constantly search for better, more rational ways of intervention.

The General Charles Chandler Krulak, the thirty-first U.S. Marine Corps commander in the 1995–1999 period, clearly presented the complexity of the spectrum of challenges that the soldier already does and will continue to confront in the modern battlefield, especially in urban areas. This was in the late 1990s and based on lessons learnt from Somalia, Haiti, and Bosnia from the so-called "The Three Block War". According to him, the concept of "The Three Block War" is to be understood as an operative requirement, which will demand soldiers, non-commissioned officers and officers simultaneously to efficiently carry out the classic operations, peacekeeping and humanitarian operations, and this – he specially emphasized – only within the three blocks of the same town. This causes the need for simultaneous training of individuals and units to operate in various operations, the necessity to take the initiative from independent actions to key decision-making on the basic tactical levels and the idea of the role of the "Strategic Corporal".

18.11 'Let Us Abolish the Slovenian Armed Forces. We Have a Good Army. We Have the Army, Such as We Built It'

In mid May 2010, numerous celebrations took place on occasion of the Slovenian Armed Forces (SAF) Day, the anniversary of the Manoeuvre Structure of National Defence and celebration on occasion of the Day of veterans' organizations associated with the attainment of independence of Slovenia. Numerous experts, popular gatherings and discussions on events that occurred 20 years ago took place, aimed at building awareness of these important historical deeds and events for the Slovenian democracy. In some circles, which were bound together into the initiative for the "Abolishment of the Slovenian Army" by the weekly *Mladina*, discussions were also held in the later weeks on the abolition of the SAF and the rearrangement of its resources and infrastructure to other areas of Slovenes' lives. The SAF is a significant part and symbol of the Republic of Slovenia (RS); it is not only formally written in the Constitution and other important documents of Slovenia, but most of the opinion surveys over the 19 years of Slovenia's history show that the citizens' confidence in the SAF is extremely high. The SAF carries out important tasks in the context of North Atlantic Treaty Organization (NATO) and the EU, too; there Slovenia has been an equal SAF and important member for several years, actively solving security threats and crises in several unstable areas of the world. The social environment is very dynamic and is changing extremely rapidly, presenting a major challenge for the SAF, which has to build upon, and adapt to, these changes as well as assume new duties in addition to the traditional tasks of homeland defense. Today, the Slovenian Army must be capable of conventional warfare, when the gun's barrels are raised against the enemy. It has to demarcate two or more parties in a conflict, when the gun's barrels are lowered and different skills are used. It must perform humanitarian activity, where weapons are not needed. All this mostly happens very far from the homeland Slovenia and requires exceptional logistics.

However, it seems that in the middle of all this, the heads of the Ministry of Defense and the SAF frequently lose contact with their environment and do not know how to explicate their role and act in the best way. These constant changes in the environment and operational needs pose a problem in the transformation of the Defence System, which is unfortunately sometimes very inert and too slow. In these conditions, initiatives such as the one concerning the abolition of the SAF in the weekly *Mladina* also find their place. The initiative seems legitimate, as it would likely open a discussion on the future development of the very important defence field. Unfortunately, the head of the Ministry of Defense did not immediately respond in the best way. In fact, even now there is no real debate among the dissenters, but different speakers at the celebrations only express views on the current role and the need for the SAF. There is a need for more open discussion on the SAF: then the Government would also realize that the budget cuts have its limits in SAF, too.

This can only be changed by a RH re-engineering, i.e. by seeking an organization, where all the tasks will be carried out in the same way and better with reduced resources. Yet now there is a lack of such a readiness in society and its environment, but the SAF abolition initiators' admonitions are directed at the beginning of these processes. Slovenia had the opportunity to prepare a plan of transition from the conscript to the professional army for the late Janez Drnovšek, who was Prime Minister at the time. Ever since it was believed that it is necessary to change thoroughly the Slovenian National Security System: to adapt and prepare it for these rapid non-transparent changes in its environment and to prepare a faster, better, higher-quality response to these challenges.

Perhaps a few sentences on RH re-engineering of the National Security System can help explanation and understanding. Today about 21,500 employed civil servants are working in the Ministry of Defence, the SAF, Security and Rescue Forces; the Ministry of Internal Affairs and the Slovenian Police; the Ministry of Foreign Affairs and Slovenian diplomacy; the Ministry of Justice and its Bureau for enforcement of the criminal-law sanctions and Intelligence services. For their functioning, including the salaries, €1.35 billion from the budget are spent per year. Slovenia is facing a non-transparent threat that may never even develop into a crisis. Yet this huge machine is uncoordinated, an enormous amount of time is spent to determine who among them will be the bearer of the problem solving process and who will be the participant.

To address the crisis, a rapid and RH response is required, and it should be immediately clear who owns the solution to the problem. Authors' thinking on RH re-engineering, as a comprehensive and radical change of the aforementioned subjects of the National Security System, goes towards greater integration and greater efficiency of all. Authors are confident that with greater integration and better skills, it will be possible to reduce this multitude of 21,500 civil servants to 12,000 professionals and the financial means from €1.35 billion to €750 million. Slovenia will obtain a National Security Force that will own and solve all non-transparent threats and problems at any moment. At the same time, Slovenia will have about 11,000 people with solid multi-disciplinary skills who are free on the labour market with the remaining €750 million at hand for them. If Slovenia spends part of this money for their re-integration into other areas of the labour market, there are at least €400 million left for the development of new technologies and other development opportunities. One may talk about the fact that this would require fewer ministers in the Slovenian Government some other time, although these changes can result also in reduction of the staffing levels in the Government of the RS.

Thus, RH re-engineering, in short words, is a fundamental innovation in the National Security System, which will certainly occur in the coming years. Many will find the aforementioned suggestions unrealistic, but let things run their course. Likewise the role of the SAF and its tasks change constantly, as does the role and use of the *Poček* Training Area. It is completely illusory to think that Slovenia will no longer need it in the coming years; therefore, it is necessary to look for synergies and the best possible solution all the time. The current arrangements allow for short-term solutions and compensation for the use of this land, but they don't regulate the strategic, ecological and common long-term use of this territory. It is also necessary to establish an undisturbed communication between all parties involved and for these discussions to not be a matter of the current elites on positions and raise distrust among people. The territory of the Training Area *Poček* needs to be open for the people of Slovenia to use when the SAF does not need it. Worldwide, for example in Eastern Europe, programs are run where such areas are ecologically arranged with considerable means, and attractive tourist amenities programs are created, offering much to the local community as well. Last but not least, the tourism industry is a priority. Such complexes may provide support to industry, which has the potential of inclusion in the military industrial complex of Slovenia. In short, there are many possibilities – decision-makers only need to know the substantive issue and bring some experience from other environments to their areas.

Without a doubt all the discussions and different views and opinions are useful, if humans only know how to listen to them. The brief critical comments are intended to foster communication between each other and seek the best possible solutions to improve coexistence of SAF and civilians. There is far too little of that in Slovenia now, and consequently Slovenia is not taking advantage of all the opportunities at hand. The authors hope and wish that in the future this will be different.

18.12 Conclusions

The innovative society of today does not attain success, once criteria of sustainability and SR are added to the one-sided economic criteria, even if the crisis showing up in 2008 is not the focus. Even if the "West" considers itself successful, research and the public press report on increasing numbers of humans feeling unhappy and hence abusing drugs, from alcohol to marijuana, etc., and doing so at an increasingly young age. This is a sign that there is a lack of incentive for creation, for the Fromm's transition from 'owner to creator', as the most human attribute (James 2007). Such processes have been around before. The Roman and other empires have faced ruin once their people entered affluence and became complacent. Hopefully, SR reaching beyond CSR to SR of all, and incentives such as happiness based on creativity, can be a way out of the blind alley towards RH. Technology alone does not make it. Innovation of VCEN is necessary for technology to play a supportive rather than destructive role for human survival on the Planet Earth, at least in the form of the current civilization. The process of making the idea an innovation needs consideration as well; but this is a topic requiring another contribution (Ženko et al. 2008).

References

Agnew E (2009) Interpreting peace: Jane Adams, the cathedral of humanity and pacifism. In: Martin G, Murphy P (eds) 12th International conference of international philosophers for peace, July 2–8, 2009, Nainital, India. International Philosophers for Peace Society, Nainital

Antonio Lopez Garzia MP (2008) Systems thinking in the post modernity era. Gen Syst Bull 37:14–15

Avadhuta AAA (2009) Shri Shri Anandamurtiji's concept of one universal human society. In: Martin G, Murphy P (eds) 12th International conference of international philosophers for peace, July 2–8, 2009, Nainital, India. International Philosophers for Peace Society, Nainital

Banerjee G (2009) Awakening of freedom and harmony. In: Martin G, Murphy P (eds) 12th International conference of international philosophers for peace, July 2–8, 2009, Nainital, India. International Philosophers for Peace Society, Nainital

Baumol WJ, Litan RE, Schramm CJ (2007) Good capitalism, bad capitalism, and the economics of growth and prosperity. Yale University Press, New Haven/London

Bourg D (2007) Toward a planet-wide ethic – special report climate change. Research*eu 52:16–17. Resource document. http://ec.europa.eu/research/research-eu/pdf/research_eu_52_en.pdf. Accessed 17 May 2010

18 Solutions to Threats and Risks for the National Security of Slovenia 343

Božičnik S (2007) Dialektično sistemski model inoviranja krmiljenja sonaravnega razvoja cestnega prometa (Eng. Dialectical control system model of innovation for sustainable development of road traffic). University of Maribor, Faculty of Economics and Business, Maribor

Božičnik S, Ećimović T, Mulej M (eds) (2008) Sustainable future, requisite holism, and social responsibility. ANSTED University, Penang in co-operation with SEM Institute for climate change, Korte, and IRDO Institute for Development of Social Responsibility, Maribor

Branson R (2009) Richard Branson: personal interview by David Sheff. Playboy 56(1):39–46

Brglez J (1999) Razvojni potenciali majhnih gospodarstev v razmerah evropskega integracijskega procesa (Eng. Developmental potentials of small economies in the circumstances of the European integration process). University of Maribor, Faculty of Economics and Business, Maribor

Brown LR (2008) Plan B 3.0: mobilizing to save civilization. Earth Policy Institute, W. W. Norton, New York/London

Chatopadhyay SN (2009) Freedom, socio-political justice and peace and harmony. In: Martin G, Murphy P (eds) 12th International conference of international philosophers for peace, July 2–8, 2009, Nainital, India. International Philosophers for Peace Society, Nainital

Chesbrough HW (2003) Open innovation. The new imperative for creating and profiting from technology. Harvard Business School, Boston

Churchman CW (1991) Education management. Pitman, London

Collins J (2001) Why some companies make the leap … and others don't. Good to Great. Random House Business Books, Sidney

Collins J, Porras J (1994) Built to last. Successful habits of visionary companies. HarperBusiness, New York

Cordoba J-R, Campbell T (eds) (2008) System thinking and corporate social responsibility. Syst Res Behav Sci 25(3):359–437

Creech B (1994) The five pillars of TQM. How to make total quality work for you. Truman Talley Books, Dutton

Crowther D, Caliyurt KT (eds) (2004) Stakeholders and social responsibility. ANSTED University, Penang

Dyck R, Mulej M, Desai N (1998) Self-transformation of the forgotten four-fifths. Kendall/Hunt, Dubuque

Ećimović T (ed) (2008) Proceedings of the 20th WACRA conference. http//:www.institut-climatechange.si. Accessed 10 June 2010

Ećimović T, Mulej M, Mayur R (2002) System thinking and climate change system. SEM, Korte

Ećimović T, Mulej M, Martin G (2009) The philosophy of past, present and future of mankind. In: Martin G, Murphy P (eds) 12th International conference of international philosophers for peace, July 2–8, 2009, Nainital, India. International Philosophers for Peace Society, Nainital

EU (2000a) Communication from the commission to the council and the European parliament innovation in a knowledge-driven economy. Commission of the European Communities, Brussels

EU (2000b) Communication concerning corporate social responsibility: a business contribution to sustainable development. Commission of the European Communities, Brussels

EU (2006a) Commission of the European communities: implementing the partnership for growth and jobs: making Europe a pole of excellence on corporate social responsibility. Commission of the European Communities, Brussels

EU (2006b) CSR Europe (2006): a European roadmap for business towards sustainable and competitive enterprise. Commission of the European Communities, Brussels

EuroCommerce (2001) Green paper on promoting a European framework for corporate social *responsibility*. http://www.eurocommerce.be/content.aspx?PageId=40125. Accessed 10 June 2010

Florida R (2005) Vzpon ustvarjalnega razreda (Eng. The rise of the creative class). IPAK, Velenje

François C (ed) (2004) International encyclopedia of systems and cybernetics, 2nd edn. K. G. Saur Verlag, München

Galtung J (2009) 12 approaches to global warming. PROUT 20(4):10–11

Goerner S, Dyck RG, Lagerroos D (2008) The new science of sustainability. Building a foundation for great change. Triangle Center for Complex Systems, Chapel Hill

Gorenak S, Mulej M (2010) Upravljanje popolne odgovornosti kot vir primerne celovitosti vodenja poslovanja in dolgoročnih konkurenčnih prednosti inovativnih podjetij. (Eng. Management of complete responsibility as a source of the appropriate integrity and long-term competitive advantage of innovative companies). Naše gospodarstvo 56(1–2):22–31

Harris EE (2008) Twenty-first century democratic renaissance. From Plato to neoliberalism to planetary democracy. The Institute for Economic Democracy Press, Sun City

Hilton B (2008) The global way. The integral economics of the post-modern world. Trafford, Canada

Hrast A (2007) Družbena odgovornost je v trendu (Eng. Social responsibility is the trend). Glas gospodarstva 1:44–45

Hrast A, Mulej M (eds) (2008) Social responsibility and current challenges 2008. Contributions of social responsibility to long-term success of all market stakeholders. Proceedings of the 3rd international conference on social responsibility. IRDO Institute for Development of Social Responsibility, Maribor

Hrast A, Mulej M (eds) (2009) Social responsibility and current challenges 2009, work – a bridge to cooperation and coexistence of generations. The 4th international conference. IRDO Institute for development of Social Responsibility, Maribor

Hrast A, Mulej M (eds) (2010) Družbena odgovornost: Narava in človek (Eng. Social responsibility: nature and human). IRDO Inštitut za razvoj družbene odgovornosti, Maribor

Hrast A, Mulej M, Knez-Riedl J (eds) (2006) Družbena odgovornost in izzivi časa 2006 (Eng. Social responsibility and challenges of the time in 2006). IRDO Institute for Development of Social Responsibility, Maribor

Hrast A, Mulej M, Knez-Riedl J (eds) (2007) Družbena odgovornost 2007 (Eng. Social responsibility 2007). Proceedings of the second IRDO conference on social responsibility. IRDO Institute for Development of Social Responsibility, Maribor

Ing D (2008) how much have we learned about the 'new economy' associated with 'the information revolution' or 'services revolution', and what we do not know yet? In: ISSS 'Systems that make a Difference', University of Wisconsin, Madison, 13–18 July 2008

IRDO (2006) Leaflet. IRDO Institute for Development of Social Responsibility, Maribor

ISSS (2008) Systems that make a difference, University of Wisconsin, Madison, 13–18 July 2008

James O (2007) Affluenza – a contagious middle class virus causing depression, anxiety, addiction and ennui. Ebury, Random House UK, Vermillion

Knez-Riedl J (2003a) Korporacijska društvena odgovornost i komuniciranje sa vanjskim okruženjem (Eng. Corporate social responsibility and communication with external community). Informatologia 36(3):166–172

Knez-Riedl J (2003b) Social responsibility of a family business. MER Rev manag razvoj 5(2):90–99

Knez-Riedl J (2003c) Corporate social responsibility and holistic analysis. In: Chroust G, Hofer Ch (eds) IDIMT-2003: proceedings. Universitätsverlag R. Trauner, Linz, pp 187–198

Knez-Riedl J (2006) Družbena odgovornost in univerza. (Eng. Social responsibility and the university). In: Hrast A, Mulej M, Knez-Riedl J (eds) Družbena odgovornost in izzivi časa 2006. IRDO Institute for Development of Social Responsibility, Maribor

Knez-Riedl J (2007a) Kako DOP povečuje konkurenčnost - projekt CSR – Code to Smart Reality (Eng. How CSR increases competitiveness – project CSR – code to smart reality). Gospodarska zbornica Slovenije – Območna zbornica Maribor, Maribor

Knez-Riedl J (2007b) Družbena odgovornost podjetja in evropski strateški dokumenti - projekt CSR – Code to Smart Reality (Eng. Social responsibility and the European strategic documents – project CSR – code to smart reality). Gospodarska zbornica Slovenije – Območna zbornica Maribo, Maribor

18 Solutions to Threats and Risks for the National Security of Slovenia

Knez-Riedl J, Hrast A (2006) Managing Corporate Social Responsibility (CSR): a case of multiple benefits of socially responsible behaviour of a firm. In: Trappl R (ed) Cybernetics and systems 2006: Proceedings of the eighteenth European meeting on cybernetics and systems research. Austrian Society for Cybernetic Studies, Vienna, pp 405–409

Knez-Riedl J, Mulej M, Ženko Z (2001) Approaching sustainable enterprise. In: Lasker GE, Hiwaki K (eds) Sustainable development and global community. International Institute for Advanced Studies in Systems Research and Cybernetics, Baden-Baden, pp 57–68

Knez-Riedl J, Mulej M, Dyck RG (2006) Corporate social responsibility from the viewpoint of systems thinking. Kybernetes 35(3/4):441–460

Korten DS (2009) Agenda for a new economy: from phantom wealth to real wealth. Berrett-Koehler, San Francisco

Lester G (2005) Researchers define who we are when we work together and evolutionary origins of the "Wait and See" approach. http://www.upenn.edu/pennnews/article.php?id=738. Accessed 10 June 2010

Letnar-Černič J (2009) Fundamental human rights obligations of corporations. In: Hrast A, Mulej M (eds) Social responsibility and current challenges 2009, work – a bridge to cooperation and coexistence of generations. The 4th International Conference. IRDO Institute for development of Social Responsibility, Maribor

Logan R (2009) A call for new economic power. PROUT 20(4):6–8

Malačič J, Drnovšek M, Jaklič M, Kotnik P, Mrak Jamnik S, Pahor M et al (2006) Študija o kazalcih ustvarjalnosti slovenskih regij (Eng. The study on indicators of creativity of Slovenian regions). Služba za regionalni razvoj RS, Ljubljana

Martin G (2006) World revolution through world law: basic document of the emerging earth federation. World Constitution and Parliament Association, Redford, The Institute on World Problems, Radford, Colombo, Sri Lanka, and Kara, Togo

Martin G, Murphy P (2009) 12th International conference of international philosophers for peace, Nainital, India. International Philosophers for Peace Society, Nainital

Menon EP (2009) The inevitability of world government. In: Martin G, Murphy P (eds) 12th International conference of international philosophers for peace, July 2–8, 2009, Nainital, India. International Philosophers for Peace Society, Nainital

Meško-Štok Z (2008) Management znanja kot temelj odličnosti v gospodarskih družbah (Eng. Management of knowledge as the foundation of excellence in companies). Univerza na Primorskem, Fakulteta za management, Koper

Metcalf G (2008) Incoming presidential address. Gen Syst Bull 37:5–12

Misra AP (2009) Foundation of freedom, harmony and peace. In: Martin G, Murphy P (eds) 12th International conference of international philosophers for peace, July 2–8, 2009, Nainital, India. International Philosophers for Peace Society, Nainital

Mulej M (1974) Dialektična teorija sistemov v ljudski reki (Eng. Dialectical systems theory and the peoples' sayings). Naše gospodarstvo 21(3–4):207–212

Mulej M (1979) Ustvarjalno delo in dialektična teorija sistemov (Eng. Creative work and the dialectical systems theory). Razvojni center, Center

Mulej M (2007a) Systems theory – a worldview and/or a methodology aimed at requisite holism/realism of humans' thinking, decisions and action. Syst Res Behav Sci 24(3):347–357

Mulej M (2007b) Inoviranje navad države in manjših podjetij (Eng. Habits innovation of the country and the smaller businesses). Univerza na Primorskem, Fakulteta za management, Koper

Mulej M (2009) Lack of requisitely holistic thinking and action – a reason for products to not become winners. In: Rebernik M, Bradač B, Rus M (eds) The winning products. Challenges of developing profitable products and services. Proceedings of the international conference on entrepreneurship and innovation PODIM, pp 161–176, IRP Institute for Entrepreneurship research, Maribor, 25th and 26th Mar 2009

Mulej M, Kajzer S (1998) Ethic of Interdependence and the Law of Requisite Holism. In: Rebernik M, Mulej M (eds) Proceedings of STIQE. ISR, Maribor, pp 56–67

Mulej M, Mulej N (2006) Innovation and/by systemic thinking by synergy of methodologies: "Six Thinking Hats" and "USOMID". In: Trappl R (ed) Cybernetics and systems 2006: Proceedings of the eighteenth European meeting on cybernetics and systems research, held at the University of Vienna, Austria, 18–21 April 2006. Austrian Society for Cybernetic Studies, Vienna, pp 416–421

Mulej M, Prosenak D (2007) Society and economy of social responsibility – the fifth phase of socio-economic development? In: Hrast A, Mulej M, Knez-Riedl J (eds) Družbena odgovornost 2007. Zbornik povzetkov 2. IRDO conference o družbeni odgovornosti (Eng. Social responsibility 2007. Proceedings of the second IRDO conference on social responsibility). IRDO Institute for Development of Social Responsibility, Maribor

Mulej M, Espejo R, Jackson M, Kajzer S, Mingers J, Mlakar P et al (2000) Dialektična in druge mehkosistemske teorije (podlaga za uspešen management) (Eng. Dialectical and other soft-system theories (The basis for successful management)). University of Maribor, Faculty of Economics and Business, Maribor

Mulej M, Kajzer Š, Potočan V, Rosi B, Knez-Riedl J (2006) Interdependence of systems theories – potential innovation supporting innovation. Kybernetes 35(7/8):942–954

Mulej M, Potočan V, Ženko Z, Prosenak D, Hrast A (2007) What will come after the investment, innovation and affluence phases of social development? In: Sheffield J, Fielden K (eds) Systemic development: local solutions in a global environment: proceedings of the 13th Annual Australia and New Zeeland Systems (ANZSYS) Conference, Auckland, New Zealand ISCE, Goodyear, 2–5 Dec 2007

Mulej M, Fatur P, Knez-Riedl J, Kokol A, Mulej N, Potočan V et al (2008) Invencijsko- inovacijski management z uporabo dialektične teorije sistemov (podlaga za uresničitev ciljev Evropske unije glede inoviranja) (Eng. Invention-innovation management, using the dialectical systems theory (The basis for achieving the objectives of the European Union's innovation)). Korona plus. D.o.o. Inštitut za inovativnost in tehnologijo, Ljubljana

Mulej M, Potočan V, Ženko Z (2009a) Kriza 2008 – posledica neoliberalncga zancmarjanja primerno celovitega obnašanja glede inoviranja (Eng. The crisis of 2008 – the result of neoliberal neglecting of complete appropriate behaviour concerning innovation performance). University of Maribor, Faculty of Economics and Business, Maribor

Mulej M, Ženko Z, Potočan V, Božičnik S, Hrast A, Štrukelj T (2009b) Nujnost in zapletenost ustvarjalnega sodelovanja za inoviranje kot pot iz krize 2008 (Eng. The urgency and complexity of the creative collaboration for innovation as the way out of crisis 2008). In: Hrast A, Mulej M (eds) Social responsibility and current challenges 2009, work – a bridge to cooperation and coexistence of generations The 4th International Conference. IRDO Institute for development of Social Responsibility, Maribor

Mulej M, Ženko Z, Potočan V, Božičnik S, Hrast A, Štrukelj T (2009c) Kriza – povod za inoviranje planiranja in vodenja v smeri k družbeni odgovornosti (podjetij) ter zadostni in potrebni celovitosti obnašanja ljudi (Eng. The Crisis – the cause for innovation of planning and managing into the direction of social responsibility (of companies) and the sufficient and necessary integrity of people's behaviour). Društvo ekonomistov Maribor, Maribor

Mulej M, Božičnik S, Potočan V, Ženko Z, Hrast A (2009d) Social responsibility as a way of systemic behaviour and innovation leading out of the current socio-economic crisis. In: Allen TFH (ed) 53rd Meeting of the international society for the system sciences, Program and abstracts: making liveable, sustainable systems unremarkable, pp 94–95. Brisbane, AUS, Hosted by the School of Integrative Systems and ANZSYS. University of Queensland, Brisbane, 12–17 July 2009

Mulej M, Božičnik S, Čančer V, Hrast A, Jere T, Jurše K et al (forthcoming) Dialectical systems thinking and the law of requisite holism. ISCE Publishing, Goodyear, Arizona

Nixon B (2004) Speaking plainly – a new agenda for the 21st Century. In: Crowther D, Caliyurt KT (eds) Stakeholders and social responsibility. ANSTED University, Penang

Nussbaum B, Berner R, Brady D (2005) Special report. Get creative! How to build innovative companies. And: a creative corporation toolbox. Business Week 8/15(August):51–68

Nystrom M (2009) The coming golden age. PROUT 20(4):19–22

18 Solutions to Threats and Risks for the National Security of Slovenia

Pavkov M, Kulić S (2009) Science in the culture of peace and development of freedom and democracy. In: Martin G, Murphy P (eds) 12th International conference of international philosophers for peace, July 2–8, 2009, Nainital, India. International Philosophers for Peace Society, Nainital

Pivka M, Mulej M (2004) Requisitely holistic ISO 9000 audit leads to continuous innovation/improvement. Cybern Syst 35(4):363–378

Porter M (1990) The competitive advantage of nations. Basics Books, New York

Porter ME, Kramer MR (2006) Strategy & society. Harv Bus Rev 84(12):1–17

Potočan V, Mulej M (eds) (2007) Toward an innovative enterprise. University of Maribor, Faculty of Economics and Business, Maribor

Prosenak D, Mulej M (2008) O celovitosti in uporabnosti obstojecega koncepta druzbene odgovornosti poslovanja (Eng. About holism and applicability of the existing concept of corporate social responsibility). Naše gospodarstvo 54(3/4):10–21

Prosenak D, Mulej M, Snoj B (2008) A requisitely holistic approach to marketing in terms of social well-being. Kybernetes 37(9/10):1508–1529

Quinn F (2006) Crowning the customer. How to become customer-driven. The O'Brien Press, Dublin

Šarotar-Žižek S, Maučec M (2009) Družbena odgovornost podjetja in osebni razvoj posameznika. (Eng. Social responsibility of the company and personal development of an individual). In: Hrast A, Mulej M (eds) Social responsibility and current challenges 2009, work – a bridge to cooperation and coexistence of generations. The 4th International conference. IRDO Institute for development of Social Responsibility, Maribor

Shankar P (2009) The earth as a female deity. In: Martin G, Murphy P (eds) 12th International conference of international philosophers for peace, July 2–8, 2009, Nainital, India. International Philosophers for Peace Society, Nainital

Shaz R (2009) Calling for an alternative world system. In: Martin G, Murphy P (eds) 12th International conference of international philosophers for peace, July 2–8, 2009, Nainital, India. International Philosophers for Peace Society, Nainital

Singh RP (2009) Managing economic upheaval in world governance. In: Martin G, Murphy P (eds) 12th International conference of international philosophers for peace, July 2–8, 2009, Nainital, India. International Philosophers for Peace Society, Nainital

Sinha RR (2009) Basics of world harmony and peace. In: Martin G, Murphy P (eds) 12th International conference of international philosophers for peace, July 2–8, 2009, Nainital, India. International Philosophers for Peace Society, Nainital

Škafar B (2006) Inovativnost kot pogoj za poslovno odličnost v komunalnem podjetju (Eng. Innovation as a precondition for business excellence in the sanitation company). University of Maribor, Faculty of Economics and Business, Maribor

Stern N (2006) The stern review. The economics of climate change. www.hmtreasury.gov.uk/independent_reviews/stern_review_economics_climate_change/sternreview_index.cfm. Accessed 11 Mar 2007

Stern N (2007) Economics of climate change. Cambridge University Press, Cambridge

Štibler F (2008) Svetovna kriza in Slovenci Kako jo preživeti? (Eng. World crisis and slovenes. How to survive it?). Založba ZRC, Ljubljana

SZK (2007) 16. konferenca Slovenskega združenja za kakovost: Kakovost, inovativnost in odgovornost. Zbornik (Eng. 16th Conference of the Slovenian society for quality: quality innovation and accountability. Conference proceedings). Slovensko združenje za kakovost, Ljubljana

Targowski A (2009) How to transform the information infrastructure of enterprises into sustainable, global-oriented and to monitor and predict the sustainability of civilization. In: Cruz-Cunha MM, Qintela Varajào JE, Martins do Amaral LA (eds) Proceedings of the CENTERIS 2009, Conference on ENTERprise Information Systems, pp 17–28., IPCA – Instituto Politécnico do Cavado e do Ave, and UTAD – Universidade de Tras-os-Montes e Alto Douro, in Ofir, Portugal,7–9 Oct 2009

Taylor G (2008) Evolution's edge: The coming collapse and transformation of our world. New Society, Gabriola Island

Toth G (2008) Resnično odgovorno podjetje (Eng. Truly responsible company). GV Založba and Društvo komunikatorjev Slovenije, Ljubljana

Wilby J (ed) (2009) ISSS 2009. Making liveable, sustainable systems unremarkable. In: Proceedings of the 53rd annual meeting. International Society for Systems Sciences and ANZSYS at University of Queensland, Brisbane

Ženko Z, Mulej M, Potočan V, Rosi B, Mlakar T (2008) A model of making theory as invention to become an innovation. In: Mulej M, Rebernik M, Bradač B(eds) STIQE 2008. Proceedings of the 9th international conference on linking systems thinking, innovation, quality, entrepreneurship and environment, University of Maribor, Faculty of Economics and Business, Maribor – Institute for Entrepreneurship and mall Business Management, and SDSR, Maribor, pp 145–152

Chapter 19
Uncertainty in Quantitative Analysis of Risks Impacting Human Security in Relation to Environmental Threats

Katarína Kampová and Tomáš Loveček

Abstract There is a close interdependency of nature and society, which can be perceived as aspects of different risks impacting human security. The purpose of this chapter is to present a systematic approach to the process of risk analysis associated with human security, focusing on the processing of uncertainty within this process. The risks related to human security are generally referred to as social risks. Various types of uncertainty present within the process of quantitative analyses of social risks are identified in this chapter. Furthermore, it defines the sources of the uncertainty and describes an approach to perception of the uncertainty with regards to the nature of social risks related to environmental threats. This approach allows the creation of a fundamental theoretical framework of a quantitative model of social risk, which is a precondition of the social risks' analysis describing and processing the existing uncertainties. There are also different quantitative methods described, which supplement the outlined framework and prove its usability. Such methods facilitate the use of a designed approach within practical applications of the social risks research and contribute to increasing levels of human security.

19.1 Environmental Aspects of Human Security

Environmental security can affect people and their lives anywhere and at anytime. The relationships between the environment and human security are certainly close and complex. A great deal of human security is tied to peoples' access to natural resources and vulnerabilities to environmental change; and a great deal of environmental change is directly and indirectly affected by human activities and conflicts

---

K. Kampová (✉)
Faculty of Special Engineering, University of Žilina, Žilina, Slovakia
e-mail: katarina.kampova@fsi.uniza.sk

T. Loveček
Faculty of Special Engineering, University of Žilina, Žilina, Slovakia

G. Meško et al. (eds.), *Understanding and Managing Threats to the Environment in South Eastern Europe*, NATO Science for Peace and Security Series C: Environmental Security, DOI 10.1007/978-94-007-0611-8_19, © Springer Science+Business Media B.V. 2011

(Clark et al. 2003). The interdependencies of nature and society originate in the globe's life-supporting eco-systems generating water, food, and air, which are necessary for human survival, well-being and productivity (Mayers 2004); in other words, aspects of human security. In this sense, people and their social relationships become the subjects that are to be secured from impacts of the risks present within the environment, in which people live.

Environmental security has been recognised as the key factor for social security, economic growth, and prosperity (Ganoulis 2007) and thus environmental preservation (water, air, soil, ecosystems, and biodiversity) is one of the main preconditions to ensure human security. Therefore, the efforts to protect nature and to better the lives of people are mutually dependent and will fail unless they are carried out simultaneously and systematically.

The interdependency of people and their environment can be illustrated by recent floods in Central Europe. This example shows the extent of human vulnerability and the intensity of risks impacting people's everyday life. The relation between nature and society can be seen not only in aspect of threats, but also as an opportunity for positive change (Clark et al. 2003), which leads to prevention or at least reduction of the adverse consequences on human society. Such change, however, requires the risk management framework based on conceptual approach to the human security.

19.2 Concept of Human Security

Besides national security and technical security, the complex understanding of security has recently emphasised the threats impacting humans (Škvrnda 2003). This dimension of the conceptual understanding of security is commonly referred to as a human security. The concept of human security is derived from the globalization thinking school, which recognizes the whole range of new problems associated with security as a result of complex globalization processes (Procházková and Šesták 2007). The increasing interconnectivity within trade, finance, technology, communications, and population mobility has created impacts on citizens across the globe, which may be difficult or impossible for states to regulate.

Human security introduces a number of new elements to the traditional security paradigm. The basic objective of human security is to provide protection to people rather than protection of territory. Thus, human security focuses on individuals and not only on nations (Kirdar and Silk 1995). Human security considers the dynamic system with processes and situations that are subject to change during certain periods of time and as such is uncertain. On the contrary, traditional security draws attention to structural interpretation of threats by executive branches of state.

The concept of human security was first systematically elaborated and gained wide recognition from the UNDP's Human Development Report in 1994, where human security has been explained as an individual feeling of safety without fear and deficiency. The report has characterized human security by four essential attributes (UNDP Human Development Report 1994):

19 Uncertainty in Quantitative Analysis of Risks Impacting Human Security 351

- *Human security is a universal concern.* This means that there are many threats (among which the environmental threats play important role) impacting people regardless of the place they live. Even though the intensity of these threats may differ depending on various factors, they are real and present all around the world.
- *The components of human security are interdependent.* The threats associated with human security are not isolated. They are related to various types of social processes within the existing environment and their consequences are often not confined within national borders.
- *Human security is easier to ensure through early prevention rather than later intervention.* When reducing the intensity of threats and their impacts, it is more costly to utilize repressive measures rather than preventive ones.
- *Human security is people centred.* Human security is assessed by people's perception of how they live and feel about their safety, freedom and possibilities of social opportunities.

However, the definition of human security has proven to be elusive since the concept of human security was first introduced. The definitional problem lies mainly in determining the boundaries of human security. It is not clear where human security begins and ends (Michel 2005). Different organizations have adopted different narrow or broad definitions of human security depending on their purpose (Human Security Report 2005). The basic difference of various definitions is apparent in the way they interpret and include different threats received by people. There are two main aspects noticeable when describing human security. The first one focuses on chronic threats such as hunger, diseases or poverty. The second one outlines human security as a way of protection from sudden disruptions in the pattern of daily life (Human Development Report 1994). The latter is the main reason of introducing the uncertainty considerations into the concept of human security. Thus, for the purpose of this chapter, human security will be considered in context of sudden and unpredictable events impacting everyday lives of people.

19.3 Social Risks Perception

In general, the term "security" often comes along with the term "risk". Security is commonly interpreted as a situation with risk minimized to an acceptable level (Šimák 2006). This relation, however, does not provide an explanation as to what risk actually means.

Risk is a part of almost every human activity and hence it is frequently understood intuitively. Situations in which people perceive risk have certain common elements. The first one is that people do not know what will happen. The second one is that personal interests are exposed to consequences in such situations (Holton 2004). Thus, there are essentially two components needed for risk to exist – uncertainty and consequences.

Risks associated with human security can be generally called social risks. The specific characteristic of these risks is that their consequences harm individual people of society as well as the society as a whole (Kampová 2008a). Thus, impacts as an important dimension of social risk define two different forms of social risk – individual and societal. The environmental threats represent a typical example, in which the consequences impact people personally, but they also have the societal dimension.

The individual form of social risk contains elements of subjectivity based on perception of social risk through the harm potentially implied to people personally. The level of individual form of risk might be as different for each individual as it is specific. The intensity of individual form is determined by different factors, which can be generally classified into three categories: personal, institutional and cultural (Jaaber and Dasgupta 2006). The individual form provides insight into complexities of public perception of social risks, but it does not allow the analysis of these risks, as the meaningful quantification of social risk can be only made within the societal form.

Societal form of social risk represents the public perception of impacts on individual interests exposed to the risk. It provides capabilities for quantification of social risk, as a basic precondition of effective social risk analysis. However, the process of social risk analysis is affected by the uncertainty complicating this process.

19.4 Uncertainty and Social Risk Analysis

Uncertainty is an essential ingredient of every risk including social risks. Uncertainty expresses the simple truth that people do not know the future. One might be aware of specific threats, but he or she does not know what will happen, when it will happen and what actual size of impact it will have. Social risk naturally contains uncertainty. This uncertainty chiefly comes from the understanding of human security in respect to sudden disruptions to pattern of people's daily life. In other words, the disruption represents the impact of social risk and the suddenness represents its uncertainty.

Probability is often used as a metric of uncertainty, but its usefulness is limited. At best, probability quantifies perceived uncertainty (Holton 2004). Therefore, the social risk cannot be assessed statically. The correct and balanced evaluation of the social risk requires studying particularly the dynamics of social risk trends and changes as well as the relations and dependencies between various factors and individual conditions. All of these circumstances comprise the environment of social risk, which is the main subject of social risk analysis.

Social risk analysis is based on the societal form of social risk and represents public perception and societal dimensions of individual factors (Kampová 2008a). Results of the analysis are explicable at societal level and thus the social risk analysis can be utilized as a support tool of the decision-making process, which is necessary to ensure a desired level of human security. The critical part of decision-making is to have sufficient and valid information. In order to provide required information, the social risk analysis depicts the context of social risk and its environment. At a particular level of abstraction, the environment of social risk is described by individual quantities that influence the existence and intensity of analyzed risk. In general, the risk environment can be designated as a set of parameters $x = \{x_1, x_2, \dots, x_n\}$. The social risks r is then described by its determining factors in form of function $r = f(x)$ (Aven 2003).

The basic problem of solving this functional relationship is that the true values of parameters x are unknown at the time of the analysis. This uncertainty is caused by insufficient knowledge, inaccurate information or by natural variability of the parameters. To understand the possibilities of quantification of this uncertainty, it can be divided into two groups. The first group is represented by stochastic uncertainty and the second one by knowledge uncertainty (Tichý 2006). Stochastic uncertainty denotes the natural variability and random character of parameters and it is represented by randomness of values in the statistic sample. Most uncertainty of environmental aspects of human security can be expressed by this type of uncertainty. On the other hand, the knowledge uncertainty is caused by insufficient knowledge or lack of information about an analyzed phenomenon or object. The concept of probability can be used to express uncertainty in both cases. Difference is in the way of its interpretation.

Stochastic uncertainty of parameter x_n can be interpreted through the random variable X_n. For instance, the parameter x_n represents a number of events n occurring in a fixed period of time distributed by Poisson distribution with mean λ (for example the number of thunderstorms of specific size in selected territory). Then, in a theoretical case of an infinite number of observations of actual parameter values, the distribution of statistical sample will follow the theoretical probability distribution of random variable X_n and arithmetic mean of the sample will equal λ. Although it is obvious that event n is occurring randomly, it might not be clear how randomly it occurs. In other words, the parameters the distribution of variable X_n follows are unknown and so is the true value of number of times the event n will occur. Such uncertainty is called knowledge uncertainty. This type of uncertainty can be expressed through subjective evaluation of the probability by different expert methods or methods of determination of probabilistic behaviour of uncertain parameter.

19.5 Social Risk Model and Uncertainty Propagation

In order to explain the relation of environmental threats and human security quantitatively within the process of social risk analysis it is necessary to introduce a quantitative model of social risk. This model is represented by the functional relationship between the analyzed risk and its determining factors, which have already been mentioned. This approach considers three domains that have to be dealt with – the risk, input parameters and function itself. All three categories are sources of specific types of uncertainty. Therefore, it is possible to identify three main groups of uncertainty in the process of social risk analysis (Abrahamsson 2002): (1) parametric uncertainty; (2) model uncertainty; and (3) risk uncertainty.

The parametric and risk uncertainties are based on unknown true values of analyzed quantities. This type of uncertainty is a combination of knowledge based and stochastic uncertainty. Individual quantities representing the input parameters and resulting risk naturally contain a certain degree of randomness (National Research Council 2000). This randomness is a source of stochastic uncertainty and is chiefly determined by the individual perception of environmental threats associated with human security and various factors influencing the size of these threats interpreted at societal level. On the other hand, knowledge uncertainty is caused by an approach implemented in quantitative risk analysis, which is based on the use of historical data. These data can be inaccurate or inappropriate. It seldom happens that a sufficient volume of data has been collected for a complete analysis of a particular area. The parametric uncertainty is commonly reduced by estimated values of parameters. The estimation is commonly prepared by statistic methods based on available data sample and respects the probabilistic behaviour of an estimated parameter.

Model uncertainty is a result of a specific feature of every model – an abstraction. Each model, whether it is an abstract model or a mathematical one, is a simplification of reality and does not cover all relations and circumstances of modelled environment. The model abstracts from particular properties of analyzed risk environment with accordance to specified objectives of analysis. The model uncertainty can be generally reduced by using several parallel models and thus increase the reliability of results (Abrahamsson 2002).

The connection between the uncertainty of risk factors and social risk itself is referred to as uncertainty propagation. The uncertainty propagation describes how the uncertain values of input parameters influence the uncertain value of analyzed social risk through the associated risk model. The uncertainty propagation can be depicted as follows (Fig. 19.1).

In general, the uncertain parameter x_n is represented by random variable X_n, which value $P[X_n < x_n]$ is defined by probability distribution $f_n(x_n)$. This distribution defines the stochastic behaviour of parameter x_n and quantifies its uncertainty. The model uncertainty is represented by function $f(x)$, which models the propagation of uncertainty from input parameters to a resulting risk. The risk uncertainty is defined by probability distribution $f_r(r)$. There are a number of quantitative methods helping to process uncertainty within the social risk analysis (Vose 2008).

19.6 Quantitative Methods of Social Risk Analysis

The safety of people is not only the object of practical steps, but also of theoretical elaboration of these issues and scientific research of the essence of security itself and its methods and tools by which it is being achieved (Šimák and Ristvej 2009).

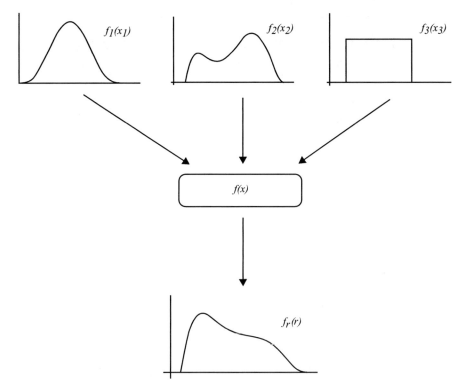

Fig. 19.1 Uncertainty propagation in social risk model (Kampová 2008a)

Therefore, in order to reveal the environmental threats and opportunities related to human security, it is necessary to introduce a systematic approach to quantitative analysis of social risk, together with methods applicable within this analysis.

The process of risk analysis typologically belongs to the scope of risk management. In general, the risk management can be perceived as a logical and methodological way of defining the context of any activities, functions or processes, identification risks, their analysis, assessment, reduction and continuous monitoring, which allows the minimizing losses and maximizing opportunities (Šimák 2006). The quantitative analysis of social risks is thus part of the risk management framework, which can be used to identify and assess the environmental threats associated with human security, to minimize their impacts and to utilize the opportunities provided by the nature.

In more detail, the quantitative analysis of social risk consists of several steps, which are based on the model of social risk. Technically, the objective of the analysis is to create and process such a model. Therefore, the following steps are necessary to be carried out (Fig. 19.2).

Firstly, a social risk to be analysed has to be selected. This step is highly dependent on purpose of the analysis and requires a complex comprehension of the analysis'

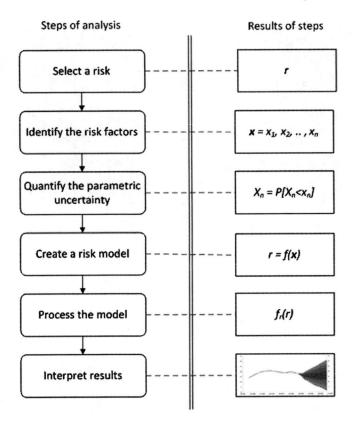

Fig. 19.2 Steps of social risk analysis

context and the risk environment. Then the appropriate factors representing the environment of the risk have to be identified. These factors become input parameters of social risk model. When the analysis is focused on the environmental threats related to human security, naturally more attention is paid to environmental factors of the analysed risk. In particular cases, this can lead to increase of the risk uncertainty, which cannot be explained by any of selected factors and thus to inaccurate results of risk analysis. Therefore, it is necessary to understand the whole context (e.g. social and technical aspects) of the social risk and to incorporate it into the analysis. The next step is defining probabilistic behaviour of the selected input model parameters. In this step, the parametric uncertainty is quantified by describing the randomness of parameters by means of probability. In the next step, the model of social risk itself is created in a form of functional relationship between input parameters and the risk. Then the model is processed and the results acquired. The results describe the uncertainty of analysed risk explained by input model parameters. The final step is to interpret the results with accordance to the purpose of the analysis.

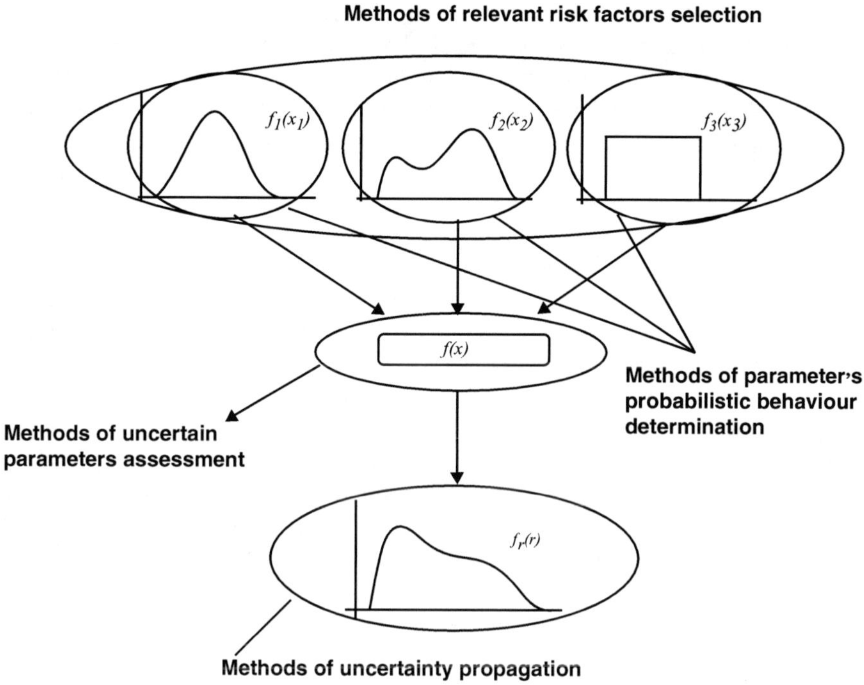

Methods of relevant risk factors selection

Fig. 19.3 Methods of social risk analysis

The outlined approach to the quantitative analysis of social risks provides several groups of methods, which help to understand the context of human security. There are basically four categories (Hnilica and Fotr 2009):

1. Methods of relevant risk factors selection
2. Methods of parameter's probabilistic behaviour determination
3. Methods of uncertain parameters assessment
4. Methods of uncertainty propagation

These categories can be employed in different stages of social risk analysis. The scope of the methods and sequence of their utilization are depicted on Fig. 19.3.

19.6.1 *Methods of Relevant Risk Factors Selection*

The purpose of the relevant risk factors selection methods is to determine the significance of input model parameters. In general, there are two ways to specify the relevance of parameters defining the model – expert assessment or sensitivity analysis (Šimák 2006).

Significance of social risks factors defined by the expert assessment is based on a subjective judgment of various social risk factors through the estimation of factor's occurrence probability and evaluation of factor's contribution to negative impacts of modelled social risk. There is no general expert technique, which can be used to acquire a definite set of relevant parameters. However, the environmental aspects of social risk commonly have the nature of risk and thus it is possible to describe the randomness of their occurrence and the size of their impacts. In such cases, the modified risk matrix method is applicable in order to determine the priority factors of social risk covering various environmental threats. The basic principle of this method is to locate considered factors into two-dimensional matrix. The position of a factor in the matrix depends on the expert evaluation of both factor's dimensions – the probability and the consequences.

Sensitivity analysis is based on the explicit representation of the effect various changes in the input parameters might have on the social risk (Marek 2004). Sensitivity analysis is often performed by changing the value of one input parameter at a time, while maintaining all others at their nominal value, and then assessing the relative impact each change has on the model output. In this way, sensitivity can be regarded as a simple measure of uncertainty importance (Abrahamsson 2002). The downside of this approach is that the sensitivity analysis does not provide accurate results, when it is performed on related and interdependent factors.

19.6.2 Methods of Parameter's Probabilistic Behaviour Determination

Methods of parameter's probabilistic behaviour determination are focused on defining the stochastic behaviour of each selected risks factor. As it has been mentioned, the true values of input parameters of a social risk model are unknown. In order to be able to deal with this uncertainty in the process of social risk analysis, it is necessary to describe probabilistic aspects of each parameter. In simple cases, the parameter's stochasticity can be described by probability distribution. There are various statistical tests to determine the distribution of probability of available model parameter's data sample (Vose 2008). The precondition to employ such approach to describe parametric uncertainty is that the values in the data sample are independent. Technically, the observable quantities of the parameter must be independent in space and in time of observation (Gregorich 2001), which is the case of parameters modelling events characterized by natural randomness (e.g. floods, storms, cyclones, etc.). In order to determine the correlation of the data sample values, it is possible to check the autocorrelation coefficient implying the interdependencies of values (Schabenberger and Pierce 2002).

However, many parameters modelling environmental threats cannot be described by simple function of probability distribution, because the values of modelled quantities are mutually dependent. A typical example is time series modelling the trend of particular risk factor (e.g. pollution of air, water or soil). In such cases, it is

19 Uncertainty in Quantitative Analysis of Risks Impacting Human Security 359

necessary to define the probabilistic behaviour of parameter in a more complex manner. For example, the central prediction function can be determined for particular parameter from its actual historical data and the uncertainty of predicted parameter's values is modelled by probability distributions of estimated standard deviations representing the volatility of the central prediction function.

19.6.3 Methods of Uncertain Parameters Assessment

The methods of uncertain parameters assessment are utilized for the design and creation of a model itself (Kampová 2008b). These methods help to quantify the contribution of each selected social risk factor to the intensity of resulting social risk. Among others, the correlation analysis and regression analysis (Vose 2008) belong to this category of methods. By using these methods, the quantitative relationship between input parameters and social risk can be defined. Besides the functional relationship between social risk and its factors, it is also important to assess the measure of how well the social risk is likely to be predicted by defined model. Such measure is provided by the coefficient of determination, which shows the proportion of variability in a data set that is accounted for by the statistical model (Steel and Torrie 1960). Thus, the coefficient of determination implies the reliability of the social risk model and determines what portion of social risk uncertainty can be explained by parameters selected in regression relationship, which effectively influence the way the uncertainty of social risk factors is propagated to social risk.

19.6.4 Methods of Uncertainty Propagation

The methods of uncertainty propagation are the key category of methods within the quantitative analysis of social risks. By employing these methods in the process of social risk analysis, the way in which the present uncertainty of input parameters is transmitted to the modelled social risk can be defined. These methods describe and quantify the uncertainty of social risk and determine its estimated values. In general, there are two types of methods of uncertainty propagation (Abrahamsson 2002): (1) exact methods; and (2) numerical methods.

The basic principle of both exact and numerical methods is that they describe a specific problem by a certain functional relationship or its approximation. The difference is in the way that the function is described. The exact methods use the description by symbolic algebra and numerical methods uses enumeration of functional values.

The exact methods manipulate with mathematical representation of the problem through symbolic algebra, in which the symbols represent the numerical values of quantities presented in a model. The use of the exact methods for problems of probability propagation commonly employs the approximation of function based on

Taylor series expansion, which provides a way of expressing the uncertainty in the function output in terms of uncertainty of inputs (Henrion et al. 1990). However, the exact methods are constrained by the computational complexity and therefore the practical application of the exact methods within the social risk analysis is narrowed to specific simple cases, such as linear combination of variables with normal probability distribution.

The goal of the numerical methods is to provide techniques capable of solving complex problems by determining the approximate solution. The main idea behind the numerical methods is that they are focused on the numerical values of observable quantities rather than symbolic algebraic description of function. Thus, the principle of the numerical methods is that the functional values of particular quantity is not defined by the function itself (as the exact methods do), but they provide an accurate estimation of functional values, which are further processed by the computational procedure of the numerical method. One very popular numerical method applicable also within the quantitative analysis of social risks is called the Monte Carlo method.

The main strengths of the Monte Carlo method are its robustness and simplicity (Fotr 2006). This method is based on generating random numbers following determined probability distributions of input model parameters. The generated numbers represent the values of quantities. The result of the Monte Carlo procedure is a data sample with estimated output values. These values define a random variable, which represents the modelled social risk and quantifies its uncertainty.

19.7 Representation of Uncertainty

The result of quantitative approach to social risk analysis and the research of environmental threats impacting human security is the enumeration of values of the risk estimated by the model covering the substantial context of social risk. The essential part of risk management framework focusing on ensuring human security is an appropriate interpretation of the analysis result, which is the main precondition of eligible decision-making. Such interpretation, however, requires the understating of uncertainty of the result. Therefore, it is necessary to suitably represent this uncertainty.

There are many ways of how to represent and interpret the uncertainty. One very simple, but effective method to represent an uncertainty is a graphical depiction in the form of a fan chart. The fan chart method was first introduced by Bank of England and the National Institute of Economic and Social Research in 1996 (Wallis 2004). The purpose of this method was to interpret the uncertainty surrounding the central projections of inflation they published. Thanks to its graphical appeal and easily understandable form, the fan charts can be effectively employed to explain the result of quantitative social risk analysis.

Fan chart technically describes the estimation of a confidence interval in which the resulting measure of risk will be found. The graph represents an extension of

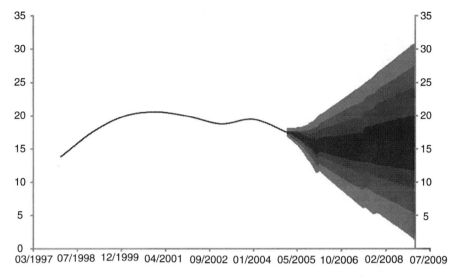

Fig. 19.4 Social risk uncertainty represented by a fan chart (Kampová 2008a)

the point forecast by an assessment of the forecast error, this being represented by the width of the interval (Gavura and Oparty 2005). The basis of the graph is a central projection represented by the line in the middle of the graph. This is the single most likely course of the risk for the given period based on current knowledge and judgment. The uncertainty of the analysis results is illustrated by gradually spreading fan surrounding the central projection (Fig. 19.4).

The fan chart is defined by the resulting data set of social risk, which is formulated by the quantitative model of the risk. The different coloured bands represent the measure of social risk uncertainty. The width of the bands depicts how likely the true values of the risk will be found in particular interval corresponding to specific confidence level. These intervals can be determined as the percentile of the resulting data set.

19.8 Conclusion

Human security is a main precondition of satisfactory quality of human life. Nevertheless, there are many environmental threats, which potentially disrupt the pattern of an individual's everyday life. A chance of exposure of individual and societal interests to the adverse consequences of future events is referred to as social risk. In order to analytically support the decision-making process, which is necessary to establish the required level of human security, the quantitative analysis of social risk can be employed. The serious problem present within this analysis is the problem of uncertainty. This chapter has introduced a conceptual approach, which describes the probabilistic background of uncertainty, identifies its sources and points out the brief overview of methods, which can be used to deal with uncertainty. The outlined approach helps to address the challenge of solving the issues of uncertainty in quantitative analysis of social risks in relation to environmental threats and thus improves our attempts at ensuring human security.

References

Abrahamsson M (2002) Uncertainty in quantitative risk analysis – characterisation and *methods of treatment*. Department of Fire Safety Engineering, Lund University, Lund. http://luur.lub.lu.se/luur?func=downloadFile&fileOId=642162. Accessed 25 Apr 2010

Aven T (2003) Foundation of risk analysis: a knowledge and decision-oriented perspective. Wiley, Chichester

Clark CW, Raad FD, Sanjeev K (2003) From the environment and human security to sustainable security and development. J Hum Dev 4(2):289–313

Fotr J (2006) Manažerské rozhodování. Postupy, mctody a nástroje (Eng. Decision-making in management. Procedures, methods and tools). Ecopress, Prague

Ganoulis J (2007) Environmental and human security in the Mediterranean: new threats and policy for reducing vulnerability. In: Linkov I, Wenning RJ, Kiker GA (eds) Managing critical infrastructure risks. Springer, Dordrecht, pp 51–61

Gavura M, Oparty T (2005) Estimating the probability distribution of inflation forecast. Biatec 8:19–21. http://www.nbs.sk/_img/Documents/biatec/bia05_05/19_21.pdf. Accessed 20 Mar 2010

Gregorich EG (2001) Soil and environmental science dictionary. CRC Press, Boca Raton

Henrion M, Morgan MG, Small M (1990) Uncertainty: a guide to dealing with uncertainty in quantitative risk and policy analysis. Cambridge University Press, Cambridge, GB

Hnilica J, Fotr J (2009) Applied risk analysis in financial management and investment decision making. Grada Publishing, Prague

Holton GA (2004) Defining risk. Financ Analysts J 60(6):19–25

Jaaber RA, Dasgupta SD (2006) Assessing social risks of battered women. Praxis International. http://praxisinternational.org/files/praxis/files/AssessingSocialRisk.pdf. Accessed 25 Apr 2010

Kampová K (2008a) Identifikácia a analýza rizík v spoločenskej praxi (Eng. Identification and analysis of risks in society). Dissertation thesis, University of Žilina, Žilina

Kampová K (2008b) Uncertain factors in process of social risk analysis. In: Fifteenth international scientific conference devoted to crisis situations solution in specific environment, University of Žilina, Faculty of Special Engineering, Žilina, 2–3 June 2010

Kirdar U, Silk L (eds) (1995) From impoverishment to empowerment. NYU Press, New York, 1995

Marek J (2004) Sensistivity analysis in project risk management. Risk Management CZ. http://www.risk-management.cz/index.php?clanek=45&cat2=1. Accessed 23 Apr 2010

Mayers N (2004) Environmental security: what's new and different? The Hague Conference on Environment, Security and Sustainable Development. Resource document. http://www.envirosecurity.org/conference/working/newanddifferent.pdf. Accessed 28 Apr 2010

Michel J (2005) Human security and social development: comparative research in four asian countries. Arusha Conference: 'New Frontiers of Social Policy'. http://siteresources.worldbank.org/INTRANETSOCIALDEVELOPMENT/Resources/24432-1133801177129/Michel.rev.pdf. Accessed 27 Apr 2010

National Research Council (Committee on Risk-Based Analysis for Flood Damage Reduction) (2000) Risk analysis and uncertainty in flood damage reduction studies. http://www.nap.edu/catalog.php?record_id=9971. Accessed 23 Apr 2010

Procházková D, Šesták B (2007) Lidská bezpečnost (Eng. Human security). KKR, Police Academy of Czech Republic, Prague

19 Uncertainty in Quantitative Analysis of Risks Impacting Human Security 363

Schabenberger O, Pierce FJ (2002) Contemporary statistical models for the plant and soil sciences. CRC Press, Boca Raton

Šimák L (2006) Manažment rizík (Eng. Risk management). Department of Crisis Management, Faculty of special engineering, University of Žilina, Žilina. http://fsi.uniza.sk/kkm/old/publikacie/mn_rizik.pdf. Accessed 20 Mar 2010

Šimák L, Ristvej J (2009) The present status of creating the security system of the Slovak Republic after entering the European union. J Homel Secur Emerg Manage 6(1), Article 20. doi: 10.2202/1547-7355.1443

Škvrnda F (2003) O sociologických aspektoch vytvárania teórie bezpečnosti (Eng. Social phenomenon of security in contemporary society). Police Theory Pract 1:81–89. http://fmv.euba.sk/files/Almanach_2-2007.pdf. Accessed 25 Apr 2010

Steel RGD, Torrie JH (1960) Principles and procedures of statistics. McGraw, New York

Tichý M (2006) Ovládání rizika (Eng. Controlling the risk). C.H. Beck, Prague

United Nations Development Programme (1994) Human Development Report 1994. Oxford University Press, New York. http://hdr.undp.org/en/reports/global/hdr1994/chapters/. Accessed 25 Apr 2010

Vose D (2008) Risk analysis: a quantitative guide, 3rd edn. Wiley, New York

Wallis KF (2004) An assessment of Bank of England and National Institute Inflation Forecast Uncertainties. Natl Inst Econ Rev 189(1):64–71. doi:10.1177/002795010418900107

Chapter 20
Environmental Conflict Analysis

Katarína Jelšovská

Abstract The purpose of this chapter is to analyze environmental conflicts. Environmental conflicts with their character, stakeholders, and especially long-term impacts, may result in serious crises. A precondition for an effective prevention and conflict solution is an examination of its nature and causes, as well as possible consequences of every single environmental conflict. This is utilized in the process of the environmental conflict analysis.

20.1 Introduction

The problem of environmental conflict while very topical today is still not well known. Environmental conflicts are often viewed as a social conflict; a link between environment and security proposes a new view on security and emphasises on treats resulting from environmental degradation. The environment conflict nexus is a subset of "environmental security" – a field of inquiry that seeks to determine whether or not traditional notions of security (which emphasize countering military threats with military power) should be adapted to include threats posed by population growth and diminishing quantity and quality of environmental goods and services (Schwartz and Singh 1999). On the other hand Gleick emphasize that what is required is not a redefinition of international or national security, as some have called for, but a better understanding of the nature of certain threats to security, specifically the links between environmental and resource problems and international behaviour (Gleick 1991).

An examination of the nature, causes, and possible consequences of any conflict is based on the fact that there is an effective prevention and solution for every conflict. This is why we cannot omit a process of the analysis of the given environmental conflict. It is important to point out that many of these environmental conflicts may

---

K. Jelšovská (✉)
Mathematical Institute, Silesian University, Opava, Czech Republic,
Faculty of Special Engineering, University of Žilina, Žilina, Slovakia
e-mail: katarina.Jelsovska@math.slu.cz

result in serious crises. To avoid these crises and to reduce their eventual consequences, the prevention of environmental conflicts or their effective solution requires and leads to the origin of the conflict.

In general, prevention and problem solving tools for crises and crisis events are measured to the level of knowledge and technical possibilities available at the time. It is useful to emphasize the fact that focusing on crisis and crisis events prevention through environmental conflicts elimination is a question for crisis management strategy in the future.

20.2 Environmental Conflict

Many of the theoretical and empirical work regarding the links between environmental change and conflict have been conducted by the Peace and Conflict Studies Program at the University of Toronto (the Toronto Group). This Institute has conducted two studies under the leadership of Thomas H. Dixon: (1) the project on Environment Population and Security, and, (2) the project on Environmental Scarcities, State Capacities and Civil Violence. Another institution doing work in this area is the Centre for Security Studies and Conflict Research in Zurich (the Swiss group). They produced research under the Environment and Conflict Project (ENCOP). Two other institutions involved in research in this area are: the Peace Research Institute in Oslo and the Environmental Change and Security Project at the Woodrow Wilson Centre, in Washington, DC.

Most of the research in the field has been done in the form of case studies, exploring the particular contribution of scarcity as a generating factor of conflicts; not only at a national level but also at subnational levels. So far, the two research teams that stand out as the main contributors, are the group at the University of Toronto under the leadership of Thomas Homer-Dixon, and the ENCOP group directed by Günter Bächer and Kurt Spillmann at the Swiss Federal Institute of Technology in Zurich and the Swiss Peace Foundation in Bern. While both research groups focus on the causal linkage between the environment and violent conflicts, the difference between the two definitions lie in the terms degeneration versus scarcity.

In the words of one of the Swiss group participants, environmental conflict: "is a conflict caused by the environmental scarcity of resource, that means: a human-made disturbance of the normal regeneration rate of a renewable resource" (Libiszewski 1992). Thus, a conflict over agricultural land is an environmental conflict if the land becomes an object of contention as a result of soil erosion or climate change, but not in the case of ordinary territorial or colonial conflict or an anti-regime civil war aiming at the redistribution of land (López 2000). The ENCOP definition only considers environmental disturbance caused by human activities, while excluding natural events in their analysis. The problem is that it is sometime difficult to draw a line between the two.

For the purpose of this contribution environmental conflict as is "conflict of interests caused by environmental scarcity of renewable resources as a result of

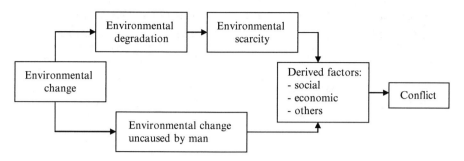

Fig. 20.1 Cause of the environmental conflict (Jelšovská 2008)

environmental degradation or another environmental change in interaction with variety of derived factors" (Jelšovská 2008).

The cause and effect of the environmental conflict may be represented as follows (Fig. 20.1):

The primary difference between the ENCOP definition and the definition mentioned above is in the cause of the environmental conflict. *Environmental change* means change caused either by human activities or by natural events. Environmental change may contribute to conflicts as diverse as war, terrorism, or diplomatic and trade disputes (Homer-Dixon 1991). The term *environmental degradation* is understood as a human-made environmental change having a negative impact on human society (Libiszewski 1992). *Environmental scarcity* is defined by Libiszewski as a: "type of scarcity concerning resources that have traditionally been regarded as plentiful and naturally renewable but are becoming scarce now because of the failure of human beings to adopt sustainable methods of their management" (Libiszewski 1992). Research indicated that environmental scarcities are already contributing to violent conflicts in many parts of the developing world (Homer-Dixon 1994).

The most common environmental factors that lead to conflicts are deforestation, soil erosion, desertification, flooding, and pollution. These environmental *factors*, in turn, combine with population growth and an uneven distribution of resources, to engender and interact with a series of social effects. The most common of those are poverty, relative deprivation, migration, shifts in political power, and state failure or loss of state legitimacy (Schwartz and Singh 1999). These social effects create and reinforce instability. Under certain circumstances, this leads to collective violent action (Hagmann 2005).

Based on the above mentioned definition of environmental conflict, and referring to Spillmann (1995), the following environmental conflict classification is proposed (Table 20.1):

The first type of conflict is caused by the consequences of natural disasters. These environmental changes are not caused by human activities. They are independent of human planning or decision making. The second type of conflict is caused by a conscious government decision in order to benefit the state. The third

Table 20.1 Environmental conflict classification according to environmental changes sources (Spillmann 1995)

Environmental changes type	Uncaused by man	Caused by man – planned, proposed, received	Caused by man – unplanned, undesirable
Environmental changes source	Environmental change caused by a known source which is not a society	Environmental change caused by known source which is a society	Environmental change caused by interconnected sources or unknown man source
	Natural disasters	Large-scale engineering	Cumulative consequences of multitude of smaller measures which are severally useful and suitable
Types of conflicts arising from environmental change	Changes can probably lead to a conflict between impacted groups of society	Changes can probably lead to a conflict between groups which caused environmental damage and groups which suffer a damage	Changes can probably lead to conflicts between groups which fight with effects elimination

type of conflict is also caused by man, but the source of this environmental change cannot be identified, therefore it is very difficult to predict it. A linkage between causes and consequences of those conflicts is still unclear, so a solution to those types of conflicts is the most complicated.

20.3 Instruments of the Environmental Conflict Analysis

The prevention of environmental conflicts seems to be a more effective way of precluding of the above mentioned crises. The instruments for dealing with these crises underline the main difference between prevention and solving environmental conflicts. Nowadays, participation and public ventilation seem to be the most suitable instruments for prevention. For environmental conflict solving, some alternatives as well as authoritative methods can be used. However, currently there exists an opinion that environmental conflict solving by authoritative methods is not effective from the long-term point of view. This is because, very often, at least one stakeholder is not satisfied with the outcome. The environmental conflict then may still exist in some modified form.

Various recommendations, methods, and procedures for prevention and conflict solving (both general and specific) offer certain possibilities for environmental conflict analysis. Considering their particularities, we cannot utilize them completely for prevention and environmental conflict solving. For example, utilization of participation and public ventilation as suitable instruments for environmental conflict prevention is embodied in Aarhus Convention and Environmental Impact Assessment, among others.

20.3.1 Environmental Impact Assessment

The Directive 85/337/EEC – Environmental Impact Assessment (EIA) represents an effective prevention system emerging from the examination and assessment of expected impacts of planned actions on the environment. The aim of the impact assessment is to secure a high standard of environment protection with active public participation. The impact evaluation is addressed by professionals from various fields. Figure 20.2 illustrates the process of Environmental Impact Assessment.

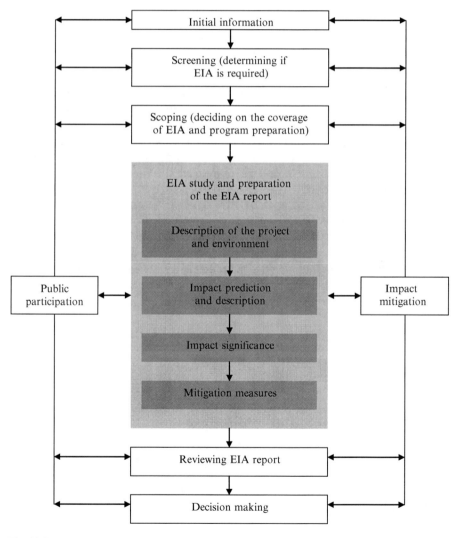

Fig. 20.2 The Environmental Impact Assessment Process (Environmental Impact Assessment 2004)

The EIA process consists of three stages:

1. Screening
2. Scoping – it is made with an aim to identify issues requiring a detailed examination, potential direct or indirect impacts and alternatives which will be further investigated in the EIA Study and measures to avoid or mitigate the effects of important negative impacts
3. Report – The EIA Study and preparation of the EIA Report. The EIA study is a systematic, reproducible and interdisciplinary prediction, identification and evaluation of the impact of the proposed development. Information that is generated during the Study is presented in the EIA Report. This phase of the EIA focuses on several primary tasks: baseline environmental information, identification and forecasting of potential impacts, elaboration on the nature of these impacts, extensive analysis of these impacts, and an assessment of the significance of the impacts regarding their acceptability and the needs for mitigation measures

The environmental impact assessment of projects and concepts is based on a systematic investigation and examination of their possible impacts on the environment. The purpose of the environmental impact assessment is to provide a comprehensive identification, description and evaluation of the expected impact of upcoming projects and concepts on the environment and on public health in all important aspects. The objective of the process is to mitigate the environmental impacts the implementation might have.

One of the reasons for executing the EIA Directive was the arrangement of conditions for the Czech Republic to meet regarding the Convention on Access to Information, Public Participation in Decision-making and Access to Justice in Environmental Matters – i.e., the Aarhus Convention. Its aim is to provide people with the access to information on the environment and give them an opportunity to participate in decision-making in this field (Aarhus Convention 1998).

20.3.2 Aarhus Convention

The Aarhus Convention (1998) is a kind of environmental agreement. The Convention:

- Links environmental rights and human rights
- Acknowledges that we owe an obligation to future generations
- Establishes that a sustainable development can be achieved only through the involvement of all stakeholders
- Links government accountability and environmental protection
- Focuses on interactions between the public and public authorities in a democratic context

The content of the Convention is structured around three pillars (Aarhus Convention 1998):

- Public access to information about the environment. The general public should be entitled, without needing to prove a special interest, to access to information about the state of the environment, of public health and of other factors affecting the environment.
- Public participation in certain environmentally relevant decisions. The Convention provides the mechanisms of public participation, particularly with respect to its time, form and scope.
- Access to courts of law or tribunals in environmental matters. The Convention states that members of the public, meeting the national criteria, shall have access to administrative or judicial procedures to challenge acts and/or omissions by private persons and public authorities which contravene national law relating to the environment.

The Aarhus Convention grants the public rights and imposes on parties and public authorities obligations regarding access to information, public participation and access to justice. It is also planning a new process for public participation in the negotiation and implementation of international agreements.

20.3.3 Risk Management

The creation of environmental conflicts brings various risks for the environment as well as for the society. When analyzing environmental conflicts, the 2004 Australian/New Zealand Risk Management Standard 4360 was used. It provides a generic framework for establishing the context, identifying, analysing, evaluating, treating, monitoring and communicating risk. The standard says that risk analysis can be applied in a variety of sectors and range of subject areas.

Risk management involves managing to achieve an appropriate balance between realizing opportunities for gains while minimizing losses. It is an integral part of good management practice and an iterative process consisting of steps, when undertaken in sequence, enable continuous improvement in decision-making and facilitate continuous improvement in performance (Risk Management Standard AS/NZS 4360/2004).

The main elements of the risk management process, as shown in Fig. 20.3, are the following (Risk Management Standard AS/NZS4360/2004):

- Communicate and consult with internal and external stakeholders as appropriate at each stage of the risk management process and concerning the process as a whole.
- Establish the context (the external, internal and risk management context) in which the rest of the process will take place. Criteria against which risk will be evaluated should be established and the structure of the analysis defined.

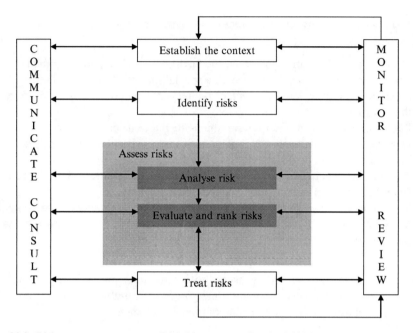

Fig. 20.3 Risk management process (Risk Management Standard 2004)

- Identify risks: where, when, why and how events could prevent, degrade, delay or enhance the achievement of the objectives.
- Analyse risks. Identify and evaluate existing controls. Determine the consequences, likelihood and the level of risk. This analysis should consider the range of potential consequences and how these could occur.
- Evaluate risks. Compare estimated levels of risk against the pre-established criteria and consider the balance between potential benefits and adverse outcomes. This enables decisions to be made about the extent and nature of treatments required and about priorities.
- Treat risks. Develop and implement specific cost-effective strategies and action plans for increasing potential benefits and reducing potential costs.
- Monitor and review. It is necessary to monitor the effectiveness of all steps of the risk management process. This is important for continuous improvement. Risks and the effectiveness of treatment measures need to be monitored to ensure changing circumstances do not alter priorities.

The individual phases of risk management are one of the points of origin for creating the proposal of environment conflict analysis. The standard itself is quite general and does not deal with particularity of the judged system. Due to the sequence chain, the standard is essential for the environment conflict analysis.

20.4 Project of the Environmental Conflict Analysis

In accordance with the above mentioned standards Risk Management, Aarhus Convention and EIA, the following structure of the environmental conflict analysis and a frame content of its individual phases is presented. See Fig. 20.4.

20.4.1 Nature of the Environmental Conflict

To detect the nature of the environmental conflict, it is important to identify:

- A potential conflict (or an already evident conflict if the phase of prevention had been underestimated and the potential conflict became evident)
- The background of the environmental conflict (mutual dependence of the judged background)
- The particularities of the environmental conflict and determination of its characteristic features
- The criteria in comparison with which the consequences of the environmental conflict will be evaluated (to what extent is the public able to accept the environmental degradation)

It is characteristic mainly of the first phase of the environmental conflict analysis that searching for, and identification of, the nature of accrued situation looks to be the most effective, if there is teamwork and a participative approach. An incorrect examination of the accrued problem can influence the prevention or solving of the environmental conflict in a negative way.

20.4.2 Causes of the Environmental Conflict

Identification of the environmental conflict cause is necessary to avoid crises or other environmental conflicts with similar causes. In general, we can divide the causes of the environmental conflict into the following:

- An objective cause – environmental change resulting in a lack of renewable resources
- A subjective cause – interests of the individuals, groups, countries; it is much more complicated from the wider international and world-wide point of view than within a municipality, region, and a country

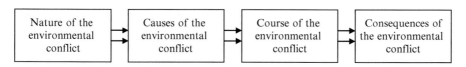

Fig. 20.4 Structure of the environmental conflict analysis (Jelšovská 2008)

Objective and subjective causes of the conflict are not sharply separated and what is more, they are interrelated.

Detecting the cause of an environmental conflict also means to identify the interests or the approaches of the stakeholders. The stakeholders may be the representatives of:

– Countries (if the conflict or its consequences cross the borders)
– State administration
– Municipality
– Non-governmental organizations
– Scientific and research institutions
– Local entrepreneurs
– Public

Stakeholders can be divided into:

– Subjects initiating the conflict
– Subjects impacted by the conflict

Then, these are followed by:

– An identification of interests and positions of the stakeholders
– An interaction between them
– Their mutual perception
– Their perception of the environment

20.4.3 Course of the Environmental Conflict

In this phase, it is the most important to distinguish if it is an environmental conflict prevention or crisis prevention, in case that environmental conflict has been already incurred. Here we talk about these actions:

– Formulating possible scripts how the occurred situation can be developed in time (see Fig. 20.5)
– Choosing methods suitable for determination of the scripts (with regard to the uniqueness of every single environmental conflict it is considered the What If Analysis to be the most suitable. This method has no exact structure, but to apply it, it is necessary to fit the basic concept for a particular purpose – it means for mitigation of the environmental degradation. Another suitable method is the Causal and Consequence Analysis. This method, as for environmental conflicts, has an indispensable role and it is used in almost each phase of the analysis of environmental conflicts. The Event Three Analysis is a usable method as well.)
– Identification of the influencing factors which concur with the environmental degradation and make the conditions for creation of an environmental conflict

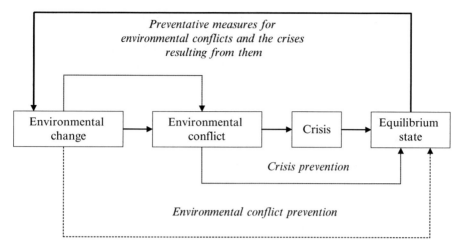

Fig. 20.5 Scripts of environmental conflicts (Jelšovská 2008)

- Identification of an impulse (indicator) changing a potential environmental conflict into an open conflict demonstration (Šimonová and Danihelka 2007)
- Creation of conditions suitable for cooperation between the stakeholders

In the case of an origin of a conflict, this evolves into a discussion which may have a constructive course (the parties negotiate and look for the solutions) or a destructive course (the parties blame and attack each other, use violence, and refuse to find a solution).

20.4.4 Consequences of the Environmental Conflict

When assessing potential consequences it is necessary to do the following:

- Identification of potential environmental, social, economic, political, and other consequences impacting each other and contributing to the tension between the stakeholders. The conflict may be manifested by other consequences and often by other symptoms as well
- Solving of the occurred environmental conflict requires focusing on the conflict solution first, then on the elimination of the cause (Mason and Spillmann 2002). To solve an already existing conflict it is necessary to choose suitable tactics, recognition and elimination of the environmental conflict cause is a matter of strategy
- Proposals of preventative measures (Brainstorming is a suitable method)
- Return to the original (objective) cause of the environmental conflict and a proposal of measures for its elimination

It is typical of environmental conflicts that an environmental change is an objective cause of their origin. Therefore it is necessary to mitigate environmental degradation to prevent environmental conflicts. Although solving the subjective causes results in a conflict solution, it does not form a background for an effective environmental conflict prevention. It is considered the main idea of environmental conflicts and crisis prevention which result from their non-solving.

In Fig. 20.5 there is a schematic representation of the scripts of the environmental conflicts which can be formed as a consequence of an environmental change. These scripts are, inter alia, influenced by the sources of the environmental change and by the speed of the environmental change formation (if it were formed suddenly or gradually).

The given proposal of the structure of the environmental conflict analysis provides us with the basic resources for their examination as well as the management itself. It is a part of crisis prevention, as consequences of unsolved environmental conflicts, focusing on prevention of the origin of the environmental conflicts.

20.5 Conclusion

Environmental conflict analysis is, together with a schematic representation of environmental conflict scripts, usable by potential as well as real stakeholders. It should be useful for experts to reduce undesirable effects on the society as well as general public to increase the environmental awareness and create an opportunity to participate in the environmental conflict management.

References

Aarhus Convention (1998) Resource document. UNECE. http://www.unece.org/env/efe/history%20of%20EfE/Aarhus.E.pdf. Accessed 25 Jul 2010

EIA Directive 85/337/EEC (2010) Resource document. AERS. http://www.aers.rs/FILES/MedjunarodniPropisiISporazumi/Directive%2085-337-EEC%20as%20amended%20by%20Dir%2097-11-EC.pdf. Accessed 25 Jul 2010

Environmental Impact Assessment (2004) Resource document. http://www.le.lt/en/main/atom/PAV_nuclear. Accessed 25 Jul 2010

Gleick P (1991) Environment and security: the clear connection. Bull At Sci 47(3):17–21

Hagmann T (2005) Confronting the concept of environmentally induced conflict. Peace Conflict Dev 6(January):1–22

Homer-Dixon T (1991) On the threshold. Environmental change as cause of acute conflict. Int Secur 16(2):76–166

Homer-Dixon T (1994) Environmental scarcities and violent conflict: evidences from the cases. Int Secur 19(1):5–40

Jelšovská K (2008) Analýza a manažment environmentálnych konfliktov (Eng. Environmental conflict analysis and management). Dissertation thesis, University of Žilina, Faculty of Special Engineering, Žilina

20 Environmental Conflict Analysis

Libiszewski S (1992) What is an environmental conflict? Occasional Paper no. 1. Environment and Conflict Project (ENCOP). Centre for Security Policy and Conflict Research/Swiss Peace Foundation, Zürich/Berne

López A (2000) Environmental change, social conflicts and security in the Brazilian Amazon: exploring the links. Dissertation thesis, Department of Political Science – University of Oslo, Oslo

Mason SA, Spillmann KR (2002) Environmental conflicts and regional conflict management. Centre for Security Studies and Conflict Research, Swiss Federal Institute of Technology ETH, Zürich

Risk Management Standard AS/NZS 4360:2004 (2004) Standards Australia, Australia

Schwartz D, Singh A (1999) Environmental conditions, resources, and conflicts. United Nations Environment Programme (UNEP), Nairobi

Šimonová M, Danihelka P (2007) Typology of combined natural and technological risks and case studies. In: Ochrana území postihnutých ničivými prírodnými pohromami (Eng. Protection of areas affected by destructive natural disasters). Conference Proceedings, Štrbské Pleso, pp 125–131, doi: 978-80-228-1803-2

Spillmann KR (1995) From environmental change to environmental conflict. Centre for Security Studies and Conflict Research. Swiss Federal Institute of Technology ETH, Zürich

About Editors and Authors

About the Editors

Gorazd Meško, PhD, is Professor of Criminology and Dean at the Faculty of Criminal Justice and Security, University of Maribor, Slovenia. He was a visiting scholar at the University of Cambridge (1995) and Oxford (1996, 1999), as well as a visiting professor at Grand Valley State University, Michigan, USA (2000). He conducted a post-doctoral research (OSI-HESP) on crime prevention at the Institute of Criminology, University of Cambridge, UK in 2001. Currently, he is heading a national basic research project entitled "Crimes against the environment – criminological, victimological, crime-prevention, psychological and legal aspects" (2009–2012). He is also involved in international projects on preparation of a WEB portal of colleges of security studies in Europe (eSEC), and a European master programme in urban safety (EFUS, Paris). In addition, Gorazd Meško is a member of the scientific board of the International PhD in Criminology at the Catholic University in Milan, Italy. He also serves as the editor in chief of the Journal of Criminal Investigation and Criminology (orig. Revija za kriminalistiko in kriminologijo) and a member of the editorial board of the Policing – An International Journal of Police Strategies and Management. His research fields are crime prevention and provision of safety/security, policing, fear of crime, and crimes against the environment.

Dejana Dimitrijević, PhD, is an Assistant Professor in environmental security as well as environmental monitoring and analysis at the Department of Environmental Protection, Faculty of Security Studies, University of Belgrade, Serbia. She has a MSc in Clinical Biochemistry (University of Belgrade, 1992) and holds a PhD in novel drug delivery systems and toxicology (University of London, 1998). Her post-doctoral research at the School of Pharmacy, University of London focused at chemical and biological assessment and characterization of nanomaterials (1998–2001). She was involved in several research projects receiving international funding at the University of London. Her main research area is today environmental and health risk assessment of nanomaterials, regulation, and standardization of nanomaterials. She has several scientific publications and conference presentations and is actively involved in the Serbian Standardization Workgroup on

G. Meško et al. (eds.), *Understanding and Managing Threats to the Environment in South Eastern Europe*, NATO Science for Peace and Security Series C: Environmental Security, DOI 10.1007/978-94-007-0611-8, © Springer Science+Business Media B.V. 2011

Nanotechnology. She has recently published a monograph entitled "Trends of Environmental Security in the 21st century".

Charles B. Fields, PhD, is Professor of Criminal Justice and Police Studies in the College of Justice and Safety at Eastern Kentucky University, Kentucky, USA. He has a BA (1980) and MA (1981) in Political Science from Appalachian State University, and a PhD in Criminal Justice Theory (1984) from Sam Houston State University, USA. He has taught for the past 26 years at several universities in the United States, and was Department Chair of Criminal Justice at the California State University, San Bernardino, and Corrections and Juvenile Justice Studies at Eastern Kentucky University. He has been a visiting lecturer at several institutions in Finland. He is a member of numerous regional, national, and international professional associations and has published 11 edited books and Technical Reports, over 30 scholarly articles, chapters, encyclopaedia entries, and reviews, and presented at over 30 academic conferences. Current research interests include The "Terza Scuola" (Third School) of the Italian Criminologists: 1880–1910, Drugs and International Drug Policy, and Comparative Justice Systems.

About the Authors[1]

Klemen Bančič, BA, is working in the Unit for Officers in Slovene Army, Ministry of Defence of the Republic of Slovenia. He got a grant from the Ministry of Defence of the Republic of Slovenia for studying at the Faculty of Criminal Justice and Security, University of Maribor, Slovenia. Due to his diligence and academic performance he was selected to spend a semester at the Flinders University in Australia as an exchange student in the academic year 2006/2007. His research fields are criminology, crime prevention and environmental crime.

Avrelija Cencič, PhD, is Associate Professor a Vice Dean for research at Faculty of Agriculture and Life sciences and Faculty of Medicine, University of Maribor, and Chair of the Department of Biochemistry at the Faculty of Medicine, University of Maribor, Slovenia. She is a founder of Biochemistry, Microbiology, and Biotechnology development at the University of Maribor. At present, her area of expertise is food and health, including risk assessment, and food safety. In this regards, she is also a chair of the master study programme on food safety in Slovenia developed by Faculty of Agriculture and Life Sciences and Faculty of Medicine, University of Maribor. She is involved in several international and national projects, members of national and international bodies and acts as an expert or evaluator in several programmes globally.

---

[1] Editors of this monograph are also (co)authors of chapters in this book. As they are presented in a section About the Editors they do not reappear in About the Authors section.

About Editors and Authors

Bojan Dobovšek, PhD, is Associate Professor of Criminal Investigation and Vice Dean at the Faculty of Criminal Justice, University of Maribor, Slovenia. He is a partner of Institute fur Politikwissenschaft und Socialforschunf, University of Wurzburg, Germany, and a visiting professor at the University of Ghent, University of Sarajevo, University of Belgrade and University of Zagreb. His areas of research are organized crime, corruption, art crime, environmental crime, criminal investigation and investigative journalism.

Tijana Đorđević, MA, is PhD student at the Faculty of Geography, University of Belgrade, Serbia. During the studies in 2009, she received scholarship from Greek Ministry of Education for Greek language and Culture Summer Course at the University of Ioannina, Greece, and Scholarship for participating in the study programme "International Security & NATO VI", Prague, Poland. She participated in several Geography congresses, held in Greece (2005), Russia (2006), Poland (2007) and Serbia (2009).

Katja Eman, MA, Junior Researcher, Teaching Assistant and PhD student at the Faculty of Criminal Justice and Security, University of Maribor, Ljubljana, Slovenia. During her MA study, she became especially interested on crimes against the environment; therefore her Master thesis was entitled Environmental Crime in Slovenia. She became junior researcher in the field of criminology and is currently conducting a PhD research in the field of green criminology, entitled Crimes against the Environment – Comparative Criminological and Criminal Justice Perspectives. In the academic year 2009/2010 she spent 2 months at the College of Justice and Safety at Eastern Kentucky University, Kentucky, USA, as visiting student working on her thesis. Her research fields of interest are green criminology, environmental crime, environmental justice and organised crime.

Sabina Fijan, PhD, is Assistant Professor in the field of Textile chemistry and Applicative microbiology at the Faculty of Mechanical Engineering, University of Maribor and at the Faculty of Health Sciences, University of Maribor, Slovenia. Her professional training includes research work at the Forschunginstitut Hohenstein, Germany and at the Institute for Microbiology and Immunology at the Medical Faculty of the University of Ljubljana. Her areas of research include textile hygiene, hospital textiles and wastewater treatment.

Mario Gorenjak, MA, is PhD student at the Faculty of Medicine, University of Maribor and junior researcher at Faculty of Agriculture and Life sciences, University of Maribor, Slovenia. From 2003 to 2007, he was a student at Faculty of Health sciences at the University of Maribor and has a degree in health sciences. From 2008 to 2009 he was an intern at University Clinic Maribor. In 2010 he finished a master's degree of bioinformatics and he is specializing in biochemistry, molecular biology, genetics and genomics. He is also involved in many projects and has recently discovered new genetic risk factors contributing inflammatory bowel disease. His expertise are also IT, multivariate statistic analysis and data management.

382 About Editors and Authors

Haris Guso, BA, is MA candidate on the Faculty of Economy, University of Sarajevo, Bosnia and Herzegovina. His research interest consists of economic related crime and ecology.

Teodora Ivanuša, PhD, is Assistant Professor for Crisis Management at the Faculty of Criminal Justice and Security and at the Faculty of Logistics, University of Maribor, Slovenia. The subject Terrorism and Weapons of Mass Destructions, which she is lecturing at the Faculty of Logistic, is adapted for Security/Armed Forces by which the Force Protection as a highest philosophy of NATO has been fulfilled in the Republic of Slovenia at the University level as well. She holds a PhD in Diagnostic Imaging from the University of Ljubljana since 2000. She is an OF-5 Military Specialist, former National Representative in NATO/CNAD/AC225/JCGCBRN, and former Advisor for Education and Special Tasks. Her main activities for the Slovenian Armed Forces covered Doctrine, Development, Training and Education Command. She holds the Chair of TIEMS International Program Committee (IPC), and is a member of Research Institute for European and American Studies – RIEAS.

Midhat Jašić, PhD, is Associate Professor of Food technology at the Faculty of Pharmacy, University of Tuzla, Bosnia and Herzegovina. Among other things, he deals with health and food safety, in the teaching process and advising the food industry. He advised more than 20 food companies in introducing the system of food safety management. He published six books and about 50 professional and scientific papers.

Katarína Jelšovská, PhD, is Assistant Professor at the Faculty of Special Engineering, University of Žilina, Slovakia. She works at the Mathematical Institute in Opava at the Silesian University in Opava, Czech Republic. She finished BA and PhD studies at the Department of Crisis Management, Faculty of Special Engineering, University of Žilina, Slovak Republic. Her main research field of interest is crisis management.

Katarína Kampová, PhD, is Assistant Professor at the Department of Security Management, Faculty of Special Engineering, University of Žilina, Slovakia. Her professional qualification is focused on studying and analyzing statistical methods applicable in processes of risk identification and analysis. Simultaneously, she deals with issues of company security and the effectiveness of security projects. She is a member of working group of the Leonardo da Vinci project – EIT portal and eSEC portal. She is author of several publications in international journals and conferences and co-author of publication "Econometric" published in Slovakia.

Zoran Keković, PhD, is Professor of Systems of security and Corporative security at the Faculty of Security Studies, University of Belgrade, Serbia. Since 2006, he is the founder and headmaster of the Postgraduate studies of Crisis Management at the Faculty of Security Studies, University of Belgrade aiming to

About Editors and Authors 383

develop managing competencies in the area of security, defence, but also other emergency organizations and educational institutions connected with risk management and crisis management. He is president of Serbian's Committee for standards and related documents ISO/TC 223 – Societal Security on incident preparedness and operational continuity management and author of many scientific and expert articles and publications and participant in a lot international conferences in the field of security. He is president of an NGO "Risk analysis and crisis management centre – Belgrade".

Ana Klenovšek, BA. During her study at the Faculty of Criminal Justice and Security, University of Maribor, Slovenia, she was selected due to her academic performance and diligence for the international student exchange programme between Europe and Australia. She studied one semester at the Griffith University, Queensland, Australia and participated in the Summer School: *Security: local, global and supranational* which was held at the Institut d'Etudes Politiques de Grenoble, in Grenoble, France. At the moment she is a Climate Advocate in Challenge Europe project, supported by the British Council, its aim being to reduce carbon emissions.

Damir Kulišić, MA, is Senior Lecturer in the field of Safety and Protection Engineering at the Fire and Explosion Protection Engineering and Investigations Department of the Fire Protection Centre at the Fire Fighting School in Zagreb, and in the field of fire and protection engineering, environmental criminality, sabotage and modus operandi of modern terrorism investigations at the Police College in Zagreb, Croatia. Since 2002 he has been teaching Fire and Explosion Engineering and Forensics at the University of Karlovac and the University in Split, Croatia. He is author of 45 professional and scientific papers and a member of temporary state commissions for major fire and explosion accidents investigations.

Tomáš Loveček, PhD, is Associate Professor at the Department of Security Management, Faculty of Special Engineering, University of Žilina, Slovakia. His professional qualification is focused on evaluation a designing of physical protection systems and information security management systems. Simultaneously, he deals with issues of company security and the effectiveness of security projects. He is a member of editorial board at a Journal of Crisis Management and Journal of Criminal Justice and Security. He is a member of working group of the Leonardo da Vinci project – EIT portal and eSEC portal. He is author of several publications in international journals and conferences and author or co-author of publications "Security Systems – Surveillance Monitoring Systems" and "Security Systems – Security of Information Systems" published in Slovakia.

Marina Malis Sazdovska, PhD, is Professor of Environmental Criminology and Vice Dean of Faculty of Security, University of Saint Kliment Ohridski, Macedonia. She worked as an inspector in MIA in Skopje and as Assistant Professor in the field of criminology at the Police Academy in Skopje. She participated in courses for

384 About Editors and Authors

training of criminologists organized by the OSCE as part of the ATA-programme (combat against terrorism) and courses in operative work, organized by the USA government. She is member of the Association of Criminologists, the Association for Penal Law and Criminology, the Association Anticrime-Veles and the Association 8ka – for ecology and education.

Elizabeta Mičović, MA, is Undersecretary at the Ministry of Agriculture, Forestry and Food, Food Safety Directorate in Food and Feed Safety and Quality Division. From 1984 to 2002, she worked in food industry as food technologist, after that she worked in Health inspectorate of Republic of Slovenia at Ministry of Health as Health Inspector. In the beginning of 2005 she started to work in Directorate for Public Health as an advisor on areas of nutrition, food safety and environmental impacts on Health. She was a Member of the Consumer Protection Council under the Ministry of the Economy, as representative of Ministry of Health, from 2005 to 2010. At present she is Undersecretary at the Ministry of Agriculture, Forestry and Food and a PhD student at the Faculty of Criminal Justice and Security, University of Maribor, Slovenia. Her research areas are of food safety, invisible threats and consumer protection.

Miroljub Milinčić, PhD, is Assistant Professor and Vice-Dean at the Faculty of Geography, University of Belgrade. He participated in realization of several national scientific and applied projects. He is Planet Earth's national Committee member and member of the editorial board of a number of scientific and professional publications. Main fields of his research are: ecology, environment protection, geo-ecology, demo-ecology, applied geography, human and social ecology, geopolitics and speleology. He has published five scientific monographs and over 80 scientific papers in several national and international journals, bulletins and proceedings.

Matjaž Mulej, PhD, is Professor Emeritus of Systems and Innovation Theory at the Faculty of Economics and Business, University of Maribor, Slovenia. He is member of three International Academies of Science and Arts and president of the International Federation for Systems Research (with 35 member associations and membership on all continents). He has published over thousand publications in more than 40 countries and spent 15 semesters as a Visiting Professor at universities abroad.

Elmedin Muratbegović, PhD, is Assistant Professor of Criminology at the Faculty of Criminal Justice Sciences, University of Sarajevo, Bosnia and Herzegovina. His research interest consists of crime control and prevention, criminological prognosis, spatial analysis of crime and fear of crime.

Radmilo V. Pešić, PhD, is Professor of Economics at the Department of Agricultural Economics, Faculty of Agriculture, University of Belgrade, Serbia. He was Fulbright Visiting Fellow at the Department of Economics TEXAS A&M University, College Station, TX, USA in 1990–1991. During 1994 he contributed

About Editors and Authors

as a reviewer to the Second Assessment Report of the Working Group III of the IPCC. He is Environmental Expert Team Leader for the Republic of Serbia National Sustainable Development Strategy drafting and implementation. He is member of the European Economic Association and the European Association of Environmental and Resource Economists. His main research fields of interests are environmental and natural resource economics, energy economics and post-communist transition, and economic theory.

Iztok Podbregar, PhD, is Associate Professor and Head of Management and Police Administration Department at the Faculty of Criminal Justice and Security, University of Maribor, Slovenia. He also lectures Crisis management at the Faculty of Organizational Sciences and the Head of Military Logistics Department at the Faculty of Logistics, University of Maribor. Prior to that he was an Advisor of National Security to the President of the Republic of Slovenia, Director of the Slovene Intelligence and Security Agency, National Counter-terrorism Coordinator, Minister-Counsellor in the Prime Minister's Office, and State Secretary at the Ministry of Defence. He was Deputy Chief and afterwards a Chief of the Slovene Armed Forces General Staff. He was also a member of the Slovene delegation appointed to start the talk on the Republic of Slovenia's accession to NATO Alliance, International Inspector for Weapons Supervision within the OSCE, and Chief of the Air Force Section in the Republic Territorial Defence Staff.

Robert Praček, BA, is Chief Criminal Police Inspector at Criminal Technician Unit in Police Directorate Ljubljana, Slovenia. He has been working in the field of crime scene investigation since 1994. He is Lecturer at Police College and Instructor of Criminal investigators at Criminal Police Directorate, Ljubljana.

Ivica Ristović, PhD, is Senior Lecturer in the field of Engineering of Safety at Work and Environmental Protection and Mining Engineering at the Faculty of Mining and Geology, University of Belgrade, Serbia, and at the Faculty of Technical Science in Kosovska Mitrovica, University in Pristina, Kosovo. As an author and co-author he published 130 scientific and professional papers in international and national scientific journals and at symposia in the country and abroad. He is author and co-author of four monographs, he participated in the preparations of 15 projects and ten studies conducted for the industry, was leader of and participated in eight scientific-research, strategic and national projects and two international project financed by the Ministry of Science and Technological Development of the Republic of Serbia. He is president of the Group for Power Mining of the Serbian Chamber of Commerce, member of the Group for Environmental Protection of the Serbian Chamber of Commerce and chairman of the International Symposium on Energy Mining.

Andrej Sotlar, PhD, earned his BA, MA and PhD in defence studies at the Faculty of Social Sciences (University of Ljubljana). He is Assistant Professor, Vice Dean for international cooperation and Head of Chair of Intelligence and Security Studies at the Faculty of Criminal Justice and Security, University of Maribor, Slovenia. He is a leading national expert in the fields of national security, intelligence and security services and private security.

Sonja Šostar-Turk, PhD, is Professor of Environmental engineering, Vice Dean for postgraduate studies and Chair of general sciences at the Faculty of Health Science, University of Maribor. Her professional training includes research work and education at the Institut fur Textilchemie, Denkendorf, Germany; Curtin University of Technology, Perth, Western Australia; Forschunginstitut Hohenstein, Germany; Institute for Product Development, Copenhagen, Denmark and University of Aalborg, Denmark. Her areas of research include textile printing, textile care, textile hygiene, hospital textiles, wastewater treatment and the environmental influence on health protection.

Bojan Tičar, PhD, holds a Doctorate in Law from The Law Faculty, University of Ljubljana (2001). He is Associate Professor for public law and Vice Dean for economic affairs and development at the Faculty of Criminal Justice and Security, University of Maribor, and Associate Professor at the Faculty of Management, University of Primorska, Slovenia. He is Head of the Law Department at the University of Primorska.

Bernarda Tominc, BA, is Teaching Assistant and a PhD student at the Faculty of Criminal Justice and Security, University of Maribor, Slovenia. Her research interests include international political and security integrations, security threats in contemporary society and national security.

Igor Winkler, PhD, is Associate Professor at the Department of Physical and Environmental Chemistry, Chernivtsi National University, Ukraine. His main research fields of interest are physical chemistry, electrochemistry, environmental control and monitoring, water ecology, oil products control and monitoring, reuse of the coal refining wastes. He has published articles in 21 refereed journals, 43 communications to scientific meetings, and three books.

Grygoriy Zharykov, PhD, is Head of Environmental Control and Licensing Section, State Department of Ecology in the Region of Chernivtsi, Ukraine. His main research fields of interest are problems of the solid waste management and mitigation of the landfill influence on the environment.

Index

A

Acceptable daily intake (ADI), 7, 299–302, 305–308, 313–314, 318–319
Accident, 4–5, 19, 27, 31–32, 75, 86, 137, 151–182, 187–201
Accumulated water, 5, 227, 231, 233, 238, 239, 241–247, 275
Accumulation, 5, 102, 108, 226, 231, 233, 234, 238–239, 242, 244, 245, 314, 321
ADI. *See* Acceptable daily intake
Africa, 55, 57, 58, 62, 82, 83, 85, 90, 93, 95, 200, 228, 232
Agriculture, 22, 49–50, 81, 85, 105, 109, 118, 193, 199, 229, 231, 235, 236, 238, 239, 244, 246, 257, 260, 262, 266, 294, 296, 313, 315–316, 318, 323, 366
Air, 1–2, 5, 8, 14–15, 22, 31, 46, 75, 102–108, 114–115, 119, 146, 176, 179, 180, 189, 190, 251–262, 270, 314, 350
  pollution, 3, 7, 49–50, 105–108, 148, 194, 195, 252, 255, 257, 263, 267, 314, 358
Albania, 84, 208, 209, 234–239, 253
Ammunition storage, 5, 187–203
Animal, 16, 17, 22, 23, 29, 35, 42, 44, 47–50, 54–62, 80, 81, 103, 104, 108, 112–116, 125–130, 132, 133, 141, 146–147, 219, 267, 299, 301, 313–316, 318, 321, 322
  rights, 3, 61, 102, 112–114
Asia, 58, 84–86, 92, 93, 95, 228, 232, 267

B

Bamako Convention, 3, 90
Basel Convention, 3, 80, 86–92
Basin, 231, 233, 236, 237, 244, 245, 247, 254, 268

Biodiversity, 22, 24, 29, 108, 136, 350
Biogas, 267, 268, 276
Bjelašnica, 105
Bleach, 6, 280, 282–284, 287–289
Bosnia and Herzegovina, 3, 101–119, 208, 209, 234, 238, 253, 254
Brazil, 132
Brčko, 106, 109
Bulgaria, 8, 208, 209, 211, 218, 234–239, 253

C

Cairo, 88, 266
Calabria, 84
California, 125–126
Camorra, 84
Canada, 53, 54, 81, 89, 126, 127, 268
Chemical, 6, 16, 49, 83, 104, 143, 152, 194, 212, 271, 280, 293, 313
  accident, 199, 200
  pollutants, 104, 105, 200
Chemo-thermal laundering, 6, 281, 286, 289
Chernivtsi, 6, 267–272, 275
China, 81, 83, 85, 86, 92, 93, 126
Civil protection, 23, 27, 49, 82, 102, 115, 191–193, 195, 197
Clark's model of situational crime prevention, 2, 41–64
Climate change, 12, 17, 18, 21–22, 24, 26, 36, 225, 235, 246, 294, 366
Coal, 105, 106, 251–263
  combustion, 5, 257, 258, 261
  excavation, 5, 6, 252, 255–258
  mining, 6, 252, 255, 257, 258, 261, 263
  mining process, 251–263
Colour, 7, 20, 288, 295, 300, 302–305, 307, 308, 361
Consumer protection, 296, 297, 317

387

388 Index

Contamination, 6, 7, 31, 50, 53–59, 61, 81, 85,
86, 92, 104, 105, 145, 146, 148,
149, 168, 192, 193, 213, 236, 238,
262, 266, 268, 270–273, 275, 276,
281, 313–315, 319, 320
Contemporary criminology, 102, 103
Continuity process, 188
Control theory, 43, 46
Country, 3, 12, 47, 74, 80, 102, 124, 136,
175, 190, 206, 233, 252, 267, 280,
312, 331, 373
Crime prevention measure, 2, 41–64
Crime prevention methods, 2, 43, 46, 51–63
Crimes against nature, 2, 44, 62
Crimes against the environment, 3–4, 8, 20,
34, 44, 46–51, 53–62, 130
Crime scene, 137, 139, 141, 143, 146, 147,
153, 175
investigation team, 4, 137–139, 142,
151–182
Criminal justice, 9, 34, 45, 47, 103
Criminal law, 28, 34–35, 42, 46, 49, 50, 341
Criminal-legal regime, 1–2
Criminology, 1, 3, 8, 42–46, 63, 79, 102, 103,
119, 128, 138, 143, 147, 295
Crisis management, 5, 196, 201, 366
Croatia, 4, 8–9, 53, 60, 118, 152, 158, 159,
169, 171, 174–180, 208, 209, 211,
234–239
Czech Republic, 8, 86, 370

D

Danger, 2, 16, 44, 72, 80, 113, 124, 138, 152,
188, 212, 267, 315
Danube River, 169, 237, 241, 268
Defence industry, 202–203
Deforestation, 3, 108–112, 367
Desertification, 108, 124, 367
Detergent, 6, 105, 177, 280–284, 288, 289,
299, 316
Dialectical systems theory, 1, 7, 330, 337, 338
Disinfectant, 6, 280–289, 316
Domino effect, 152, 165, 177
Dump area, 6, 262, 265–276

E

East Europe, 85
EC. *See* European Commission
Eco-crime, 2, 3, 69–77, 102, 124, 125, 127
prevention, 2–3, 69–77
Eco-global criminology, 49, 124
Ecological justice, 34

Ecological system, 102
Ecomafia, 84
Economic benefit, 3, 70–72
Economic instrument, 1, 69–77
Eco-risk, 1, 16, 194, 195, 202, 225–247
Ecosystem, 24, 81, 105, 108, 189, 212, 226,
229, 236, 247, 257, 294, 350
EDI. *See* Estimated daily intake
EIA. *See* Environmental impact assessment
Emergency, 4, 27, 61, 106, 156, 158–160, 163,
166, 171–173, 181, 188, 189, 196–200
plan, 158–160, 171
preparedness, 4, 188
process, 188
Emery–Watson model, 2–3
Energy conservation, 205–206
Energy-mining industry, 6
Energy source, 103
Enterococcus faecium, 283, 285, 287
Environment, 1, 11, 41, 70, 79, 102, 123, 135,
152, 187, 205, 225, 251, 265, 279,
294, 313, 328, 349, 365
Environmental
change, 294, 349–350, 366–368, 373,
375, 376
conflict analysis, 1, 7–8, 365–376
crime, 1–4, 8, 20–21, 26, 35–36, 42–50, 53,
56, 62–64, 79–80, 84, 92, 102–114,
119, 123–133, 135–149, 174
criminality, 42–47, 51–63, 136, 137, 148
criminology, 8, 295
degradation, 3, 42, 50, 53–62, 255,
365–367, 373–376
destruction, 12
disaster, 1, 3, 18, 164
harm, 5, 34, 42, 61, 102, 103, 280
hazards, 62, 314
issue, 1, 5, 8, 22, 139, 196
policy, 1–3, 11–36, 70, 72, 73, 76, 77, 82, 124
protection, 2, 13, 17, 18, 24, 25, 28–33,
35, 43, 47–50, 53, 72, 75, 92,
102, 103, 114, 116, 118, 119,
124–129, 136, 171, 213, 217, 252,
258, 333, 370
Protection Act, 29, 48–49, 74, 75, 103,
106, 119, 197, 198, 252
remediation, 205–206
risk factor, 1, 5, 6, 31, 193, 225–231, 233,
236, 240, 243–244, 247, 251–263,
279–290
security system, 2, 12–13, 20–27, 36,
187–203, 258, 349, 350, 365
threat, 2, 5, 7, 8, 13–20, 23, 27, 35–36,
41–64, 124, 201, 265–276, 349–362

Index 389

Environmental impact assessment (EIA), 106, 368–370, 373
Erosion, 105, 111–112, 238, 261, 262, 366, 367
Estimated daily intake (EDI), 7, 300, 302, 304–308
European Commission (EC), 24, 31, 86–87, 89, 91, 94, 154, 162–163, 212, 215–217, 219, 267, 280, 313, 316
European Union (EU), 2, 5, 22, 23, 28, 31, 34, 36, 53, 59, 62, 80, 82, 84, 86–87, 89–96, 104, 107, 163, 164, 199, 200, 205–220, 261, 267, 280, 300, 328–330, 332–333, 340
Europol Convention, 132
Evidence, 4, 5, 51, 125, 127, 128, 137–140, 145, 152–155, 157, 169, 172–174, 181, 211, 216, 219, 227, 260, 273, 275, 306, 373
E-waste, 80, 81, 83–86, 91–94, 115, 267
Explosion, 1, 5, 132, 152, 153, 158, 159, 164–168, 174, 176–178, 180, 187–203, 211

F
Finland, 108
Fire, 17, 22, 26, 27, 49, 60, 112, 133, 137, 145, 146, 153, 158, 159, 164–168, 173, 174, 177–180, 188, 191, 192, 197, 200
Food, 1, 16, 46, 105, 169, 202, 206, 229, 251, 271, 285, 292, 311, 350
  additives, 7, 169, 295–7, 299–308, 313–314, 318
  chain, 7, 202, 213, 312, 314–319, 321–324
  consumption, 296, 302–305, 314, 316
  contaminant, 7, 297, 299, 300, 313–316
  industry, 285, 295–297, 315, 317
  intake, 7, 295, 300–305, 307, 315, 321, 322
  production, 7, 294, 296, 310–324
  safety, 1, 7, 294–302, 307, 312–313, 315–317, 319, 321–324
  scarcity, 294
Food safety management (FSM), 7, 312, 313, 316, 317, 319, 323, 324
Forensic, 4, 27, 125, 138, 152–154, 156–165, 167, 170–175, 177, 180–182
Forest, 24, 29, 35, 49, 50, 54, 56–58, 60, 108–109, 111–112, 114, 118, 193, 199, 238, 257, 269–270
  mafia, 130, 131, 133
FramingNano Report, 207, 216
France, 88, 132, 253

Fresh water, 226–228, 230, 231, 246, 266, 273, 280, 282, 283, 289
FSM. *See* Food safety management

G
Geography of thirst, 228
Germany, 84, 86–89, 111, 124, 180, 206, 266, 281
Ghana, 85, 90, 228
Global warming, 17, 124
Government, 5, 12, 14, 16–17, 21, 26, 32, 45, 56, 57, 82–83, 87, 108, 111, 116–119, 125–127, 132, 153, 158, 189, 193, 196, 197, 199, 201, 211, 218, 220, 266, 296, 297, 319, 334–337, 340, 341, 367, 370
Great Pacific Garbage Patch, 266
Greece, 87, 96, 208, 209, 211, 227, 234–239
Green crime, 3, 80, 101–119, 280
Green criminology, 1, 3, 43–45, 79, 102, 103, 119
Groundwater, 6, 105, 179, 213, 236, 238, 241, 242, 257, 261–263, 267–276

H
Harmful substances, 50, 54, 55, 58–60, 105, 169, 263, 318, 323–324
Hazardous material, 4, 5, 86, 142, 151–182, 187, 201
Hazardous waste, 3, 18, 49–50, 62, 75, 80–82, 84–85, 88–93, 115–116, 148, 179
Hazards, 3, 18, 48, 75, 80, 105, 142, 151, 187, 212, 294, 312
Health, 1, 12, 43, 81, 103, 125, 141, 152, 191, 205, 227, 263, 286, 293, 311, 329, 370
  protection, 47, 119, 169, 171, 293–308
Helsinki Convention, 104
Hoax case, 125–129
Holism, 7, 329–331, 337, 338
HOQ. *See* House of quality
Hospital bed linen, 6, 281–284, 286–289
Hospital wastewater, 1, 6, 288–289
House of quality (HOQ), 138, 140–147
Human security, 1, 7, 14, 36, 349–362
Hungary, 208, 209, 211

I
Illegal logging, 111, 118, 124
India, 86, 88, 89, 93, 132, 169, 231
Industry, 5, 17, 42, 76, 86, 102, 153, 190, 206, 226, 252, 266, 280, 295, 314, 328

Index

International, 1, 12, 42, 79, 104, 123, 136,
  163, 166–167, 169, 171, 189, 206,
  237, 295, 318, 330, 365
International environmental protection,
  124–129, 136
International waste trafficking, 1, 3, 79–97
Interpol, 81, 83, 88, 92, 124–129, 132, 174
Invention-innovation-diffusion, 330, 331, 335
Italy, 31, 54, 81, 84, 87, 88, 92, 95, 96, 132,
  168–169, 266

J
Japan, 58, 85, 89, 331
Judge, 4, 139, 141–142, 298, 372, 373

K
Karadorde Hill, 5, 191
Kolubara, 240, 254, 255, 258–260
Kosovo and Metohija, 234, 238, 244,
  254, 255

L
Landfill, 6, 81, 82, 84, 87, 91, 114, 115, 257,
  266–273, 275–276
Laundry, 279–290
Lazarevac, 260
Legal regulation, 2, 11–36, 71, 80, 106, 201
Legislation, 3, 5, 13, 23, 24, 26, 28–33, 42–43,
  47, 48, 60, 70–72, 80–82, 85, 86,
  88–92, 94, 96, 103, 107, 108, 112,
  114, 116, 118, 119, 136, 171, 190,
  197–201, 215–217, 219–220, 242,
  296, 300, 312
Lifestyle theory, 43, 46
Lignite, 253–259
Ljubljana, 8, 9, 18
Local, 5, 6, 21–24, 26, 42–43, 48, 55, 57–59,
  62, 81, 95, 96, 109, 115–117, 124,
  125, 132, 146, 159, 171, 176, 179,
  188, 191, 193, 196, 197, 199,
  201–202, 230, 233, 234, 236, 238,
  242, 243, 246, 255, 260, 266,
  269–273, 275, 333, 336–337, 339,
  341, 374
Lomé IV Convention, 3, 90

M
Macedonia, 3, 123–133, 208, 209, 234, 253
Madagascar, 126, 228
Mainstream criminology, 83
Malaysia, 125, 126, 128

Management, 1, 2, 5–8, 11, 14, 21, 22, 24,
  26, 27, 29, 36, 42, 45, 49–50, 84,
  86, 88, 91, 92, 107–109, 112, 114,
  116, 118, 119, 156, 160, 161,
  172–174, 181, 188–190, 196–203,
  217–219, 225–247, 263, 265–276,
  280, 295, 297, 298, 307, 312, 313,
  316, 317, 319, 320, 323, 324,
  350, 355, 360, 366, 367,
  371–373, 376
Mexico, 58, 93, 125–127, 129, 164,
  168, 173
Microorganism, 104, 280, 283, 285, 287, 297,
  315, 319, 322
Mining, 1, 6, 105, 244, 251–263, 316
Mining-energy process, 255–263
Modelled environment, 2, 7, 44, 213, 354
Monte Carlo method, 360
Municipal waste, 6, 241, 265–276, 281

N
Nanoscience, 5, 208, 217, 218
Nanotechnology, 1, 5, 205–220
Naples, 266
National, 2, 3, 5, 8, 12–29, 31, 34, 42, 49,
  53–55, 57, 75, 82, 89, 92–96, 124,
  125, 131, 132, 136, 152, 159, 171,
  197–201, 211–213, 218, 226, 230,
  233, 237, 246, 266, 268, 280, 295,
  324, 327–342, 350, 351, 354, 360,
  365, 366, 371
  environmental policy, 1
  security policy, 12, 13, 21–23, 36
  security strategy, 15–17, 21, 25
NATO. *See* North Atlantic Treaty
  Organisation
Natural water, 5, 104, 226, 231, 233,
  242, 246
Nature, 2, 7, 8, 16, 21, 24–26, 28, 33, 34,
  43, 44, 47–50, 54, 57, 61–63,
  102–104, 106, 108, 117–119,
  127, 128, 153, 156, 161, 171,
  181, 190, 196, 207, 226, 231,
  263, 299, 315, 320–322, 327,
  329, 339, 350, 355, 358, 365,
  370, 372, 373
The Netherlands, 87, 88, 95
NGO. *See* Non-governmental organisation
Nigeria, 85, 88, 90, 228
Non-governmental organisation (NGO), 14,
  22, 24, 116, 117
North Atlantic Treaty Organisation (NATO), 8,
  14, 189, 199, 340
Nutrition, 294, 296, 302, 307, 315

Index

391

O

Obrenovac, 259–260
Oceania, 267
Offender, 2, 4, 44–47, 51, 52, 54, 61, 63, 73–76, 128, 130–131, 137, 295
Organization for Economic Co-operation and Development (OECD), 86, 89, 91, 211, 215, 218
Organization of African States (OAS), 90
Organized crime, 3, 16, 17, 127–129, 131–133

P

Pakistan, 86, 89, 231, 233
Pan European Corridor 10, 189, 191, 200
Paraćinske Utrine, 5, 187, 190–192
Patent data generation, 207–211
PEN. *See* Project of Emerging Nanotechnologies
Penal code, 20, 26, 34, 35, 50, 53, 57, 143, 147
Perpetrator, 2, 44, 45, 50, 124, 127, 130, 132, 136, 141, 147, 153, 162, 175
Pesticide, 84, 105, 169, 296, 297, 299, 300, 313–316, 318–320
The Philippines, 89, 126, 132
Plant, 17, 30–33, 35, 42, 44, 47, 49–50, 54, 55, 58, 59, 61, 62, 74, 80, 103–108, 111, 118, 128, 133, 161, 166, 212–214, 219, 241, 262, 266, 280, 281, 286, 289, 290, 313–316
Police, 3–4, 20, 23, 26, 27, 45, 51, 55, 56, 94–96, 119, 124, 127, 128, 130–131, 133, 136–139, 142, 146–149, 171–172, 174, 175, 181, 188, 191, 197, 340–341
Policy making, 5, 12, 16, 24–25, 27, 77, 187–203
Pollutant, 2, 6, 36, 104–107, 200, 268, 269, 271, 273, 275, 314
Pollution, 2, 3, 6, 7, 16–18, 20, 22, 24, 27, 30, 31, 35, 36, 42, 46, 49–50, 53–56, 59, 61, 62, 70, 72, 74–76, 103–107, 118–119, 124, 125, 130, 136, 137, 139, 141–143, 146–148, 194, 195, 198, 200–202, 213, 226, 238, 240, 241, 252, 255, 257, 260–263, 266, 267, 272, 273, 275, 280, 294, 316, 367
Precautionary principle, 29, 34, 215, 219
Preservative, 7, 295, 300, 302–308
Prevention, 2, 4, 5, 8, 17, 18, 20, 22, 23, 25, 26, 31, 32, 36, 41–64, 69–77, 91, 103, 107–108, 124, 125, 130, 131, 133, 152, 159, 160, 168, 170–173, 197, 199, 201, 213, 225–247, 275,

295, 350, 351, 365, 366, 368, 369, 373–376
Proactive green criminology, 43
Project of Emerging Nanotechnologies (PEN), 206
Prosecutor, 4, 27, 125, 139, 141, 142
Protection, 1–2, 12, 43, 72, 89, 102, 124, 136, 152, 188, 207, 227, 252, 266, 286, 293, 318–319, 333, 350, 369
Public, 5, 12, 42, 73, 82, 107, 136, 152, 188, 207, 226, 273, 280, 293, 318, 329, 352, 368
health, 27, 47, 238, 293–308, 318–320, 370, 371

Q

Quality function deployment (QFD), 138–145

R

RAL-GZ 992 criteria, 281, 285, 287, 289
RAT. *See* Routine activity theory
Rational choice theory, 43, 44, 46
Rational polluter model, 70–74, 76
Region, 6, 18, 44, 63, 84, 93, 148, 227, 234–239, 242, 247, 252–254, 259, 263, 266–269, 373
Regional, 5, 6, 42, 90, 93, 107, 129, 152, 159, 195, 197, 200, 201, 226, 230, 231, 233, 236, 237, 240, 242–246, 253–255, 263, 333, 337
Remediation, 2, 11, 36, 205, 213, 332
Republic of Srpska, 106, 109, 115, 118
Residue level, 180, 316, 318
Risk analysis, 7, 158, 160, 200, 294–295, 297–302, 307, 312–313, 352–360, 371
Risk assessment, 4, 6–7, 62, 75, 158–161, 163, 167–168, 172, 181, 187, 207, 211–218, 293–308, 312–313, 316–324
Risk quantification, 310–324
River, 49, 58, 71, 81, 85, 86, 107, 114, 118, 152, 169, 227, 231, 235–238, 240, 241, 243–247, 262, 268–269, 315
RKI. *See* Robert-Koch Institute
Robert-Koch Institute (RKI), 281, 283
Romania, 169, 208–209, 211, 234–239, 253
Routine activity theory (RAT), 43, 46, 51, 295, 297

S

Safety, 4, 14–15, 47, 75, 92, 115, 124, 136, 152, 197, 206, 229, 266, 294, 312, 329, 350
report, 31–32, 154, 157–161, 164–166, 171, 173, 215–216

Sanitation, 238, 294, 313–314, 316
Sarajevo, 104, 106, 107, 114, 115
Sardinia, 266
Save-for-health food production, 7, 311–324
Secret agent, 127, 129
Security, 1, 12, 42, 141, 152, 187, 227, 258, 294, 327, 349, 365
    policy, 1, 2, 12–13, 15–16, 20–27, 35, 36, 196–197
    system, 1, 2, 13, 14, 20, 25–27, 36, 158, 164–165, 338, 340–341
SEE. *See* South Eastern Europe
Serbia, 3, 5, 6, 8, 74–76, 188, 189, 193, 194, 197–201, 208, 209, 211, 225–247, 253–256, 259, 262, 263
Situational crime prevention, 2, 41–64
Slovenia, 1–2, 7–9, 11–36, 43, 44, 46–55, 57, 59, 63, 138, 139, 148, 208, 209, 211, 234–239, 289, 302–307, 327–342
SMW. *See* Solid municipal waste
Social control, 3, 4, 64, 119
Social efficiency, 329
Social justice, 329
Social responsibility (SR), 252, 327–338, 342
Social risk, 17, 226, 352–354, 358–361
    analysis, 7, 352–360
    model, 353–356, 358, 359
    perception, 351–352
Social science, 7, 18, 42, 44, 47
Soil, 2, 5–8, 22, 24, 31, 36, 42, 47, 49–50, 75, 81, 84, 104, 114–115, 143, 169, 173, 179, 194, 195, 198, 199, 202, 213, 214, 229, 238, 240, 251–263, 271–273, 275, 288, 294, 314, 328, 329, 350, 358, 366, 367
Solid municipal waste (SMW), 6, 265–276
South Eastern Europe (SEE), 1–9, 206–211, 218, 220, 225–247, 252–254, 259, 263
South Korea, 85
Spring zone, 225–247
SR. *See* Social responsibility
Staphylococcus aureus, 283, 285, 287
Surface water, 5, 6, 225–247, 255, 257, 261, 268, 271–273, 275, 276
Sustainable development, 6, 22, 117, 205–206, 247, 333, 370
Sweden, 108
Sweetener, 7, 295, 301–308
Symbiotic green crime, 3, 80, 101–119
System theory, 7, 330, 333, 337, 338

T
Terrorism, 16–19, 125, 132, 169, 182, 196, 200, 329, 367
Thermal laundering, 281, 289
Thermal power plant (TPP), 107, 244, 246, 252–256, 258–263
Threat, 1, 12, 41, 82, 103, 124, 136, 162, 188, 218, 226, 265, 294, 327, 349, 365
Tourism, 57, 62, 118, 234, 236, 245, 246, 341
Toxicological effect, 280
Trace, 52, 128, 137, 141, 143, 145, 147, 148, 153, 154, 156, 157, 312
Transportation, 45, 82, 88–89, 95, 128, 155–159, 163, 165, 167, 169, 175, 181, 200, 252, 255–257, 259–261, 263, 266, 313

U
Ukraine, 6, 265–276
Uncertainty, 7, 17, 196, 212, 217, 219, 301–302, 313, 317, 318, 323, 324, 329, 349–362
United Kingdom (UK), 8, 45, 81, 83, 87, 89, 124, 206, 212, 214, 217
United Nations (UN), 22, 36, 163, 190, 193, 229, 234, 235, 336
United States of America (USA), 1, 8, 9, 45, 74, 81, 88, 89, 91, 92, 113, 125, 126, 216, 219–220, 230, 231, 267, 268, 335, 336

V
Victim, 19, 44, 45, 51, 62, 130, 136, 137, 167, 168, 295–296, 308

W
Washing agent, 6, 289
Washing procedure, 283
Waste, 1, 13, 42, 71, 79, 102, 125, 148, 153, 195, 212, 236, 257, 265, 280, 304, 314
    dumping, 3, 61, 71–72, 81, 115, 267–268
    management, 6, 24, 49–50, 84, 86, 88, 91, 92, 114, 116, 119, 266
    recycling, 18, 267
    water, 1, 6, 49–50, 71, 72, 104, 105, 179, 205–206, 213, 236, 240–242, 257, 271–273, 275, 276, 280–282, 285–286, 288–290, 314
Water, 2, 14–15, 46, 71, 81, 102, 130, 143, 159, 189, 205–206, 225, 251, 266, 280, 294, 314, 329, 350

management infrastructure, 225, 246
management system, 226, 229, 245, 246
pollution, 3, 35, 49–50, 54–56, 59, 61, 104–106, 194, 257, 275
power plant, 244, 246
regime, 26, 228, 230, 231, 233, 236, 242, 246, 247, 255, 257
resource, 5, 104, 107, 226–242, 244, 246, 247, 280

supply, 164–165, 227, 230, 231, 233, 234, 236, 239, 241–247
use, 104, 233, 240, 263, 280, 281, 289
Weapon, 17, 18, 51, 54–55, 81, 114, 128, 129, 169, 178, 182, 187–196, 332, 336, 337, 340
Welfare, 12, 18, 113, 114, 205–206, 233
World Health Organization (WHO), 81–82, 168, 294, 298, 300–302, 312–314, 319